Collins

Cambridge IGCSE™

Maths

TEACHER'S GUIDE

Also for Cambridge IGCSE (9–1)

AF834467

Isabel Marsden, Chris Pearce

Contents

Key: [E] = Extended only content

All Answers, Homework, Consolidation, Extension and Resource sheets, and Lesson plans, can be found online at:
www.collins.co.uk/pages/cambridge-international-downloads

Introduction

Welcome to the Teacher's Guide for Collins Cambridge IGCSE™ Maths, which has been updated for the revised Cambridge IGCSE and IGCSE (9-1) Mathematics syllabuses (0580/0980) for examination from 2025.

With this syllabus, students will be expected to calculate with and without a calculator. It is important that students are confident working without a calculator and understand the theory and steps in calculations. Straightforward non-calculator questions are flagged in the Student's Book so they become familiar with the requirement. Most topics can be tackled without a calculator, so it is good practice to encourage students to attempt all questions by working through the steps, only using a calculator for complex numbers and checking their work. This approach is encouraged in the lesson plans, and we suggest attempting homework, consolidation and practice in this way to build skills.

Contents of the Teacher's Guide

This Teacher's Guide contains lesson plans to accompany all the topics in the Student's Book. Accompanying resources are available to download in editable Word format:

- Lesson plans for every topic in the Student's Book (1.2, 1.2, 1.3…).
- Homework for every lesson.
- Consolidation questions for every chapter for students who need more support.
- Extension questions for every chapter for students who need more challenge – these do not extend beyond the syllabus but provide further practice and challenge on the syllabus content.
- Questions progress from basic to challenging in each exercise with extended syllabus content flagged.
- Answers to all Student's Book, homework, consolidation and extension exercises. Answers to the past paper questions in the Student's Book – these answers have been written by the author.

Downloads are available at www.collins.co.uk/pages/cambridge-international-downloads

Using the lesson plans

Each lesson plan follows the same format making it easy to use and to prepare lessons.

- Syllabus statements show the syllabus reference numbers either at Core level for content that both Core and Extended learners need to know, or at Extended level for students studying the Extended syllabus. The statements are based on the syllabus wording but have been broken down into parts where a statement is covered across more than one lesson.
- Learning aims indicate clearly what the students should master during the lesson, and are a useful way of measuring its success.
- Key words highlight important mathematical vocabulary.
- Prior knowledge highlights the underpinning maths that students need for the lesson.
- Common mistakes and remediation are pinpointed so they can be quickly recognised and rectified.
- Useful tips help students remember key concepts easily.
- These are followed by suggestions for the structure of the lesson:
    - Starter ideas involve the whole class to capture the attention and interest of students.
    - Main lesson activities help you lead students into exercise questions.
    - Plenaries offer guidance on how to round off the three-part lessons.

Planning a course

The Student's Book has topics arranged in four sections: Number; Algebra; Geometry and Trigonometry; Statistics and Probability. The chapters within each section can be taught in order. However, the structure gives flexibility to put together a course that best suits the needs of your students or timetabling by choosing one or more chapters from each section in turn to provide a sequence for teaching.

1.1 Square numbers and cube numbers

Syllabus references

Statement	Core
C1.1/E1.1	Identify and use: square numbers, cube numbers.

Collins references

- Student's Book pages 8–12

Learning aims

- Find and recognise square numbers and square roots.
- Find and recognise cube numbers.

Key words

- square number, square root, cube number

Prior knowledge

Students will need to know the multiplication tables to 10 × 10.

Common mistakes and remediation

A common mistake is to multiply the number by 2, rather than by itself when finding its square. Similarly, one should not multiply by 3 when finding the cube of a number, or divide by 2 when finding the square root.

Useful tips

Encourage students to draw some squares of dots to reinforce that square numbers can be arranged to for a square pattern.

Starter

- Show a 10 × 10 number square on the board (or give each student a paper copy of one).
- Draw a ring around the leading diagonal, enclosing the numbers 1, 4, 9, …, 100.
- Ask students what is special about those numbers. (They are the square numbers.)
- Now draw a square around 1 and around 1, 2, 2, 4 and around 1, 2, 3, 2, 4, 6, 3, 6, 9. Ask students to add up the numbers in each square (1, 9, 16) and then continue drawing the squares. Ask them what they notice. (You get the square numbers 1, 9, 36, 100, 225, …)
- Now consider the back-to-front L shapes formed by drawing round first of all just 1, then 2, 4, 2; 3, 6, 9, 6, 3; 4, 8, 12, 16, 12, 8, 4. Ask students to add those up (8, 27, 64). These are the cube numbers.

Main lesson activity

- Remind students that square numbers are formed by multiplying a number by itself. They should write $3^2 = 3 \times 3 = 9$.
- Also remind them that the square root of a number is the inverse of the square of a number. It is equivalent to the number of dots on one side of a square pattern of dots. They should write $\sqrt{9} = 3$.
- Make sure students can use their calculators to find squares and square roots.
- Students should now work through Exercise 1A.
- Students might have met square numbers before, but not cube numbers, so explain that to find the cube of a number you multiply (not add) the number by itself and then by itself again.
- Ask students to find a number that is both a square and a cube (1, 64, 729, ...).
- They can then do Exercises 1B and 1C.

Plenary

- Write the number 99 066 on the board. Ask the students to square it on their calculators (it is 9 814 072 356) and say what they notice (it has rotational symmetry and uses all the digits from 0 to 1 exactly once).

Homework and answers

Homework for this lesson is available in the homework folder. Consolidation and extension exercises are available in the further practice folder. Answers to the exercises can be found in the answers folder.

1.2 Multiples of whole numbers

Syllabus references

Statement	Core
C1.1/E1.1	Identify and use: natural numbers, integers, common multiples.

Collins references

- Student's Book pages 12–13

Learning aims

- Find multiples of whole numbers.
- Recognise multiples of numbers.

Key words

- multiple

Prior knowledge

Students will need to know the multiplication tables to 10 × 10.

Common mistakes and remediation

A common mistake is to fail to interpret the functional aspect of the topic correctly, e.g. if there are 10 people getting in taxis each holding four people, students may give answers of two and a half taxis required or two taxis with people left behind. Explain to students that they need to think carefully about their answers.

Useful tips

Encourage students to remember some simple multiplication rules: multiples of 2 always end in 0, 2, 4, 6, or 8; multiples of 3 have digits that add up to a multiple of 3; multiples of 5 always end in 0 or 5; multiples of 9 have digits that add up to a multiple of 9; and multiples of 10 always end in 0.

Starter

- Ask students, around the class, to take turns to increase an amount by a given number, e.g. the first person may say '6', the next '12', the next '18', and so on. Continue until someone makes a mistake.
- Go beyond the 10 × 10 multiplication tables for students who need more of a **challenge**.
- Move on to counting down, e.g. start at 60 and count down in sixes.
- Draw 'multipillars' of different tables. These also make good displays of the multiples in the tables.

Main lesson activity

- Remind students that multiples are the answers that appear in multiplication or times tables.
- Encourage students to look at the numbers in various multiplication tables and try to identify patterns or rules.
- For the 10 times table, the answer always ends in 0, numbers in the five times table always end in either 0 or 5, numbers in the two times table are always even. The digits in multiples of 3 always add up to a multiple of 3, the digits in multiples of 9 always add up to a multiple of 9. Multiples of 6 are always even numbers and the digits add up to a multiple of 3. Multiples of 4 are even when divided by 2.
- Make sure that students know how to use a calculator to find multiples; e.g. for multiples of 5 press 5 then = then +5 then repeatedly press =, and the multiples will be displayed.
- Now display this table of numbers and, as students give answers, write the answers on the board.

34	48	55	102
123	470	501	630
876	991	989	1000

- Ask students to give you the multiples of 2. (34, 48, 102, 470, 630, 876, 1000)
- Ask students to give you the multiples of 3. (48, 102, 123, 501, 630, 876)
- Ask students to pick out the multiples of 6 from the lists above. (48, 102, 630, 876)
- Ask students to give you the multiples of 5. (55, 470, 630, 1000)
- Ask students to give you the multiples of 10. (470, 630, 1000). Point out they could use the multiples of 2 and 5.
- Repeat for multiples of 4 and 9.
- Students who need more of a **challenge** could identify the multiple of 18 as multiples of both 2 and 9 (630).
- They can then do Exercise 1D.

Plenary

- Write the number 392 on the board. Ask the students which of the numbers, from one to ten, this number is a multiple of. Clearly 2 is one answer. Half of 392 is 146, so 4 is also a multiple, but what about other numbers? 3 and 9 are not answers as 3 + 9 + 2 = 14, which is not a multiple of 3. Also 5 and 10 are not answers, as the number does not end in 0 or 5.
- Use calculators to test whether 7 is a multiple (it is).
- Repeat with further examples, such as 640 or 720.
- Use smaller numbers within the 10 × 10 multiplication table for students who need more **support**, e.g. 48, 63 and 72.

Homework and answers

Homework for this lesson is available in the homework folder. Consolidation and extension exercises are available in the further practice folder. Answers to the exercises can be found in the answers folder.

1.3 Factors of whole numbers

Syllabus references

Statement	Core
C1.1/E1.1	Identify and use: common factors and common multiples.

Collins references

- Student's Book pages 13–15

Learning aims

- Identify the factors of a number.
- Identify common factors and common multiples.

Key words

- factor, factor pair, lowest common multiple, highest common factor

Prior knowledge

Students will need to know the multiplication tables to 10 × 10.

Common mistakes and remediation

When listing factors of a number, students often miss out 1 and the number itself. Remind them always to begin with 1. Many students will mix up the terms *multiple* and *factor*. Relate the word *multiple* to multiplications.

Useful tips

Remind students that factors of any number always include 1 and the number itself.

Starter

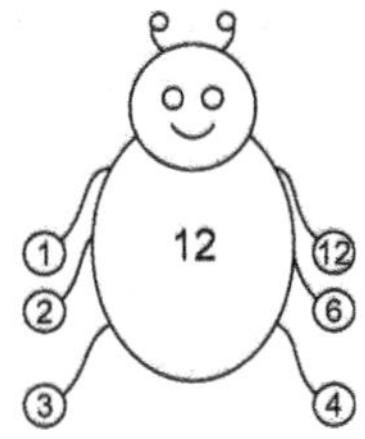

- Ask students to draw rectangles, each with an area of 12 cm^2. Look at the different possibilities and list the lengths of the sides. Repeat with other areas.
- Show students how to draw factor bugs and let them complete their own bugs for different numbers. Check that students have always included 1 and the number itself as factors.
- Now ask students to find the number between 20 and 30 that has eight factors. (24 → 1, 2, 3, 4, 6, 8, 12 and 24)
- For students who need more of a **challenge**, ask further questions, e.g. 'How many factors has 36?' (9 → 1, 2, 3, 4, 6, 9, 12, 18 and 36)

Main lesson activity

- Introduce the term 'factor pair'.
- Make sure that students realise that factors always come in pairs, so when they have found one factor they can find the other. The exception to this is when they are finding factors of square numbers, where one factor pair is the same number repeated.
- Draw out the fact that every number has 1 and itself as a factor pair.
- Explain to students that, in order not to miss any factors, it is sensible to start with 1 and work upwards, obtaining all the factor pairs. Using this method, when they have tried all the numbers up to half the number itself, students will have found all the factors.
- Work through Example 1.
- Ask students to make a table, as shown.

1 factor	2 factors	3 factors	4 factors	5 factors	6 factors

- Under each column, enter numbers from 1 to 20, e.g. 10 has four factors so is entered in that column.

1 factor	2 factors	3 factors	4 factors	5 factors	6 factors
1	2, 3, 5, 7, 11, 13, 17, 19	4, 9	6, 8, 10, 14, 15	16	12, 18, 20

- Students can now do Exercise 1E.

  When they are using a calculator to find factors, encourage students who need more **support** particularly to work systematically.
- Introduce the terms 'lowest common multiple (LCM)' and 'highest common factor (HCF)'.
- Work through Examples 2 and 3.
- Students can now do Exercise 1F.

Plenary

- Give students a number, e.g. 48. Ask them to write it as a multiplication, e.g. 6 × 8.
  Carry on breaking down the separate values as multiplications (do not allow 1×), e.g. 2 × 3 × 8, then 2 × 3 × 2 × 4, then 2 × 3 × 2 × 2 × 2.
- Ask students how they can use 2 × 3 × 2 × 2 × 2 to obtain factors, e.g. 2 × 2 × 2 × 2 = 16 so the factor pair is 3 and 16.
- Repeat with 20 (2 × 2 × 5), 60 (2 × 3 × 2 × 5), 100 (2 × 2 × 5 × 5), etc.
- Finally, write the product 1 × 2 × 3 × 5 × 7 = 210 on the board and ask students to give all the factors. (1, 2, 3, 5, 6, 7, 10, 14, 15, 21, 30, 35, 42, 70, 105 and 210)

Homework and answers

Homework for this lesson is available in the homework folder. Consolidation and extension exercises are available in the further practice folder. Answers to the exercises can be found in the answers folder.

1.4 Prime numbers

Syllabus references

Statement	Core
C1.1/E1.1	Identify and use: prime numbers.

Collins references

- Student's Book pages 15–16

Learning aims

- Identify prime numbers.

Key words

- prime number

Prior knowledge

Students will need to know the multiplication tables to 10 × 10 and the prime numbers up to 50.

Common mistakes and remediation

Sometimes students confuse prime numbers with odd numbers.
They frequently also include 1 and forget 2.
Regular practice at reciting the first few prime numbers should help to avoid these errors.

Useful tips

Stress that 1 is not defined as a prime number as it only has one factor.

Starter

- Ask students to give numbers that have only two factors.
- Explain that any number with exactly two factors, itself and 1 only, is called a prime number.
- Tell students that there are 25 prime numbers under 100.

Main lesson activity

- Tell students that they are now going to do an activity that will result in them finding all the prime numbers less than 100.
- Ask students to draw a 100 number square as shown, and to put a cross through the number 1.
- Draw a circle around the number 2 and then cross out all the multiples of 2.
- Draw a circle around the next number that hasn't been crossed out (in this case 3) and cross out all the multiples of that number.

~~1~~	(2)	(3)	~~4~~	~~5~~	~~6~~	~~7~~	~~8~~	~~9~~	~~10~~
11	~~12~~	13	~~14~~	~~15~~	~~16~~	17	~~18~~	19	~~20~~
~~21~~	~~22~~	23	~~24~~	25	~~26~~	~~27~~	~~28~~	29	~~30~~
31	~~32~~	~~33~~	~~34~~	35	~~36~~	37	~~38~~	~~39~~	~~40~~
41	~~42~~	43	~~44~~	~~45~~	~~46~~	47	~~48~~	49	~~50~~
~~51~~	~~52~~	53	~~54~~	55	~~56~~	~~57~~	~~58~~	59	~~60~~
61	~~62~~	~~63~~	~~64~~	~~65~~	~~66~~	67	~~68~~	~~69~~	~~70~~
71	~~72~~	73	~~74~~	~~75~~	~~76~~	77	~~78~~	79	~~80~~
~~81~~	~~82~~	83	~~84~~	85	~~86~~	~~87~~	~~88~~	89	~~90~~
91	~~92~~	~~93~~	~~94~~	~~95~~	~~96~~	97	~~98~~	~~99~~	~~100~~

- Repeat the process until all the numbers in the table are either circled or crossed out.

  The numbers circled are the prime numbers. (2, 3, 5, 7, 11, 13, 17, 19, 23, 29, 31, 37, 41, 43, 47, 53, 59, 61, 67, 71, 73, 79, 83, 89, 97)

- Students who need more of a **challenge** can go beyond 100 to see how many primes they can find. (Primes up to 200: 101, 103, 107, 109, 113, 127, 131, 137, 139, 149, 151, 157, 163, 167, 173, 179, 181, 191, 193, 197, 199)
- Explain that to show that a number is not prime, all students need to do is find a factor that is not the number itself (1 is a factor of all numbers). For example, 432 is not prime as it is even. The only even prime number is 2.
- Students who need more **support** may need to be reminded of the rules for divisibility by 2, 3 and 5.
- Students can now complete Exercise 1G.

Plenary

- Check students' understanding by playing 'True or false'. Say a number and ask students to put their thumbs up if it is prime and down if it is not prime.
- Ask students to use whiteboards to write down a factor of the number if it is not prime; e.g. for 49, the student has to display 7. Gradually increase the numbers to make the plenary more challenging for the students.

Homework and answers

Homework for this lesson is available in the homework folder. Consolidation and extension exercises are available in the further practice folder. Answers to the exercises can be found in the answers folder.

1.5 Prime factorisation

Syllabus references

Statement	Core
C1.1/E1.1	Identify and use: common factors.

Collins references

- Student's Book pages 16–18

Learning aims

- Express any number as a product of its prime factors.

Key words

- product of prime factors, index (indices), prime factorisation

Prior knowledge

Students will need to know what a prime number is and what a factor is.

Common mistakes and remediation

Students will use factors and not always reduce these to prime number factors.

Some pupils will need reminding that 1 is not a prime number and so it is not used in prime factorisations.

Useful tips

Students could check their answers with a calculator and multiply the factorisation out to check they get back to the original number.

Starter

- Put on the board $2^3 \times 5$.
- Ask, 'What number does this represent?'
- You may need to remind some students that 2^3 means $2 \times 2 \times 2 = 8$, so the answer here is $8 \times 5 = 40$.
- Put on the board $3^2 \times 7$.
- Ask, 'What number does this represent?' (63)
- Ask several more similar questions, e.g. $5^2 \times 3$ (75), 2×7^2 (98), $5^3 \times 3$ (375).

Main lesson activity

- Put 8×1 on the board. Ask students to give as many other multiplications with the answer 8 as they can.
- Put their responses on the board, e.g. 1×8, 4×2, 2×4.
- Continue until they give $2 \times 2 \times 2$. This is the *prime factorisation* of 8.
- Then put the number 18 on the board and ask similar questions, leading eventually to the product $2 \times 3 \times 3$. Remind the students that this prime factorisation of 18 can be written as 2×3^2.
- Highlight the two prime factorisations shown and ask what is special about the numbers used in these factorisations. Students should identify them as prime numbers.

- Explain that these are the *prime factors* of 8 and 18.
- Put 12 on the board and ask for the prime factorisation of 12. ($2 \times 2 \times 3$ or $2^2 \times 3$)
- Put 100 on the board and ask for its prime factorisation.
- If you get the answer $2 \times 2 \times 5 \times 5$, or $2^2 \times 5^2$ ask students how they found it.
- Demonstrate the process of breaking 100 into any two factors first, e.g. 4×25, then breaking each of those factors down to achieve $2 \times 2 \times 5 \times 5$.
- Now put 180 on the board and ask for the prime factorisation of this number. If necessary, suggest breaking it down first as, say, 10×18 and then from there to $2 \times 5 \times 2 \times 3 \times 3$. Show how to rearrange this as $2 \times 2 \times 3 \times 3 \times 5$ and $2^2 \times 3^2 \times 5$, which lists the prime factors in numerical order.
- Demonstrate how to use the ladder division layout to achieve this result, dividing by prime numbers starting with 2, then 3, then 5, etc. as in Example 4 in the Student's Book. This gives the factors in ascending numerical order.
- Encourage students who need more **support** to use the factor tree method to find factors, writing the number at the top and splitting it into two branches, to show two factors. Continue to add branches, splitting every non-prime factor, until there are only prime numbers at the ends of the branches. The prime factorisation can be listed from this diagram.
- Students can now do Exercise 1H.

Plenary

- Ask students to answer these questions using mental arithmetic, without pen and paper or a calculator.
- What is the prime factorisation of 36? ($2^2 \times 3^2$)
- What is the prime factorisation of 100? ($2^2 \times 5^2$)
- What is the prime factorisation of 45? ($3^2 \times 5$)
- Discuss looking for square numbers as factors.

Homework and answers

Homework for this lesson is available in the homework folder. Consolidation and extension exercises are available in the further practice folder. Answers to the exercises can be found in the answers folder.

1.6 Finding the HCF and LCM

Syllabus references

Statement	Core
C1.1/E1.1	Identify and use: common factors, common multiples.

Collins references

- Student's Book pages 18–19

Learning aims

- To use prime factorisation to find the LCM or the HCF of two numbers.

Key words

- natural number, integer, real number, rational number, irrational number

Prior knowledge

Students will need to be able to express a number as the product of prime factors.

Common mistakes and remediation

Students may confuse the terms HCF and LCM, not being sure which is which. Always use the full term alongside the abbreviation in class explanations.

Useful tips

Ensure that students write the prime factors of each number clearly, to enable them to compare the numbers.

Starter

- Ask pupils for the highest common factor (HCF) of 24 and 32. (8)
- Ask them how they calculated this. Discuss looking for factors of one number, say 24, and trying them in the other number, 32. They should find that 8 is the largest number that divides exactly into both numbers, so it is the HCF.
- Now ask for the HCF of 36 and 45. (9)
- Ask for the HCF of more pairs of numbers, such as 48 and 60 (12), 45 and 60 (15).
- Now ask for the lowest common multiple (LCM) of 6 and 8. (24)
- Ask how they calculated this.
- Encourage students to look at the times tables of 6 and of 8 to find the smallest multiple that appears in both tables.
- Repeat for 12 and 10 (60), 9 and 6 (36).

Main lesson activity

- Put the number 810 on the board.
- Ask, 'What is this as a product of primes? How can you find out?'
- Write students' answers on the board, as they suggest them, showing the breakdown into prime factors. Write the final answer as $2 \times 3 \times 3 \times 3 \times 3 \times 5$.
- Ask students for the simplest way to write this down ($2 \times 3^4 \times 5$).

- Then put 252 on the board and ask students to work this out as a product of prime factors.
- Again, as they answer, produce the prime factor breakdown $2 \times 2 \times 3 \times 3 \times 7$.
- Ask students for the simplest way to write this down ($2^2 \times 3^2 \times 7$).
- Now ask for the HCF of 810 and 252.
- Explain that they can use prime factors to find this.
- Indicate the two prime factorisations:
  $810 = 2 \times 3^4 \times 52$ $\quad$ $252 = 2^2 \times 3^2 \times 7$
- Lead students to look for the common factors of each number.
  Looking at the 2s $\quad$ there is only one 2 in each number.
  Looking at the 3s $\quad$ 3×3 occurs in each number.
- There are no other common factors, so the HCF is $2 \times 3 \times 3$ or $2 \times 3^2 = 18$.
- Now ask for the LCM of 810 and 252.
- Look again at the two prime factorisations.
- Explain that the LCM will include the highest power of each prime factor; so it will include:
  two 2s (the largest power is 2^2) $\quad$ four 3s (the largest power is 3^4) $\quad$ one 5 $\quad$ one 7.
- This gives the LCM as $2^2 \times 3^4 \times 5 \times 7 = 4 \times 81 \times 5 \times 7 = 11\,340$.
- Put 63 and 675 on the board and ask for their HCF and LCM.
- Students should suggest finding the prime factorisation of each number.
  $63 \rightarrow 3^2 \times 7$ $\quad$ $675 \rightarrow 3^3 \times 5$
- Show that the HCF is 3^2 as this is the only factor common to both.
- Show that the LCM is $3^3 \times 5 \times 7 = 945$.
- The students can now do Exercise 1I.

Plenary

- Ask students to explain why it is helpful to use prime factorisation to find the HCF and LCM of two numbers.
- Allow the discussion to use the size of the numbers to decide if that's the easiest method or just use number knowledge.

Homework and answers

Homework for this lesson is available in the homework folder. Consolidation and extension exercises are available in the further practice folder. Answers to the exercises can be found in the answers folder.

1.7 Rational and irrational numbers

Syllabus references

Statement	Core
C1.1/E1.1	Identify and use: natural numbers, rational and irrational numbers, reciprocals.

Collins references

- Student's Book pages 20–21

Learning aims

- To distinguish between rational and irrational numbers.

Key words

- natural number, integer, rational number, irrational number, reciprocal

Prior knowledge

Students will need to know how to use a calculator to convert fractions to decimals.

Common mistakes and remediation

Students often think that less familiar decimals such as 0.384 615 38... must be irrational. Remind students to look carefully to see if there is a repeating group of digits.

Useful tips

When they are converting fractions to decimals, encourage students to write down the full calculator display or they may fail to notice a group of recurring digits.

Starter

- Discuss with students how they can use a calculator to convert fractions to decimals.
- Ask them to convert these fractions to decimals: $\frac{1}{2}, \frac{4}{5}, \frac{1}{3}, \frac{2}{9}, \frac{4}{7}$.

Main lesson activity

- Ensure that students understand the definitions of 'natural numbers' and 'integers' as explained in the Student's Book.
- Now ask them to copy the table on the left below and complete it, ensuring they understand the concepts of a 'terminating decimal' and a 'recurring decimal'. The answers are shown below right.

Number	Decimal	Description of decimal
$\frac{1}{2}$	0.5	terminates
$\sqrt{2}$		
$\frac{1}{3}$		
$\frac{2}{9}$		
$\frac{4}{7}$		
π		
$-\frac{4}{5}$		

Number	Decimal	Description of decimal
$\frac{1}{2}$	0.5	terminates
$\sqrt{2}$	1.414213562...	doesn't terminate or recur
$\frac{1}{3}$	0.333333...	recurring digit
$\frac{2}{9}$	0.222222...	recurring digit
$\frac{4}{7}$	0.57142857142857...	recurring group of digits
π	3.141592653589793	doesn't terminate or recur
$-\frac{4}{5}$	–0.8	terminates

- Now group the numbers accordingly:
  - Terminates $\frac{1}{2}, -\frac{4}{5}$
  - Neither terminates or recurs $\sqrt{2}$, π
  - Recurs $\frac{1}{3}, \frac{2}{9}, \frac{4}{7}$
- Explain the distinction between *rational* and *irrational* numbers.
- Rational numbers are:
  - Numbers with decimal parts that terminate.
  - Numbers with decimal parts that go on forever but have a recurring pattern.
- Irrational numbers are:
  - Numbers with decimal parts that go on forever but have no repeating patterns.
  - Numbers that cannot be written as fractions.
- Students should now be able to identify the numbers in their tables as rational or irrational.
  - Rational $\frac{1}{2}, -\frac{4}{5}, \frac{1}{3}, \frac{2}{9}, \frac{4}{7}$
  - Irrational $\sqrt{2}$, π
- Stress that if a number is a natural number that is not a square, then the square root of that number is irrational.
- Explain what a reciprocal is (the number obtained by dividing 1 by that number, often referred to as the multiplicative inverse).
- Also show them how to use the reciprocal button (often shown as 1/n on the calculator).
- Ask them to find the reciprocals of these numbers: 2, $\frac{1}{2}$, 0.1, 100, 1.618.
- Ask them what they notice about the product of a number and its reciprocal.
- Students can now do Exercise 1J. For questions 6, 7, 8 and 9, students can work in pairs and see whether they agree with their partners' answers.

Plenary

Put some numbers on the board, and ask students to say what kind of number it is – a natural number, an integer, a rational number or an irrational number.

0.6	(rational)
$\frac{1}{4}$	(rational)
2π	(irrational)
–13	(rational)
$\sqrt{5}$	(irrational)
$\sqrt{36}$	(only consider the positive square root – rational, integer, natural number)

Homework and answers

Homework for this lesson is available in the homework folder. Consolidation and extension exercises are available in the further practice folder. Answers to the exercises can be found in the answers folder.

2.1 Equivalent fractions

Syllabus references

Statement	Core
C1.4/E1.4	Use the language and notation of the following in appropriate contexts: proper fractions, improper fractions, mixed numbers.

Collins references

- Student's Book pages 24–25

Learning aims

- Create equivalent fractions.
- Simplify fractions by cancelling.
- Change improper fractions into mixed numbers and mixed numbers into improper fractions.

Key words

- numerator, denominator, cancel, lowest terms, simplest form, proper fraction, improper fraction, top-heavy fraction, mixed number

Prior knowledge

Students must be able to recognise equivalent fractions from diagrams.

Common mistakes and remediation

A common error is that students may not leave final answers in the simplest form. Repeated practice and reminders should overcome this error. Also, students may find it much more difficult to change mixed numbers into improper fractions than vice versa. Emphasise that the type of fraction remains the same in the mixed number and the improper fraction (e.g. in $3\frac{3}{4}$, the 3 and the $\frac{1}{4}$ are both in terms of quarters).

Useful tips

Using a multiplication grid may help students to tackle the questions in this lesson.

Starter

- Ask students to give fractions that are the same size as one-half. Repeat, asking for fractions that are the same size as one-quarter.
- Now give them a standard 10 × 10 multiplication grid to complete. They will need this in the main lesson activity.

Main lesson activity

Equivalent fractions

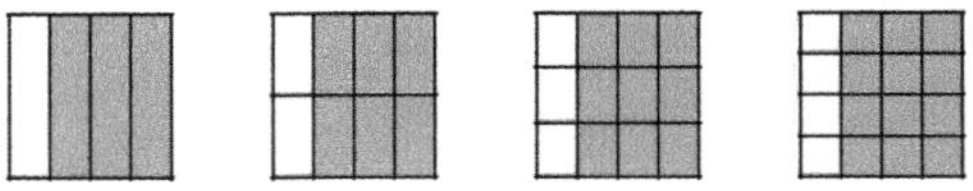

- Ask students to think about the fraction $\frac{3}{4}$. Ask them to colour in the 3 and 4 times rows in their standard multiplication grid.
- Then show them a diagram of a shape with 3 out of 4 quarter strips shaded.

- Use diagrams to demonstrate that the pairs of numbers in each vertical column are equivalent to $\frac{3}{4}$, so $\frac{3}{4}=\frac{6}{8}=\frac{9}{12}=\frac{12}{16}$ and so on.
- Explain that in each case they are multiplying the numerator and denominator by the same number to create equivalent fractions. Use Example 1 to illustrate this.

Simplifying fractions by cancelling

- Now explain that, working backwards, the numerator and denominator could be divided by the same number and that this is called 'simplifying' or 'cancelling'. Make it clear that this leaves a fraction that is in its 'lowest terms'.
- Write $\frac{24}{32}$ on the board. Ask students for numbers that divide into both 24 and 32.
- Students who need more of a **challenge** will tend to spot 8 straight away while others will usually go for division by 2. Show both methods and encourage students always to look for the highest common factor, to save time.

Converting improper fractions to mixed numbers

13	17	11	7
2	3	5	19

- Display this number board. Ask students what is special about the numbers in this board (all prime numbers).
- Ask students to give an improper fraction using these numbers (e.g. $\frac{13}{2}$) and to explain what it means (e.g. 13 halves).
- Ask them how many whole ones 12 halves is equal to (6). Explain that this leaves one-half extra so $\frac{13}{2}=6\frac{1}{2}$. This is a mixed number (a mixture of a whole number and a proper fraction). Repeat for other numbers in the table.
- Talk students through Example 2.

Converting mixed numbers to improper fractions

- Explain that converting mixed numbers to improper fractions is the reverse of the process they have just used.
- Write the mixed number $3\frac{1}{4}$ on the board. Ask students to say how many quarters there are in one (4). Then ask how many are in, for example, 3 whole ones (12).
- Show that $3=\frac{12}{4}$ and $\frac{1}{4}$ extra makes $\frac{13}{4}$. So, $3\frac{1}{4}=\frac{13}{4}$
- Now show students how to simplify fractions, using the fraction button on a scientific calculator – they should input their fraction and press the equals sign. They can use this method for checking answers.
- Students can then do Exercise 2A. Students who need more **support** may initially need to draw diagrams to help them understand the conversions.

Plenary

- Discuss different strategies for answering questions 5, 9 and 10. Students who need more of a **challenge** could attempt these questions individually and then discuss in pairs how they approached each question.

Homework and answers

Homework for this lesson is available in the homework folder. Consolidation and extension exercises are available in the further practice folder. Answers to the exercises can be found in the answers folder.

2.2 Fractions and decimals

Syllabus references

Statement	Core
C1.4/E1.4	Use the language and notation of the following in appropriate contexts: fractions and decimals. Recognise equivalence and convert between these forms.

Collins references

- Student's Book pages 26–27

Learning aims

- Change a decimal number to a fraction.
- Change a fraction to a decimal.

Key words

- decimal, fraction

Prior knowledge

Students need to know how to cancel fractions.

Common mistakes and remediation

When adding fractions, students often add the denominators as well as the numerators. Shading fractions of a shape should help students avoid this error.
Reinforce that a larger denominator does not signify a larger number.

Point out that $\frac{1}{4}$ of 24 = 6, whereas $\frac{1}{3}$ of 24 = 8.

Useful tips

Encourage students always to use a horizontal line when writing fractions so that it is very clear which number is the denominator.

Starter

- Give students the target card from Resource sheet 2.2 and ask them to match the fractions with their decimal equivalents.
- Write various unsimplified fractions on the board and ask students to rewrite each in its simplest form.

Main lesson activity

Changing a decimal into a fraction

- Write the column headings on the board as shown.

Units	.	Tenths	Hundredths	Thousandths
	.			

- Now fill in some decimals, e.g. 0.3, 0.7 and 0.8.

Units	.	Tenths	Hundredths	Thousandths
0	.	3		
0	.	7		
0	.	8		

- Tell students that this shows that $0.3 = \frac{3}{10}$, and so on.
- Repeat for hundredths, for example 0.12, 0.06, 0.97.

Units	.	Tenths	Hundredths	Thousandths
0	.	1	2	
0	.	0	6	
0	.	9	7	

- Tell students that this shows that $0.12 = \frac{12}{100}$, $0.06 = \frac{6}{100}$, and so on.
- Repeat for thousandths, for example 0.345, 0.251, 0.093.

Units	.	Tenths	Hundredths	Thousandths
0	.	3	4	5
0	.	2	5	1
0	.	0	9	3

- Tell students that this shows that $0.345 = \frac{345}{1000}$, and so on.
- Show students Example 3.
- Ask students who need more of a **challenge** for examples of decimal numbers and their equivalent fractions in tens of thousands.

Changing a fraction into a decimal

- Show students a fraction, e.g. $\frac{2}{5}$, and a divide sign ÷. Explain how the dots in the divide sign represent the numerator and the denominator. So $\frac{2}{5}$ means 2 divided by 5, which is equal to 0.4.
- Tell students that this is all they need to do to change a fraction into a decimal.
- Give more examples, e.g. $\frac{1}{2} = 1 \div 2 = 0.5$, $\frac{3}{10} = 3 \div 10 = 0.3$.
- Show students Example 4.
- Explain that changing fractions to decimals is useful for comparing the size of fractions. This will provide a good hint for question 3 of Exercise 2B.
- Students can now do Exercise 2B.

Plenary

- Use the plenary to recap the methods for the exercise covered, for example asking: 'How do I change a decimal into a fraction?'

Homework and answers

Homework for this lesson is available in the homework folder. Consolidation and extension exercises are available in the further practice folder. Answers to the exercises can be found in the answers folder.

2.3 Recurring decimals

Syllabus references

Statement	Extended
E1.4	Recognise equivalence and convert between recurring decimals and fractions and vice versa.

Collins references

- Student's Book pages 28–30

Learning aims

- Express any fraction as a decimal and change recurring decimals to fractions.

Key words

- terminating decimal, recurring decimal

Prior knowledge

Students will need to know how to change a fraction into a decimal.

Common mistakes and remediation

Students will often fail to use the recurring notation correctly. Students will often write a recurring decimal such as $0.\dot{1}5\dot{2}$ as 0.152… which would mean 0.152 222 222 2…

Useful tips

Encourage students to check their recurring decimal notation, using the correct number of dots over the recurring digits or groups of digits.

Starter

- Ask the students for the decimal equivalents of some common fractions such as: $\frac{1}{2}$ (0.5), $\frac{1}{4}$ (0.25), $\frac{3}{4}$ (0.75), $\frac{1}{5}$ (0.2), $\frac{3}{5}$ (0.6), $\frac{1}{8}$ (0.125), $\frac{3}{8}$ (0.375).
- Ask the students to tell you the fraction equivalents of decimals such as: 0.1 $\left(\frac{1}{10}\right)$, 0.7 $\left(\frac{7}{10}\right)$, 0.4 $\left(\frac{2}{5}\right)$, 0.11 $\left(\frac{11}{100}\right)$, 0.37 $\left(\frac{37}{100}\right)$.
- For students who need more **challenge**, these can all be treated mentally, with discussion. Students who need more **support** may need them to be written down, on the board.

Main lesson activity

- Put on the board the fraction $\frac{1}{8}$ and ask for its decimal equivalent (0.125).
- Explain that this is a *terminating decimal* because it terminates (finishes). This is an important term and students need to remember it. Students who need more **support** may need help in understanding the word.

- Put on the board the fraction $\frac{5}{8}$ and ask students how to find the decimal equivalent (5 ÷ 8).

  If necessary, remind them how to do this by short division.
- Now put on the board the fraction $\frac{5}{9}$ and ask students how to find the decimal equivalent (5 ÷ 9).

  Invite students to help complete this short division on the board, to see that 5 ÷ 9 = 0.555 555 55... Make sure that they realise it carries on forever.
- Explain that this is a *recurring decimal* because it does just that, 5 keeps on recurring and the calculation does not terminate.
- Show the methods for recording a recurring decimal, using the three-dot ellipsis (...) or a dot over the recurring digit. Demonstrate that if a decimal has a recurring group of digits, such as 0.545 454..., students should put a dot above the first and last digit of the recurring group, in this case the 5 and 4.
- If a fraction has recurring digits such as 0.174 174 174... then they would put a dot above the 1 and the 4 to show the recurring pattern.
- Students can now do Exercise 2C, questions 1 to 3.
- As the students who need more of a **challenge** reach questions 3c and 3d, show the class how to manipulate recurring decimals, for example, adding and subtracting:

  $\frac{1}{9} + \frac{2}{3} = 0.111... + 0.666...$

  Demonstrate the decimal addition on the board, using the conventional method of adding in columns to find the answer of 0.777.... Ask if students recognise this decimal. Students who need more of a **challenge** should have recognised the pattern for ninths and be able to identify 0.777... as $\frac{7}{9}$.
- Now put $0.2929\dot{2}\dot{9}$ on the board, pointing out the dots above the last 2 and the last 9.
- Demonstrate how to find the equivalent fraction for this decimal number.
- Write $f = 0.2929\dot{2}\dot{9}$, with the dots over the last 2 and the last 9.
- As there are two recurring digits, multiply f by 100 (two zeros) to get:

  $100f = 29.2929\dot{2}\dot{9}$ (dots above the last 2 and the last 9).
- Show that subtracting f from $100f$, in columns, gives $99f = 29$, with no decimal fraction.

  Hence $f = \frac{29}{99}$ or $0.2929\dot{2}\dot{9}$ is equivalent to the fraction $\frac{29}{99}$.
- Students can now complete Exercise 2C.

Plenary

- Ask students to give the decimal equivalence of $\frac{1}{9}$ (0.1111...), $\frac{2}{9}$ (0.2222...), etc.
- Ask if anyone remembers $\frac{1}{11}$ (from Q10a), which is $0.09\dot{0}\dot{9}$, then $\frac{2}{11}$ ($0.18\dot{1}\dot{8}$), then $\frac{3}{11}$ ($0.27\dot{2}\dot{7}$).

  Ask students to identify the pattern in the recurring digits. (Multiples of 9)
- Ask students to predict the equivalent decimal for $\frac{9}{11}$ ($0.81\dot{8}\dot{1}$) and to explain how they know. (8 is one less than the 9, and 1 is required to be added to the 8 to make 9, giving 81, which recurs.)
- If there is time, continue with the elevenths. Explain that maths is filled with such extraordinary patterns.

Homework and answers

Homework for this lesson is available in the homework folder. Consolidation and extension exercises are available in the further practice folder. Answers to the exercises can be found in the answers folder.

2.4 Converting percentages, fractions and decimals

Syllabus references

Statement	Core
C1.4/E1.4	Use the language and notation of the following in appropriate contexts: proper fractions, improper fractions, mixed numbers, decimals, percentages. Recognise equivalence and convert between these forms.

Collins references

- Student's Book pages 30–33

Learning aims

- Convert percentages to fractions and decimals and vice versa.

Key words

- percentage, decimal equivalents

Prior knowledge

Students should know how to cancel fractions and how to calculate with fractions. They must be able to multiply and divide decimals by 100. Students should know some commonly used equivalent fractions, decimals and percentages.

Common mistakes and remediation

Most of the mistakes occur when dealing with fractions, e.g. writing as 34% or $\frac{1}{3}$ as 0.3.

Students should be encouraged to learn as facts the common equivalences.

Useful tips

Reinforce common equivalent fractions, decimals and percentages, e.g. $\frac{1}{3} = 0.33 = 33\frac{1}{3}\%$.

Starter

- Ask students to give situations where percentages are used (examples are test results, in shops, taxes, bank accounts, etc.). Their examples can be used to demonstrate the value of knowing equivalences. Students who need more **support** could be shown an easier example to make the point, e.g. 'Which is better, 30 out of 50 or 59 out of 100?' (30 out of 50 is equivalent to 60 out of 100.)
- Give students a copy of the equivalence sheet, Resource 2.4. Ask them to fill in the fraction, decimal and percentage equivalents that they know. This can be a useful tool for revision as these are the basic equivalences that students need.
- Students could add extra rows to the sheet, completing other equivalent fractions, decimals and percentages that they know. This can then be used to assess the prior knowledge of the students.

Main lesson activity

- Following on from the starter, ask students for their answers to the equivalence sheet. Write the answers for the extra rows on the board.

- Suggest ways of grouping the answers, e.g. $\frac{1}{10}$, $\frac{2}{10}$, $\frac{3}{10}$ form one group and 25%, 50%, 75%, 100% form a different group.
- Check that students understand that 'per cent' means 'out of 100'. Check also that students realise that, for example, 80% = 0.8 but 8% = 0.08.
- Some students may find it easier to break the work into sections, e.g. converting from percentages and then converting from decimals and finally from fractions. If so, as you explain each section, work through the appropriate parts of Examples 8, 9 and 10 and ask students to do the relevant questions from Exercise 2D.

Converting from percentages

- Remind students that 'per cent' means 'out of 100', so 52% = $\frac{52}{100}$ = 0.52,
- Repeat with other percentages, e.g. 8% = $\frac{8}{100}$ = 0.08, 100% = $\frac{100}{100}$ = 1.
- Now ask students for equivalences to check understanding.

Converting from decimals

- Use some examples that were used previously, e.g. 0.52. Ask for the fraction out of 100 $\left(\frac{52}{100}\right)$ and now ask for the percentage (52%).
- Do not talk about multiplying by 100, as this implies something that is 100 times bigger. Multiplying by 100% is multiplying by 1, so the value is still equivalent.

Converting from fractions

- Look at fractions with denominators which are factors of 100, e.g. $\frac{3}{10}$, $\frac{1}{25}$, $\frac{7}{50}$.
- Ask students to give the fractions out of 100 and then give the decimal and percentage equivalents ($\frac{30}{100}$, 0.3, 30%, $\frac{40}{100}$, 0.04, 4%, $\frac{14}{100}$, 0.14, 14%).
- Now explain that not all fractions can be converted this way, e.g. $\frac{5}{12}$, $\frac{1}{7}$.
- Remind students that $\frac{5}{12}$ means 5 ÷ 12 and $\frac{1}{7}$ means 1 ÷ 7 and that these can be changed to decimals using a calculator or a division method.
- Demonstrate $\frac{5}{12}$ and $\frac{1}{7}$ using a calculator: $\frac{5}{12}$ = 5 ÷ 12 = 0.41666... = 41.7%; $\frac{1}{7}$ = 1 ÷ 7 = 0.142857... = 14.3%. Explain that it is sensible to give percentages to 1 decimal place.
- Show students Examples 8, 9 and 10. Students can then do Exercise 2D.

Plenary

- Say a fraction and ask a student to give the equivalent percentage, and another to give the decimal. Repeat several times, sometimes starting with a decimal and sometimes with a percentage. Start with fractions that are easy to convert and gradually increase the difficulty. Point out any mistakes and explain any errors.

Homework and answers

Homework for this lesson is available in the homework folder. Consolidation and extension exercises are available in the further practice folder. Answers to the exercises can be found in the answers folder.

2.5 Calculating a percentage

Syllabus references

Statement	Core
C1.13/E1.13	Calculate a given percentage of a quantity.

Collins references

- Student's Book pages 34–35

Learning aims

- Calculate a percentage of a quantity.

Key words

- multiplier, quantity

Prior knowledge

Students need to understand that 'per cent' means 'out of 100' and 'of' means '×'.

Common mistakes and remediation

A common error is to write, for example, 8% as 0.8. Single digit percentages should be used frequently in starters to reinforce the correct answer (0.08).

Useful tips

When asking students to calculate a percentage of an amount, always ask them which keys they will press on their calculators to ensure they learn the method.

Starter

- Give students simple quick-fire questions. Students could display answers on small whiteboards. For example: 50% of 40 (20), 10% of 30 (3), 25% of 8 (2), 30% of 100 (30), 75% of 80 (60).
- For students who need more **support**, ensure that finding 50%, 25%, 75% and 10% of amounts has been fully understood before moving on to other percentages.
- Develop this by asking linked questions. For example: 50% of 30 (15), 25% of 30 (7.5); 10% of 60 (6), 5% of 60 (3), 15% of 60 (9); 10% of 20 (2), 20% of 20 (4), 30% of 20 (6), 70% of 20 (14).
- For students who need more of a **challenge**, ask them to find 1% of 30 (0.3) and then 51% of 30 (15.3). Ask for other percentages, such as 11% of 60 (6.6), 21% of 20 (4.2), 19% of 20 (3.8).

Main lesson activity

Using a build-up method

- Using the starter examples, ask students how they worked out each answer (50% is half, 25% is one-quarter, 10% is one-tenth, 5% is half of 10% and so on).
- Show students Example 11.

Using a percentage multiplier

- Write 13% of 45 on the board.
- Ask students to 'translate' this by changing the percentage to a decimal and changing the 'of' to a 'four rules' symbol. (0.13 × 45)
- Now ask a student to work this out on a calculator (5.85).
- Show students Example 12.
- Students will now need a calculator. Give questions for students to do on their calculator, e.g. 34% of 70 (23.8), 87% of 95 (82.65).
- Students can now do Exercise 2E. Highlight the fact that many of these questions consider 'real-life' situations to do with personal and household finance.
- Remind students to take note of the hint with Question 8.

Plenary

- Go round the class asking for the equivalent percentage multiplier for various percentages, e.g. 45% = 0.45, 62% = 0.62. Include some that often cause difficulty, e.g. 3% = 0.03 but is often given as 0.3.
- Now give the multiplier and ask for the percentage.
- Move on to multipliers greater than 1, e.g. 1.25 (125%). Explain that this represents a 25% increase and is covered in the next lesson.

Homework and answers

Homework for this lesson is available in the homework folder. Consolidation and extension exercises are available in the further practice folder. Answers to the exercises can be found in the answers folder.

2.6 Increasing or decreasing quantities by a percentage

Syllabus references

Statement	Core
C1.13/E1.13	Calculate percentage increase or decrease.

Collins references

- Student's Book pages 36–40

Learning aims

- Increase and decrease quantities by a percentage.

Key words

- multiplier

Prior knowledge

Students need to know how to use a percentage multiplier and how to work out a percentage of a quantity.

Common mistakes and remediation

When students are asked to increase or decrease a quantity by a percentage they will often misread the question and simply work out the percentage. For example, when asked to increase 30 by 5%, they will work out 5% of 30 but not add it on to get the answer. Encourage students to check that they have answered the question correctly.

Useful tips

It may be necessary, with students who need more **support**, to ask some quick-fire questions on subtracting from 100.

Starter

- Revise the use of percentage multipliers by asking students questions such as, 'What is two per cent as a multiplier?' (0.02). Include some of the more difficult questions such as: 'What is 0.8% as a multiplier?' (0.008). 'What is 3.4% as a multiplier?' (0.034).
- Ask the students questions about working out a percentage of a quantity to check prior knowledge, e.g. ask how they would work out 13% of 40.
- Lead on to questions involving increasing by a percentage. For example, ask how they would increase an amount by 13%. Students will usually say work out 13% and add it on.
- Repeat for other values.
- Move on to decreasing by a percentage. Ask students, for example, how they would work out 6% of 70. Then ask them how they would decrease 70 by 6%.

Main lesson activity

- Following on from the starter, display this table of multipliers.

1.05	0.05	0.95	1.5	0.5
0.85	0.15	1.15	0.25	1.25
0.75	0.55	0.45	1.55	1.45

- Ask students questions about the multipliers, starting with simple percentages, e.g. 'What percentage is equivalent to the multiplier 0.05?' (5%), '0.45?' (45%), '0.5?' (50%) and so on.
- Now ask what percentage is equivalent to the multiplier 1.25 (125%). Check if students remember from the previous plenary that this represents a 25% increase when used as a multiplier.
- Repeat for 1.5 (50% increase), 1.15 (15% increase), 1.05 (5% increase), 1.55 (55% increase).
- Show students Example 13, pointing out that they can use the 'adding on' method but that it is good to understand the 'multiplier' method also. Now show students Example 14.
- Students who need more **support** can do Exercise 2F at this stage. Students should choose the method to use, ensuring that if they choose to use the multiplier method, they need to be able to explain why it works.
- Now ask students to identify the multiplier from the table that represents a 5% decrease (0.95). Repeat for other decreases, e.g. 0.85 (15% decrease), 0.75 (25% decrease).
- Work through Examples 15 and 16.
- Students can now do Exercises 2F and 2G. Point out that the only difference between the exercises is that Exercise 2F deals with increases and Exercise 2G deals with decreases. Also, highlight the fact that many of these questions consider 'real-life' situations to do with personal and household finance.
- Remind students to take note of the advice with Question 12 in Exercise 2G.

Plenary

- Tell students that they have $100.
- Ask one half of the class to increase the $100 by 10% and then decrease the answer by 10%. ($99)
- Ask the other half to decrease the $100 by 10% and then increase the answer by 10%. ($99)
- Then ask all the students to show their answer on a whiteboard at the same time. Establish that the result is always $99 but ask students why it is not $100. (The decrease is from a larger amount each time.)

Homework and answers

Homework for this lesson is available in the homework folder. Consolidation and extension exercises are available in the further practice folder. Answers to the exercises can be found in the answers folder.

2.7 One quantity as a percentage of another

Syllabus references

Statement	Core
C1.12/E1.12	Express one quantity as a percentage of another.

Collins references

- Student's Book pages 40–42

Learning aims

- Express one quantity as a percentage of another.
- Work out percentage change.

Key words

- percentage change, percentage increase, percentage decrease, percentage profit, percentage loss

Prior knowledge

Students need to know how to use a percentage multiplier. They should also know how to write one quantity as a fraction of another quantity.

Common mistakes and remediation

Students frequently forget to make sure the units are the same. Encourage them always to look at units carefully. They also often do not use the original value as the denominator, but the new value after the increase or decrease. Frequently repeating the formula for percentage change should help students to learn it.

Useful tips

Students will mix up increasing a quantity by a percentage and working out a percentage increase. Be careful to use the correct language to describe each one.

Starter

- Ask mental calculation questions involving increasing or decreasing quantities by a percentage. For example:

Increase \$50 by 10% (\$55)	Increase 100 cm by 2% (102 cm)
Decrease 40 kg by 25% (30 kg)	Decrease 70 grams by 50% (35 grams)

- Write a table on the board, using the first three columns of the table in the main lesson activity below. Repeat for other values.
- Increase the level of difficulty for students who need more of a **challenge** by including decimals or fractions, e.g. increase \$18.50 by 10%. (Percentage change = \$1.85, new amount = \$20.35.)

Main lesson activity

- Talk about percentage change. Say that if the change is an increase, it is called a percentage increase. Discuss the terms percentage decrease, percentage profit and percentage loss.
- Now ask students to look at the table and see if they know how to get from $50 and $55 to 10%. Students who need more **support** may need an extra column showing the change. Ask them how to get from $5 and $50 to 10%. ($5 is 10% of $50.)
- Repeat for the other rows in the table.

Original amount	Percentage change	New amount	Change
$50	+10%	$55	$5
40 kg	–25%	30 kg	10 kg
100 cm	+2%	102 cm	2 cm
70 grams	–50%	35 grams	35 grams

- Say that to express a quantity as a percentage of another, you write the first quantity as a fraction of the second quantity and then convert to a percentage. Using Examples 17 and 18, explain that this is the same as working out a percentage change; the change in Example 17 is $6 and in Example 18 it is 75 cm.

  Write the formula on the board:

  $$\text{percentage change} = \frac{\text{change}}{\text{original amount}} \times 100$$

- Show students that the formula works for each row in the table, e.g. for $50 to $55:

  $$\text{percentage change} = \frac{5}{10} \times 100 = 10\%$$

- Go through Example 19. Students who want to use the multiplier method can also go through Example 20.
- Students can now do Exercise 2H. Highlight the fact that many of these questions consider 'real-life' situations to do with personal and household finance.

Plenary

- Ask the following questions.
  - Express 40 cents as a percentage of $4 – what would the obvious mistakes be?
  - Suni buys a painting for $1500 and sells it for $2000. What is her percentage profit?
  - If Suni had sold the painting for $1200 what would the percentage loss be?
- Now set a comparison question. Two students take a test scored out of 20. One scores 5 and the other 10. They retake it. Both improve their scores by 5. Who has made better progress? Discuss the answers. Some may say both made the same progress as both know the answer to five more questions. Point out that the first student had a percentage increase of 100%, and the second student had a percentage increase of 50%.

Homework and answers

Homework for this lesson is available in the homework folder. Consolidation and extension exercises are available in the further practice folder. Answers to the exercises can be found in the answers folder.

2.8 Simple interest and compound interest

Syllabus references

Statement	Core
C1.13/E1.13	Calculate with simple and compound interest.

Collins references

- Student's Book pages 43–44

Learning aims

- Calculate compound interest.
- Solve problems involving repeated percentage change.

Key words

- simple interest, compound interest, principal, annual rate

Prior knowledge

Students should be able to increase an amount by a percentage, preferably using the multiplier method.

Common mistakes and remediation

When doing the year-by-year method, premature rounding or incorrect rounding often leads to incorrect answers. When using the compound interest formula for depreciation, the students often forget that the '+ r %' becomes a '– r %'. Another common mistake is not reading the question properly – does the question ask for the total amount in the bank after n years, or for the total interest received after n years?

Encourage students, after producing an answer, to reread the question and check that their answer is sensible.

Useful tips

Students often think compound interest only applies to savings account problems. At every opportunity, relate questions to other topics, such as population growth or car depreciation. Encourage the use of a multiplier as the most efficient method. Students who need more **support** may prefer to do their working one step at a time.

Starter

- Give the class the following problem:

  Michelle invests €100 in a savings account, which earns her 5% interest per annum.

  How much does Michelle have in the account at the end of the first year? (€105)

  How much does Michelle have in the account at the end of the second year? (€110.25)

  How much does Michelle have in the account at the end of the third year? (€115.76)

- Discuss a quicker method to find out how much Michelle had at the end of the third year. (100×1.05^3)

- Students should check the calculation on their calculators.
- Now discuss the advantage of using the multiplier for an investment of 10 years instead of calculating each year separately. (There is just one calculation, so it is much quicker to work out.)

Main lesson activity

- Now check that students can do the calculations using a calculator since the method has already been established in the starter. Write on the board: $3000 investment, 6%, 5 years.
- Ask students to give the multiplier for the calculation (1.06). Now ask students for the calculation to work out the final value of the investment (3000×1.06^5).
- Ask students to do the calculation on their calculators ($4014.68). Ensure that they give the answer in correct money notation.
- Provide a few more calculations for students to work out on their calculators. Use different contexts. The population of a village is 10 000. Each year for three years it falls by 2% of the amount at the end of the year. What is the population after three years? (9411 or 9412)
- Students might prefer to understand the method rather than learn a formula.
- Ask students who need a **challenge** to put the method into a formula:

  $$\text{Final amount} = \text{Initial investment} \times \left(1+\frac{r}{100}\right)^n$$

  r = interest rate n = number of years

- Work through Example 21 and ask students to check that the answer is correct, using their calculators.
- Students can now do Exercise 2I. Highlight the fact that these questions consider 'real-life' situations to do with personal finance.
- If students who need more **support** find the exercise difficult, advise them to work with a partner for the more difficult questions. Remind them to use the multiplier method.

Plenary

- Set these problems.
  - Mr Cash invests $1 000 000 for 5 years at an annual compound interest rate of 5%. By how much does his money grow? ($276 281.56)
  - Miss Cash buys a car for $100 000 which depreciates by 5% per year. By how much has the car depreciated after 5 years? ($22 621.91)
  - Mrs Cash invests $1 000 000 at a rate of 5% per annum, compound interest. How many years will it take for her money to reach at least $1.5 million? (9 years)
- The next task is quite difficult and may only be suitable for the students who really need more of a **challenge**.
  - Ask students to try to work out what interest rate (nearest whole number) would be needed to double the value of an investment in 5 years (15%).

Homework and answers

Homework for this lesson is available in the homework folder. Consolidation and extension exercises are available in the further practice folder. Answers to the exercises can be found in the answers folder.

2.9 A formula for compound interest

Syllabus references

Statement	Core
C1.13/E1.13	Calculate with simple and compound interest.

Collins references

- Student's Book pages 45–46

Learning aims

- To learn the compound interest formula and how to use it.

Key words

- formula

Prior knowledge

Students will need to know what compound interest is.

Students need to be familiar with the use of the power keys on their calculators.

Common mistakes and remediation

Students will be confused by what n means in different contexts, such as monthly interest or annual interest.

Useful tips

Ensure students become familiar with the terms, principal, P, rate, r and term, n.

Starter

- Put 2^4 on the board and ask students to state its value. (16)
- Ask how they worked it out. (2 × 2 × 2 × 2)
- Now ask what 3^4 is (81), relating it to 3 × 3 × 3 × 3.
- Now ask how they would work out 3^{10}.
- Through discussion, encourage students to think of using the power button on their calculators to get 59 049.
- Students who need more **support** will need a reminder about how to use a calculator to find powers, and some practice in doing it.
- Now ask students to use their calculators to find the value of 2^{16}. (65 536)
- Continue with a few other examples, such as 5^7 (781 250), 7^8 (5 764 801), 1.5^6 (11.390 625).

Main lesson activity

- Write $200 on the board and say that it will earn 5% interest per year.
- Ask what this will be worth after one year. Write down $200 × 1.05 = $210.
- Ask how much this will be worth after two years. Write down $210 × 1.05 = $220.50.

- Ask how much this will be worth after three years. Write down $220.5 × 1.05 = $231.525.
- Now ask how much an investment of $200, with 5% interest pa, would be worth after three years and show that the answer, again, is 200 × 1.05 × 1.05 × 1.05 = 231.525.
- Ask how to write 200 × 1.05 × 1.05 × 1.05 more simply. (200×1.05^3)
- Ask students how to find the value of the investment after eight years. Guide the discussion to 200×1.05^8, then use a calculator to work it out as $325.79.
- Ask students who need more of a **challenge** what the *generalisation* would be for the value of the investment after n years. (200×1.05^n)
- Explain that compound interest is used widely in banking and investments. It is calculated from a formula based on this idea. Write the compound interest formula on the board.

  $$\text{Investment} = P\left(1+\frac{r}{100}\right)^n$$

  where P is the principal amount, r is the rate of interest and n is the number of times this interest is to be applied.
- Talk through both methods for Example 22 in the Student's Book.
- Stress that students need to be clear what the principal amount is, what the rate of interest is and the number of times the interest is applied, as in the example given. Remind them to round their final answer suitably, normally to 2 dp.
- Students can now do Exercise 2J.

Plenary

- Put on the board $500 loan.
  Bank charges interest at 6% interest per year to borrow this.
  A money-lender offers monthly interest of only 0.5% per month.
  Which is the better offer? Most students will probably say the money-lender.
- Now work it out.
- The final amount to pay back at the bank is 500 × 1.06 = $530 at the end of the year.
- The final amount to pay back to the money-lender is 500×1.005^{12} = $530.84.
- The money-lender charges more, although the terms look more attractive.
- Now ask students to work it out when the money-lender's interest rate is 2% per month, which may seem quite reasonable.
- The final amount to pay the money-lender is 500×1.02^{12} = $634.12.
- This is $100 extra. The rate looks reasonable but is deceptive, because it is quoted as a low monthly rate, rather than an annual rate. Round off the lesson by reminding students of the need to take care when considering the rates offered by both banks and money-lenders.

Homework and answers

Homework for this lesson is available in the homework folder. Consolidation and extension exercises are available in the further practice folder. Answers to the exercises can be found in the answers folder.

2.10 Reverse percentage

Syllabus references

Statement	Extended
E1.13	Calculate using reverse percentages.

Collins references

- Student's Book pages 47–49

Learning aims

- Calculate the original amount, given the final amount, after a known percentage increase or decrease.

Key words

- multiplier, unitary method

Prior knowledge

Students should be able to work out a multiplier. They also need to know that the whole amount is always equivalent to 100%.

Common mistakes and remediation

A common mistake is to work out 'the percentage of', such as '10% of', rather than working out 10% off. Stress that students should always read the text carefully. The main lesson activity provides an example that could be misconstrued in this way.

Useful tips

Emphasise that this topic is not about working out a percentage, but about working out a quantity by scaling.

Starter

- Tell students that for each statement they must work out 100%. Start with simple percentages: 50% is $3 (100% is $6), 25% is 4 kg (100% is 16 kg), 20% is 5 cm (100% is 25 cm), 10% is 60 cents (100% is $6).
- Move on to more difficult percentages, using fractions and decimals for the those who need more of a **challenge**. Encourage students to give answers in two or more steps: 90% is $18 (10% is $2, so 100% is $20), 75% is 15 kg (25% is 5 kg, so 100% is 20 kg), 37.5% is 30 cm (12.5 % is 10 cm, 25% is 20 cm, 100% is 80 cm).

Main lesson activity

- Develop the questions to give a real context. Ask students to think about this problem:
  - A sale offers 10% off all prices. The sale price of a dress is $18. What was the original price of the dress?
- Point out that '10% off' does not mean '10% of', so they must not work out '10% of' at any time in this question. On the board, write: 10% *off* not 10% *of*.

- Now ask students how they would solve the problem. There are two methods. Students should be shown both methods but they should then choose the approach they prefer and stick with it.

The unitary method

- This involves scaling to obtain 1% and then 100%, as used in the starter.
- Show students the three steps for the unitary method in the Student's Book then work through the example that students have been thinking about.
- 10% off means the sale price is 90%.
- 90% is equivalent to $18.
- 1% is equivalent to 20 cents, or 10% is equivalent to $2 (dividing by 90 or 9).
- So 100% (the original price) = $20.

  Now work through Example 23.

The multiplier method

- Show students the two steps of the multiplier method in the Student's Book.
- 10% off means a multiplier of 0.9 (writing down the multiplier).
- Working backwards to find the original amount:
- $18 ÷ 0.9 gives an original price of $20 (dividing by the multiplier).

  Now work through Example 24.

- Students can then do Exercise 2K, in which questions 14–16 are problem-solving (PS) questions. Highlight the fact that many of these questions consider 'real-life' situations to do with personal and household finance. Students who need more **support** should attempt as many questions as they can, working with a partner if they find some of the questions difficult.

Plenary

Ask students to help you with a few questions.

- Sales tax: An MP3 player costs $176.25, including sales tax at 17.5%; how much would it have cost without sales tax? ($150)
- Salary: Tasnim now earns $21 840 after a 4% pay rise. How much did she earn before her pay rise? ($21 000)
- Mass: Paolo keeps a diary of his mass. He lost 6% of his body mass in one month, then gained 6% of his new body mass during the following month. Did he gain or lose mass overall?
  What was the percentage change? If he weighs 75 kg now, what was his mass to begin with? (Lose by 0.36%; 75.3 kg)

Homework and answers

Homework for this lesson is available in the homework folder. Consolidation and extension exercises are available in the further practice folder. Answers to the exercises can be found in the answers folder.

3.1 Order of operations

Syllabus references

Statement	Core
C1.6/E1.6	Use the four operations for calculations with integers, fractions and decimals, including correct ordering of operations and use of brackets.

Collins references

- Student's Book pages 52–53

Learning aims

- Work out the answer to a problem with a number of different mathematical operations.

Key words

- operation, brackets, order

Prior knowledge

Students should know the number bonds to 20 and the multiplication tables up to 10 × 10.

Common mistakes and remediation

The most common mistake that students make in the order of operations is to ignore the conventions and work from left to right; so, for example, they write 4 + 5 = 9 × 6 = 54, whereas they should write 4 + (5 × 6) = 34 (5 × 6 = 30 + 4 = 34).
Students should be encouraged to recognise questions on this topic and write down the mnemonic BIDMAS/BODMAS (as detailed in the Student's Book) before doing the calculation, as a reminder to work in the correct order.

Useful tips

Explain to students that some teachers use the mnemonic BODMAS and some use BIDMAS, but effectively they are the same.

Starter

- Ask students to work out:

  4 ÷ 4 + 4 – 4 (= 1)   (4 ÷ 4) ÷ (4 + 4) (= 1)   (4 ÷ 4) × (4 ÷ 4) (= 1)
- Make sure that the class understands that the parts in the brackets must be done first.
- Ask students to make any number other than 1, using only four 4s. Calculations could include:

  (4 ÷ 4) + (4 ÷ 4) = 2   (4 + 4 + 4) ÷ 4 = 3   (4 – 4) ÷ 4 + 4 = 4   (4 ÷ 4 + 4) = 5
- Ask students to feed back their calculations to the class and take time to explain any calculations that need adjusting, to use brackets correctly.
- Give students four different numbers (e.g. 1, 2, 3, 4) and ask them to use these numbers to make 21 (e.g. (1 + 2) × (3 + 4)). Tell students that all these calculations require them to use operations in the correct order.

Main lesson activity

- Write $2 + 3 \times 4$ on the board and ask students to work out the answer. They should write down their answers on rough paper rather than calling them out. Write the numbers 14 and 20 on the board and collect in the students' answers. Write a tally under the numbers 14 and 20 to show how many students got each answer.
- Explain that 14 is the correct answer because the convention is that multiplication is worked out before addition, so $2 + 3 \times 4 = 2 + 12$. Say that there is a convention for the order of carrying out operations in calculations and they must follow it. Show students the BIDMAS/BODMAS list in the Student's Book. Explain that all calculators are programmed to use BIDMAS/BODMAS.
- Explain that, if a calculation has brackets in it, that part must be worked out first, and so on.
- Write $2 + 3 \times 4$ and $(2 + 3) \times 4$ on the board. Remind students that the answer to the first calculation is 14. Ask students for the answer to the second calculation (20).
- Write $(1 + 2 + 3) \times 4$ and ask students to give the answer (24).
- Explain that sometimes brackets are not needed; for example, $3 \times 4 + 2$ is just as mathematically correct as $(3 \times 4) + 2$ because 3×4 will be the first calculation in both cases.
- Ask students to work in pairs, using the numbers 1, 2, 3 and 4 and a combination of brackets, division, multiplication, addition and subtraction, to create as many different answers as possible. Allow them to combine numbers, e.g. 1 and 2 to be used as 12.
- For students who need more **support** limit the number of numbers used. If the class needs more of a **challenge**, give them extra rules that they must take into account when making their calculations, e.g. all four numbers should be used exactly once, find the answers for the numbers 1 to 20, etc.
- Students can now do Exercise 3A.

Plenary

- Play a game of 'Countdown', choosing a set of numbers (such as 25, 6, 3, 10, 4, 1) and giving students 30 seconds to make 254 $(25 \times 10 + 4)$ or 60 seconds to make 287 $(25 \times 10 + (6 + 3) \times 4 + 1)$.
- Pick other target numbers of varying difficulty, depending on the ability of the class.

Homework and answers

Homework for this lesson is available in the homework folder. Consolidation and extension exercises are available in the further practice folder. Answers to the exercises can be found in the answers folder.

3.2 Choosing the correct operation

Syllabus references

Statement	Core
C1.6/E1.6	Use the four operations for calculations in practical situations.

Collins references

- Student's Book pages 54–55

Learning aims

- Solve problems set in a real-life context.

Prior knowledge

Students should know the multiplication tables and be comfortable when doing simple mental calculations. They should be able to multiply numbers such as 24 × 36, 162 × 78, 24.5 × 15, and divide numbers such as 1035 ÷ 55 without using a calculator. Students should also be familiar with short division methods.

Common mistakes and remediation

Students may fail to answer a question due to not rounding answers logically, e.g. rather than 18.81 coaches, it should be 19. Other common mistakes are generally based on careless errors in arithmetic. Students with a weak grasp of multiplication tables may make basic errors. Address this by regularly testing students' multiplication table skills.

Useful tips

Students who need more **support** can use 'carried figures' if they need practice in laying out addition and subtraction.

Make sure students know that they need to round up (question 3a) or round down (question 5), depending on the context of the question. Students who need more **support** may find it helpful to discuss each question before starting, identifying if it involves multiplication or division.

Starter

- Ask students some mental multiplications, e.g. 20 × 5 (100), 7 × 400 (2800) and 30 × 500 (15 000).
- Combine two similar multiplications, e.g. 30 × 8 (240), 2 × 8 (16) and 32 × 8 (240 + 16 = 256). Repeat for other numbers, increasing the difficulty according to ability. Now ask some mental divisions using the multiplication tables up to 10 × 10. Include examples that have a remainder, e.g. 40 ÷ 6 = 6, remainder 4.

Main lesson activity

- The focus of this lesson is to ensure that students have methods of multiplying and dividing with which they are comfortable. Depending on prior knowledge, they may be able to explain their methods. Groups who need more **support** should, at least in this lesson, concentrate on using just one method of multiplying and one method of dividing.

Multiplication

- Demonstrate the grid (or box) method, e.g. with 213 × 54. Adding each answer to find the total gives 11 502. Now demonstrate the same calculation using the column method.
- Ask students which method they prefer. If there is a preference then they should stick to that method.
- Repeat with other calculations for groups who need more **support**.

```
    2 1 3
×     5 4
    8 5 2
1 0 6 5 0
1 1 5 0 2
```

×	200	10	3
50	10 000	500	150
4	800	40	12

Division

- Display 430 ÷ 2 and ask students how they would work it out (215). They are likely to say, 'By a mental calculation'. Increase the level of difficulty, e.g. ask students how they would work out 275 ÷ 25. Expect some to say that there are four 25s in 100, so 4 (100) + 4 (100) + 3 (75) = 11. Point out that this is a chunking method. Write out the calculation:

$$\begin{aligned} 4 \times 25 &= 100 \\ 4 \times 25 &= 100 \\ 3 \times 25 &= 75 \\ \hline 11 \times 25 &= 275, \text{ so } 275 \div 25 = 11 \end{aligned}$$

- Now demonstrate a short division giving an answer with a remainder, e.g. 100 ÷ 7 (14, remainder 2) or 300 ÷ 21 (14, remainder 6). Explain that some questions will have remainders, as the numbers do not divide exactly. Repeat the same question, using long division. Explain that long division is similar to short division but requires them to show more working, which may reduce errors.
- Ask: There are 30 students in the dinner queue. They sit at tables with four seats. How many tables are needed? (8). Ask students to explain why eight tables are needed. Discuss the fact that there are two spare seats.
- Go through Example 1. Demonstrate the solution, using their method, e.g. for chunking 10 × 53 = 530, this leaves 672 – 530 = 142, 2 × 53 = 106, this leaves 142 – 106 = 36 as the remainder, so 640 ÷ 53 = 12, remainder 36. So 13 coaches are needed.
- Explain that if they have a variation on any of these methods or their own successful method, they can continue to use it. Encourage students to be precise with their calculations and to check answers to see if they seem sensible.
- Students can now do Exercise 3B. In this exercise, the problem-solving (PS) questions are 5, 6 and 7. Students who need more of a **challenge** can try more than one method.

Plenary

- Ask for the answer to 400 ÷ 24. Ask students to write this as a question, e.g. 400 people are on a train, with 24 people in every coach but the last one. How many are in the last coach? (16)
- Groups who need more **support** may find it easier to use questions involving money or smaller numbers, e.g. I have $20 for bus fares ($3 per day). How many days' bus fare do I have? (6 days with $2 over.) Repeat for other numbers.

Homework and answers

Homework for this lesson is available in the homework folder. Consolidation and extension exercises are available in the further practice folder. Answers to the exercises can be found in the answers folder.

3.3 Finding a fraction of a quantity

Syllabus references

Statement	Core
C1.6/E1.6	Use the four operations for calculations with fractions, including correct ordering of operations and use of brackets.

Collins references

- Student's Book pages 55–57

Learning aims

- Find a fraction of a given quantity.

Key words

- quantity, fraction

Prior knowledge

Students need to know how to use the fraction button on a calculator. They should also know how to use a calculator to multiply and divide.
They must know the multiplication tables to 10 × 10.

Common mistakes and remediation

A common mistake is to muddle 'of' and 'off', e.g. $\frac{1}{4}$ of 80 and $\frac{1}{4}$ off 80. Point out that students need to read questions very carefully.

Useful tips

When tackling word questions, students should first extract the information and decide the method, and then rewrite the question as a standard calculation, e.g. in question 4 students should write the profit as of $45 600 and then write it as profit = $\frac{2}{15}$ × 45 600.

In mathematics 'of' is interpreted as ×. For example, 2 lots of 3 is the same as 2 × 3 (6).

Starter

- On the board, draw a line that is exactly 60 centimetres long. Say, 'I want to mark a point that is halfway along the line. Where should I put it?' Discuss their answers and then mark the point in the correct position. Repeat with other fractions such as one-third, one-tenth, one-fifth.

Main lesson activity

Before beginning the main lesson, it is important to recognise that students who need more **support** will find it easier to use the word 'of' instead of the symbol ×. However, it is important to plant the idea that these mean the same, to help their progress with more difficult questions and to aid understanding.

- Continue from the starter by asking students to explain how they decided where the correct position for each fraction was.

- With $\frac{3}{4}$, for example, some will work out one-quarter and multiply by 3, while others will work out one-quarter and subtract their answer from 1.
- Encourage students to use one method for the calculation and the other method for checking the answer.
- Work through a few examples, firstly showing the students Example 2. If units are given, they must be included in the answers.
- Finally, check that students are able to do the same calculations on their calculators.
- Depending on the ability of the group, calculators can be used for either completing some of the calculations or for checking answers. In general, students who need more of a **challenge** should use them for checking answers.
- The students can now complete Exercise 3C.
- In the word problems, students who need more **support** may need help extracting the mathematical calculation required.

Plenary

- Check students' understanding by asking quick-fire questions such as: 'What is three-quarters of eight centimetres? What is two-thirds of 90°? What is three-fifths of 60 minutes?'
- Move on to context questions with mixed units. Give the conversions if they are not known.
- What is $\frac{2}{3}$ of a full turn in degrees?
- What is one-fifth of an hour, given in minutes?
- What is $\frac{3}{4}$ of a kilogram, in grams?

Homework and answers

Homework for this lesson is available in the homework folder. Consolidation and extension exercises are available in the further practice folder. Answers to the exercises can be found in the answers folder.

3.4 Adding and subtracting fractions

Syllabus references

Statement	Core
C1.6/E1.6	Use the four operations for calculations with fractions, including correct ordering of operations and use of brackets.

Collins references

- Student's Book pages 57–60

Learning aims

- Add and subtract two fractions with the same denominator, then simplify the result.
- Solve problems in words.

Key words

- proper fraction, improper fraction, lowest terms, simplest form, denominator, mixed number, equivalent fraction

Prior knowledge

Students should understand equivalent fractions and be able to simplify by cancelling. They should also be able to convert mixed numbers to improper fractions.

Common mistakes and remediation

A common mistake is to omit working and just write down answers. This should be discouraged. Students do not always give the answer in its lowest terms. Remind students to check whether their answers can be simplified in order to answer questions fully. Also, when adding fractions, students often add the denominators as well as the numerators. Shading fractions of a shape should help students avoid this error.

Useful tips

Whenever possible, while practising addition and subtraction of fractions, students should use a scientific calculator to check answers. Remind students that they can only add or subtract fractions if the denominators are the same. Also encourage students always to use a horizontal line when writing fractions so that it is very clear which number is the denominator.

Starter

- As a revision, ask students to convert mixed numbers to the appropriate improper fractions.
- Ask simple division questions in the form: 'How many halves are there in 6? How many quarters in 12?'
- 'How many three-quarters are there in 12?' Ask students who need more of a **challenge**: 'How many three-quarters in 13?'

Main lesson activity

- Ask students to read through the six bullet points in the Student's Book.
- Practise the methods shown and then use very simple fractions, with different denominators, to develop the method, e.g.

  $\frac{1}{2}+\frac{1}{4}\left(\frac{3}{4}\right)$, $\frac{1}{4}-\frac{1}{8}\left(\frac{1}{8}\right)$, $\frac{1}{2}+\frac{3}{8}\left(\frac{7}{8}\right)$ and $\frac{5}{8}-\frac{1}{2}\left(\frac{1}{8}\right)$

- Move on to questions that give rise to fractions that can be simplified or that can be changed to mixed numbers.
- Show students Example 3. They can then do Exercise 3D. If the class needs more **support**, cover one or two bullet points at a time, followed by corresponding questions from Exercise 3D.
- Students who need more of a **challenge** may not need to complete all questions.
- Work through Example 4a $\left(\frac{2}{3}+\frac{1}{5}\right)$. Point out that students need to make the denominators the same, by working out equivalent fractions, and then add the numerators together.
- Write $2\frac{3}{4}-1\frac{5}{6}$ on the board
- Ask students to change the fractions into twelfths $\left(2\frac{9}{12}-1\frac{10}{12}\right)$ then ask them to change the mixed numbers into improper fractions, writing their response on the board $\left(\frac{33}{12}-\frac{22}{12}\right)$.
- Ask students to subtract the second numerator from the first for the correct answer $\left(\frac{11}{12}\right)$. Now show students Example 4b, pointing out the alternative method of reaching the same answer.
- Students can now do Exercise 3E. In the case of word problems, remind students that they just need to extract the necessary information and then solve the problem.

Plenary

- Start by asking students to add fractions with an answer of 1, e.g. $\frac{2}{5}+\frac{3}{5}$, $\frac{1}{4}+\frac{3}{4}$.

  Now do some mental calculations of the complements to 1, e.g. 1 minus three-fifths.

  After a while ask students if they can see a simple rule for doing this. In this case $\frac{5}{5}-\frac{3}{5}=\frac{2}{5}$.
- Then check students' understanding of subtracting a fraction from a whole number.

  Use questions such as $3-\frac{5}{8}$.

Homework and answers

Homework for this lesson is available in the homework folder. Consolidation and extension exercises are available in the further practice folder. Answers to the exercises can be found in the answers folder.

3.5 Multiplying and dividing fractions

Syllabus references

Statement	Core
C1.6/E1.6	Use the four operations for calculations with fractions, including correct ordering of operations and use of brackets.

Collins references

- Student's Book pages 60–63

Learning aims

- Multiply a fraction by a fraction; dividing fractions.

Key words

- numerator, reciprocal

Prior knowledge

Students should know the multiplications tables up to 10 × 10.

Common mistakes and remediation

Students often forget to simplify the final answer by cancelling. Remind them always to give fractions in their simplest form unless instructed otherwise.

Useful tips

Remind students that × means 'of', and that dividing by a fraction is the same as multiplying by the fraction turned upside-down.

Starter

- Ask students, 'What is a half of a half?' (One-quarter.) Some may answer '1'.
- Ask students to draw a 2 × 2 grid and then shade half of it. Then ask them to shade half of the half they shaded before. Ask: 'What fraction of the original shape has been shaded twice?'
- Repeat with a 3 × 2 grid, first shading half then a third of the same half. Ask, 'How much is shaded now?' Then, 'What do you notice about the fractions you have shaded?' (The final answer is always smaller than the original fractions.)
- Ask students what size of grid would be needed for calculating with halves and fifths (2 × 5).
- Now ask students to use a grid to work out $\frac{2}{5}$ of $\frac{1}{2}$, ($\frac{2}{10}$ or $\frac{1}{5}$).

Main lesson activity

Multiplication of fractions

- The aim of the first part of the lesson is to show that the formal method of multiplying fractions by multiplying numerators together and then multiplying denominators together gives the correct answer, when compared to the diagram method used in the starter.
- Ensure that students understand that, when they multiply by a proper fraction, the answer will be smaller than the original number. Stress that they should cancel and simplify the answer if necessary.

- Write the questions and answers used in the starter on the board, e.g. $\frac{1}{2} \times \frac{1}{3} = \frac{1}{6}, \frac{1}{4} \times \frac{2}{3} = \frac{2}{12} = \frac{1}{6}, \frac{2}{5} \times \frac{1}{2} = \frac{2}{10} = \frac{1}{5}$
- Ask students if they can see a way of getting the unsimplified answers from the numbers in the question, using the same rule each time.
- Establish that the rule is to multiply the numerators and multiply the denominators, then cancel if possible. Use Example 5 to show this method.
- Then show examples to demonstrate that answers can be worked out and then simplified, or simplified and then worked out; e.g. $\frac{5}{6} \times \frac{3}{5}$ can be worked out as $\frac{15}{30} = \frac{1}{2}$ or $\frac{1}{2} \times \frac{1}{1} = \frac{1}{2}$.
- Show students Example 6. Then students who need more **support** may do questions 1–4 of Exercise 3F at this point before looking at Example 7.
- Discuss Example 7 with the students, highlighting the fact that in multiplications with mixed numbers, they need to be converted to improper fractions first.
- Students can now do Exercise 3F. Encourage them to simplify first.

Dividing fractions

- Ask students: 'How many quarters in 3?' (12), 'How many halves in 3?' (6), 'How many $\frac{3}{4}$ in 3?' (4).
- Students probably found the last questions difficult, so write on the board: $2 \times 3 = 6$ so $6 \div 2 = 2$ and $3 \times 4 = 12$ so $12 \div 3 = 4$.
- Show students the diagrams in the Student's Book, illustrating that three whole squares can be divided into four lots of three-quarters.
- Now write $4 \times \frac{3}{4} = 3$ so $3 \div \frac{3}{4} = 4$. Ask 'How many $\frac{3}{4}$ in 3?' (4).
- Explain that dividing by a fraction is the same as multiplying by its reciprocal.
- Give more examples, e.g. how many halves in 6? (12, $6 \div \frac{1}{2} = 6 \times 2 = 12$).
- Remind students that to divide mixed numbers they must first change the mixed numbers into improper fractions. Show students Example 8.
- Students can now do Exercise 3G.
- Ask students to simplify their answers. Students who need more **support** can work in groups.

Plenary

- Put the following multiplication on the board: $\frac{2}{15} \times \frac{5}{9} \times \frac{5}{6} \times \frac{3}{4} \times \frac{12}{25}$.
- Discuss why it may not be a good idea to multiply all the numerators and all the denominators (easy to make a mistake, difficult calculations).
- Ask students to cancel as much as possible. All the top will cancel and 5 and 9 will be left on the bottom. Make sure students understand that there are ones left on top, not zero.

  The final answer is $\frac{1}{45}$.

Homework and answers

Homework for this lesson is available in the homework folder. Consolidation and extension exercises are available in the further practice folder. Answers to the exercises can be found in the answers folder.

4.1 Introduction to negative numbers

Syllabus references

Statement	Core
C1.6/E1.6	Use negative numbers in practical situations.

Collins references

- Student's Book pages 66–67

Learning aims

- Understand how negative numbers are used in real life.

Key words

- negative, positive

Prior knowledge

Students must know how to order positive numbers. They need to understand what a negative number is: a negative number is a number less than zero.

Common mistakes and remediation

Students will sometimes forget to write down the minus sign when reading from a scale. The most common error is to assume that all scales are calibrated in ones. To prevent this, students should be encouraged to complete the whole scale and check that it makes sense, as shown in the diagram. Some students will think that, for example, –6 is bigger than –2. Encourage students to think of temperatures and this error is then often overcome, as –6 °C is clearly colder than –2 °C.

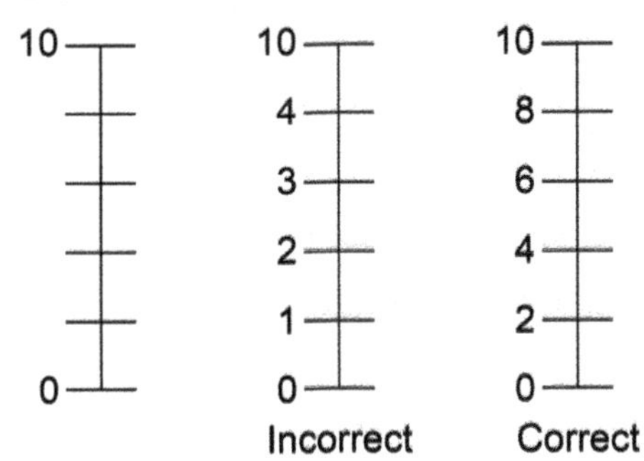

Useful tips

Remind students that, for example, 9 is **more than** 3 but –9 is **less than** –3. Be careful to use the words 'more than' and 'less than' correctly when negative numbers are involved.

Starter

- Display some vertical number lines with some numbers missing.
- Using the first number line, ask students to point out where different numbers should be positioned, for example, +8, –3, –9 and so on. Repeat with the other number lines, moving on to numbers including fractions and decimals, for example, $-\frac{1}{4}$, 0.6 and so on.

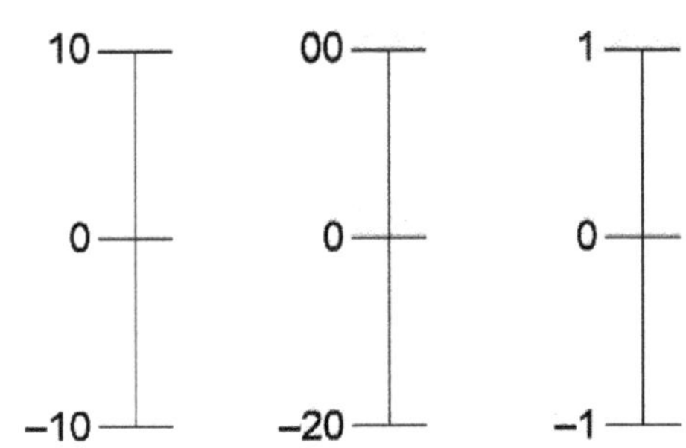

Main lesson activity

- Remind students that a negative number is a number below zero.
- Display these number scales.

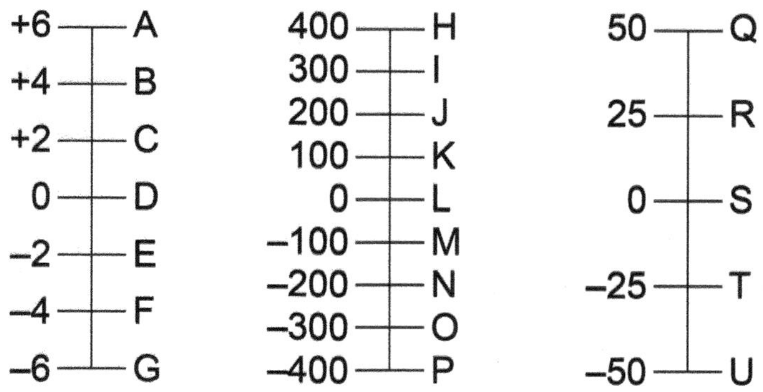

- Ask students for the difference between points on the scale, e.g. ask 'How much higher is B than G?' (10), 'What is the difference between J and O?' (500). Repeat for lots of values. Students who need more support could start by using positive numbers only. Students who need more of a **challenge** could have more complex scales.
- Students can then do Exercise 4A.

Plenary

- Check students' understanding by going through Exercise 4A.

Homework and answers

Homework for this lesson is available in the homework folder. Consolidation and extension exercises are available in the further practice folder. Answers to the exercises can be found in the answers folder.

4.2 Everyday use of negative numbers

Syllabus references

Statement	Core
C1.6/E1.6	Use negative numbers in practical situations.

Collins references

- Student's Book pages 67–68

Learning aims

- Using positive and negative numbers in everyday life.

Key words

- profit, loss

Prior knowledge

Students should understand what a negative number is and be able to order numbers.

Common mistakes and remediation

Students working on bank statements may confuse a debit with a negative balance or debt. Point out that on some statements debts are shown by marking the amounts with an asterisk (*) rather than a negative sign. Others will use red print. Emphasise that a debit is a negative and a credit is a positive.

Useful tips

Some students will cope better with negative numbers if they also insert a + sign for positive numbers.

Starter

- Use the data below about melting and boiling points to ask questions about positive and negative numbers, e.g. 'What is the melting point of oxygen? What is the difference in temperatures for nitrogen and helium? What are the melting point temperatures from coldest to hottest?'

Melting points		Boiling points	
nitrogen	–210 °C	sulfuric acid	340 °C
oxygen	–218 °C	acetic acid	118 °C
hydrogen	–259 °C	nitric acid	83 °C
helium	–272 °C		

- **Information for teachers.** Absolute zero is the lowest temperature that is theoretically possible. Its value is equivalent to –273.15 °C or zero on the Kelvin scale.

Main lesson activity

- Explain that in this lesson students continue the work from the previous lesson, looking at more applications of negative numbers in everyday contexts.
- Students will need to know where they are likely to need to use negative numbers in real life. The most common examples include temperature, sea level and below, and credit and debit. They will also have met negative scales on graphs but will only have used them to plot coordinates. It is important that students realise that negative numbers are less than positive numbers and they should ignore the value of the digit, for example, –10 is less than –2.
- Refer the students to the bank statement in the Student's Book and discuss the meaning of the negative signs. Explain the word 'balance'. Introduce the words 'credit', which is positive, and 'debit', which is negative. Ensure that students understand the language used.
- The students can now do Exercise 4B.
- Students who need more of a **challenge** could do the first part of this exercise orally, whilst students who need more **support** may need the reinforcement of looking at questions orally and then writing them out.

Plenary

- As preparation for the next lesson, revisit the melting points data in the starter and ask students to give facts using the words 'less than' or 'greater than', e.g. –218 is less than –210.
- Ensure that students understand which is the lowest and which is the highest.
- For students who need more of a **challenge** give extra data, such as:

  argon –189 °C

  krypton –157 °C

  xenon –111 °C

Homework and answers

Homework for this lesson is available in the homework folder. Consolidation and extension exercises are available in the further practice folder. Answers to the exercises can be found in the answers folder.

4.3 The number line

Syllabus references

Statement	Core
C1.6/E1.6	Use the four operations for calculations with negative numbers, including correct ordering of operations and use of brackets.

Collins references

- Student's Book pages 68–70

Learning aims

- Use a number line to represent negative numbers.
- Use inequalities with negative numbers.

Key words

- number line, less than, more than, greater than

Prior knowledge

Students will need to have a basic understanding of the significance and meaning of negative numbers.

Common mistakes and remediation

Students may miss out zero altogether. Emphasise the importance of including it.
When comparing numbers, students often look at the value of the digit and not its sign.
Encourage them to look at the sign first.

Useful tips

The 'less than' symbol (<) resembles a letter L and this is a way that students can remember it.
The 'greater than' symbol (>) is the opposite way round. Students may also find it helpful to remember that the 'closed' end always points to the smaller number.

Starter

- Go around the class, starting at 40 and counting down in threes. Encourage students to keep going when they reach negative numbers.
- Ask quick-fire questions such as, 'What is three less than seven? What is five less than one? What is four more than minus two?'
- Use larger numbers for students who need more of a **challenge**.

Main lesson activity

- Emphasise that numbers to the right of zero on a number line are positive and those to the left of zero are negative. Some students prefer a number line that goes up and down rather than across. Remind students of the inequality signs (<) and (>) and make sure that they understand them. For example, –7 is smaller than 4. Stress that it is the sign that is important with negatives, not the value of the digit.
- Give out cards with numbers on, e.g. –6, –1, –3, +4, 0, –7, +5, –11, +9, +2. The numbers could be wider-ranging or include decimals for students who need more of a **challenge**. It may be necessary to reduce the list of numbers for students who need more **support**.
- Ask students to sort the numbers from lowest to highest (–11, –7, –6, –3, –1, 0, +2, +4, +5, +9).
- Display the < symbol and ask students to use two numbers to make a correct statement, e.g. –11 < (is less than) –6.
- Repeat, using > (is greater than).
- The students can now complete Exercise 4C.

Plenary

- Draw a number line the other way from how you have been working. Check that students can fill in missing numbers on this line.
- Use different calibrations for the scales, according to the ability of the group.

Homework and answers

Homework for this lesson is available in the homework folder. Consolidation and extension exercises are available in the further practice folder. Answers to the exercises can be found in the answers folder.

4.4 Adding and subtracting with negative numbers

Syllabus references

Statement	Core
C1.6/E1.6	Use the four operations for calculations with negative numbers.

Collins references

- Student's Book pages 70–73

Learning aims

- Add and subtract positive and negative numbers to or from both positive and negative numbers.

Prior knowledge

Students need to be able to add and subtract positive numbers and to be familiar with a number line.

Common mistakes and remediation

The common mistake is to forget that – – is the same as +. Remind students about the idea of two changes taking them back to the start, i.e. – – is the same as +.

Useful tips

Emphasise that, if the signs are the same, the result is a plus, so 5 – –2 = 5 + 2

Starter

- Ask students to look at the bullet points in the Student's Book. Ensure they understand how to add and subtract positive numbers. They can then complete the addition pyramids on Resource 4.4.
- Go through Example 1 with the students.
- Students can then do Exercise 4D. In this exercise, the problem-solving (PS) question is 3.

Main lesson activity

- Give students an outline addition table.
- Ask them to complete the top right-hand corner, then the top left-hand corner following the patterns. Then ask them to continue to follow the patterns to complete the whole table.
- Pick out some answers, e.g. –3 + –3 = –6. Use the number line to show that this is the same as –3 – 3 = –6 so that + –3 is the same as –3.

			+3			
			+2			
			+1			
–3	**–2**	**–1**	**0**	**+1**	**+2**	**+3**
			–1			
			–2			
			–3			

0	1	2	**+3**	4	5	6
–1	0	1	**+2**	3	4	5
–2	–1	0	**+1**	2	3	4
–3	**–2**	**–1**	**0**	**+1**	**+2**	**+3**
			–1			
			–2			
			–3			

- Go through the rules for adding and subtracting negative numbers.
- Adding leaves the sign unchanged so may be treated as subtraction, subtracting 'changes the sign'. As a mnemonic, + (–) gives –, – (–) gives +.
- Emphasise the fact that a negative sign changes the direction of movement on a number line. To subtract 3, move down or to the left. However, to subtract –3 (– –3), changing direction twice means going up or to the right.
- Show the students how they can use a calculator to carry out calculations with negative numbers.
- Go through Example 2 with the students. They can then do Exercise 4E.
- The students can now do Exercise 4E. In this exercise, the problem-solving (PS) question is 9. Students who need more of a **challenge** should attempt questions 8 and 9 but students who need more **support** may need to limit the questions to 1–7.

0	1	2	**+3**	4	5	6
–1	0	1	**+2**	3	4	5
–2	–1	0	**+1**	2	3	4
–3	**–2**	**–1**	**0**	**+1**	**+2**	**+3**
–4	–3	–2	**–1**	0	1	2
–5	–4	–3	**–2**	–1	0	1
–6	–5	–4	**–3**	–2	–1	0

Plenary

- Ask students to draw this 4 by 4 square.

1	2	3	4
5	6	7	8
9	10	11	12
13	14	15	16

- Now reverse the numbers on each diagonal as shown here, to get a '34' magic square. If students have not met magic squares previously, you will need to explain to them that they are ones in which each row, column and diagonal adds up to the same number.

16	2	3	13
5	11	10	8
9	7	6	12
4	14	15	1

- Ask students to subtract 10 from each number and check they now have a '–6' magic square.

6	–8	–7	3
–5	1	0	–2
–1	–3	–4	2
–6	4	5	–9

Homework and answers

Homework for this lesson is available in the homework folder. Consolidation and extension exercises are available in the further practice folder. Answers to the exercises can be found in the answers folder.

4.5 Multiplying and dividing with negative numbers

Syllabus references

Statement	Core
C1.6/E1.6	Use the four operations for calculations with negative numbers.

Collins references

- Student's Book pages 73–75

Learning aims

- Multiply and divide with negative numbers.

Prior knowledge

Students need to understand what negative numbers are and where they fit on the number line. Students need to know the multiplication tables up to 10 × 10.

Common mistakes and remediation

The common mistake is that students misinterpret the rules when doing addition or subtraction questions. For example, –3 – 5 is given as +8, as students interpret the rule that two negatives make a positive incorrectly. To avoid this, it is important to stress that the rule is that two negative numbers multiplied or divided make a positive number.

Useful tips

As the rules for multiplying and dividing negative numbers are very different from the rules for adding and subtracting, do not include both at the same time until they are both understood.

Starter

- Ask students to copy the multiplication table on the right and fill in the top left-hand corner. Explain that you have already filled in the zeros as any number multiplied by zero is zero.

×	+3	+2	+1	0	–1	–2	–3
+3				0			
+2				0			
+1				0			
0	0	0	0	0	0	0	0
–1				0			
–2				0			
–3				0			

- Check that the students' tables appear as shown on the right.

×	+3	+2	+1	0	–1	–2	–3
+3	9	6	3	0			
+2	6	4	2	0			
+1	3	2	1	0			
0	0	0	0	0	0	0	0
–1				0			
–2				0			
–3				0			

- Now ask students to read the top row from left to right and continue the pattern: 9, 6, 3, 0, –3, –6, –9
- Ask students to use the pattern to complete the top row.
- Do the same for the second and third rows. It should now appear as shown on the right.

×	**+3**	**+2**	**+1**	**0**	**–1**	**–2**	**–3**
+3	9	6	3	0	–3	–6	–9
+2	6	4	2	0	–2	–4	–6
+1	3	2	1	0	–1	–2	–3
0	0	0	0	0	0	0	0
–1				0			
–2				0			
–3				0			

Main lesson activity

- Continuing from the starter, ask students to complete the whole table using patterns horizontally and vertically as shown on the right.

×	**+3**	**+2**	**+1**	**0**	**–1**	**–2**	**–3**
+3	9	6	3	0	–3	–6	–9
+2	6	4	2	0	–2	–4	–6
+1	3	2	1	0	–1	–2	–3
0	0	0	0	0	0	0	0
–1	–3	–2	–1	0	1	2	3
–2	–6	–4	–2	0	2	4	6
–3	–9	–6	–3	0	3	6	9

- Now summarise their results with one of the tables below:

×	**+**	**0**	**–**
+	+	0	–
0	0	0	0
–	–	0	+

×	**+**	**–**
+	+	–
–	–	+

- Explain that working backwards you can show that exactly the same results work for division. For example, –3 × +2 = –6 so –6 ÷ –3 = +2 and so on. Modify the table to show this as on the right.

× or ÷	**+**	**–**
+	+	–
–	–	+

- Emphasise that these results apply for multiplication and division only. Encourage students to learn that two negative numbers multiplied or divided make a positive number.
- Now work through Example 3 to show the impact of squaring a negative number.
- Students can now complete Exercise 4F.

Plenary

- Ask questions such as: 'What number would you multiply by –4 to get –20?'
- For students who need more of a **challenge**, ask some additions or subtractions to check that the common mistake above is not being made. If it is, ask the question: 'You are in a lift, you go down three floors, then you go down another five floors. How many have you gone down altogether?' Relate this to –3 – 5 = –8. Repeat for other numbers.
- For students who need more **support**, you may prefer to concentrate solely on the rules for multiplying and dividing negatives until you feel they are confident working with this topic.

Homework and answers

Homework for this lesson is available in the homework folder. Consolidation and extension exercises are available in the further practice folder. Answers to the exercises can be found in the answers folder.

5.1 Squares and square roots

Syllabus references

Statement	Core
C1.3/E1.3	Calculate with squares and square roots of numbers.

Collins references

- Student's Book pages 78–79

Learning aims

- Find a square root of a square number.
- Use a calculator to find the square root of any number.

Key words

- square, square root

Prior knowledge

Students will need to know the multiplication tables up to 10 × 10 and square numbers up to 15 × 15.

Common mistakes and remediation

Students may divide by two instead of finding the square root. Encourage students always to check their answer by squaring it to see if it gives the number in the question. The other common mistake is to square a number instead of taking the square root. Students can avoid this by using the symbol when writing down the question, rather than using words.

Useful tips

Introduce negative square roots at the same time as positive square roots to avoid it being seen as a harder concept.

Starter

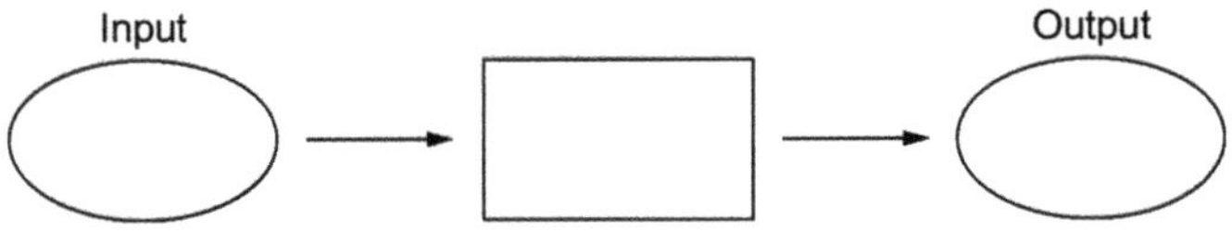

- Draw a function machine on the board, but leave the operation blank. Give the students three inputs and three outputs, e.g. (1, 3), (5, 7), (10, 12), and then ask them to give the output for a given input and the input for a given output. Once they have spotted the rule (add 2), ask them to describe it and share the inverse rule (subtract 2).
- Repeat with other rules such as 'double', 'divide by 3' and 'double and add 3'.
- Finish with the operation 'square'. Students should have an intuitive idea of what a square root is, with students who need more of a **challenge** being able to describe it.
- Use numbers such as 5 and –5 to show the same output and also to demonstrate that a square root has two possible values. Emphasise that finding the square root is the reverse operation of squaring.

Main lesson activity

- Point out that squaring a number always gives a positive answer, so it is not possible to obtain a square root of a negative number.
- Ask students to try to work out $\sqrt{-10}$ on their calculator. The display will show an error.
  Now write on the board: $3^2 = 3 \times 3 = 9$ and also: $(-3)^2 = -3 \times -3 = 9$, so $\sqrt{9} = 3$ or -3.
- Ask students to give the equivalent sentence for the number 4. ($4^2 = 4 \times 4 = 16$ and also $(-4)^2 = -4 \times -4 = 16$, so $\sqrt{16} = 4$ or -4.)
- Repeat for other numbers up to 15.
- Students who need more **support** will tend to forget negative square roots so this is worth revisiting during starters for other lessons.
- Check that students realise, for example, that because $5^2 = 25$ and $6^2 = 36$, squaring a number between 5 and 6 gives an answer between 25 and 36. Similarly, taking the square root of a number between 25 and 36 gives an answer between 5 and 6.
- Show students how to use the square and square root buttons on a calculator and tell them that calculators will show the positive square root.
- Students can now do Exercise 5A.

Plenary

- Allow students to use calculators. Ask them to think of an odd number, say 9. Square the number, add 1 to the answer and divide by 2: $(81 + 1) \div\div 2 = 41$.
- Now square the odd number, subtract 1 from it and divide by 2: $(81 - 1) \div 2 = 40$.
- Ask students to square their two results and subtract the second answer from the first answer. Then take the square root of this result.
- If they have done it correctly they will get the number they started with.
- Conclude by telling students that they will see square roots being used within mathematical formulae, possibly in science lessons.

Homework and answers

Homework for this lesson is available in the homework folder. Consolidation and extension exercises are available in the further practice folder. Answers to the exercises can be found in the answers folder.

5.2 Cubes and cube roots

Syllabus references

Statement	Core
C1.3/E1.3	Calculate with cubes and cube roots of numbers.

Collins references

- Student's Book pages 79–80

Learning aims

- Find the cube root of a cube number.
- Use a calculator to find the cube root of any number.

Key words

- cube, cube root

Prior knowledge

Students will need to know the multiplication tables up to 10 × 10 and be able to multiply a two-digit number by a single digit. They also need to know the rules for multiplying directed numbers.

Common mistakes and remediation

Students may divide by 3 instead of finding the cube root. Encourage students always to check their answer by cubing it to see if it gives the number in the question. Another common mistake is to cube a number instead of taking the cube root. Students can avoid this by using the symbol when writing down the question, rather than using words.

Useful tips

Introduce cube roots of negative numbers at the same time as cube roots of positive numbers so that students appreciate that unlike square roots, it *is* possible to find the cube root of a negative number.

Starter

- As a revision, ask students to find squares and square roots of numbers, with and without a calculator. Aim for instant recall with square numbers up to 15 × 15 and ensure that students do not forget negative square roots.
- Ask students to recall the rules for multiplying directed numbers.
- Write the following calculations on the board, and ask students to give you the answers. This exercise can be done orally.

  a 2 × 4   **b** 3 × –6   **c** –4 × 7   **d** –8 × –9

  e 2 × 3 × 5   **f** –2 × 6 × –4   **g** –4 × –2 × –3   **h** –5 × –2 × 4

Main lesson activity

- Write on the board: $2 \times 2 \times 2 = 8$.
- Explain that 8 is called the 'cube of 2'. This can be written as 2^3.
- Alternatively, we can say that '2 is the cube root of 8', which can be written as $\sqrt[3]{8}$.
- Now ask students to work out $-2 \times -2 \times -2$ (-8). This means that '-8 is the cube of -2' and that '-2 is the cube root of -8'. Emphasise that 8 and -8 each have only one cube root.
- Ask students to write similar sentences for the numbers 3 and -3.
- Repeat for 4 and -4, 5 and -5.
- Ensure that students know the symbols for cube and cube root.
- Check that students understand that, for example, because $5^3 = 125$ and $6^3 = 216$, cubing a number between 5 and 6 gives an answer between 125 and 216.
- Similarly, taking the cube root of a number between 125 and 216 gives an answer between 5 and 6.
- Show students the cube and cube root buttons on their calculators.
- Students can now do Exercise 5B.

Plenary

Discuss question 7 of Exercise 5B. Is there only one possible table? Ensure that pupils realise that some of the numbers can be negative.

Homework and answers

Homework for this lesson is available in the homework folder. Consolidation and extension exercises are available in the further practice folder. Answers to the exercises can be found in the answers folder.

5.3 More powers and roots

Syllabus references

Statement	Core
C1.3/E1.3	Calculate with other powers and roots of numbers.

Collins references

- Student's Book pages 81–82

Learning aims

- Calculate with other powers and roots of numbers.

Key words

- power, root

Prior knowledge

Students need to be familiar with the use of the power keys on their calculators.

Common mistakes and remediation

Students may multiply by the power, which corresponds to adding the base that number of times, rather than multiplying it by itself that number of times. Also, they might divide by n instead of finding the nth root. Encourage students always to check their answer by powering it up, or rooting it, to see if it gives the number in the question. Another common mistake is to cube a number instead of taking the cube root. Students can avoid this by using the symbol when writing down the question, rather than using words.

Useful tips

It is very useful if students learn the squares of numbers to 225, the cubes of numbers to 1000 and the powers of 2 to 1024 as these occur so often in the course. Ensure that students understand how to use the root button on their calculators. They should always check that their result is sensible.

Starter

- As a revision, ask students to find cubes and cube roots of numbers, with and without a calculator. Aim for instant recall with the numbers suggested in the tips above.
- Ask students to recite the powers of 2, starting with 2 and doubling: see how far they can get in one minute. Then ask them to write down the powers of 2 as far as they can go in one minute. Ask them to look at the units digit and point out the cycling pattern of 2, 4, 8, 6, 2, 4, 8, 6, …
- Repeat with the powers of 3.

Main lesson activity

- Write on the board: $2 \times 2 \times 2 \times 2 = 16$.
- Explain that 16 is called the 'fourth power of 2'. This can be written as 2^4.
- Now write on the board: $-2 \times -2 \times -2 \times -2 = 16$. So, 16 is also the 'fourth power of –2'. This can be written as $(-2)^4$. Explain that it is important to use brackets here as $-2^4 = -16$, because without brackets the power has to be worked out before the negative sign is observed.
- Therefore, alternatively, we can say that '2 and –2 are the fourth roots of 16'.

- When we write and evaluate $\sqrt[4]{16}$ however, we just write down the positive root.
- Now ask students to work out –3 × –3 × –3 × –3 × –3 (–243).
  This means that '–3 is the fifth root of –243' and that '3 is the fifth root of 243'. Emphasise that 243 and –243 each have only one fifth root.
- Ask students to write similar sentences for the numbers 32 and –32.
- Repeat for 1000 and –1000, 512 and –512.
- Ensure that students know the symbols for various powers and roots.
- Check that students understand that, for example, because $2^6 = 64$ and $3^6 = 729$, finding the sixth power of a number between 2 and 3 gives an answer between 64 and 729.
- Similarly, taking the sixth root of a number between 64 and 729 gives an answer between 2 and 3.
- Show students the power and root buttons on their calculators.
- Students can now do Exercise 5C.

Plenary

- Talk about the units digit of powers of numbers and how they cycle, as it was mentioned in the starter. Ask students what the units digit of 2^{100} will be (6). What about 5^{37}? (5) What about 3^{37}? (3)
- Move on to the last two digits of powers of 5 and 9 and ask similar questions.
- If time permits, ask students to work in pairs to find out something about the units digits of various powers of numbers.

Homework and answers

Homework for this lesson is available in the homework folder. Consolidation and extension exercises are available in the further practice folder. Answers to the exercises can be found in the answers folder.

5.4 Exponential growth and decay

Syllabus references

Statement	Extended
E1.17	Use exponential growth and decay in examples such as depreciation and population change

Collins references

- Student's Book pages 82–85

Learning aims

- To understand the concept of exponential growth and decay.

Key words

- exponential growth, exponential decay

Prior knowledge

Students need to understand how to calculate compound interest.

Students need to be familiar with the use of the power keys on their calculators.

Common mistakes and remediation

Students will sometimes use just the percentage change, rather than adding it to 1, to find the multiplier.

Useful tips

Ensure that students understand that growth means an increase, while decay or depreciation reduces the value. They should always check that their result is sensible.

Starter

- Tell the story of a wise worker who was asked to do a very difficult, dirty job. It would last for 30 days.
- He agreed he should be paid $1 for completing the first day, but then his total pay would double for every additional day he completed on this horrible job.
- Discuss with the class whether this sounds like a good offer.
- With students who need more **support**, build the sequence on the board, showing the pattern of doubling.
- The class will realise that if the worker lasted 30 days he would get $\$2^{(30-1)}$ at the end of the job.
- Ask how much they think this is. (537 million dollars)

Main lesson activity

- Refer again to the story of the worker and explain that this is an example of *exponential growth*.
- Talk through another example. 'The population of a country was 8 million in 2005. As a result of immigration, this increased by 5% every year over the next ten years. What was the population in 2015?'
- Show how to calculate this as 8 million × 1.05^{10} = 13 million.

- Ask students who need more of a **challenge** what the population would be after N years. (8 million × 1.05^N)
- Now introduce the idea of *exponential decay*. 'One year, in very hot weather, it was estimated that a reservoir lost 8% of its water every day. On 1 August the reservoir held 5 million litres of water. Estimate how much water there would be in the reservoir on 20 August.'
- Show that the reduction is 8% and so the reducing factor or multiplier is 0.92 per day. Hence, after 19 days, on 20 August, the estimated amount will be 5 million × 0.92^{19} = 1 million litres.
- Ensure that students understand the words *growth* and *decay* in these contexts. Refer to Example 3 to reinforce the idea.
- Provide an example of depreciation so that students recognise and understand the different words that describe exponential decay. Use the example, 'Jade bought a bike for $1250. She knew it would depreciate in value by about 10% each year. What is the estimated value of the bike after 5 years?'
- Show that the reduction is 10%, leading to a reducing factor of 0.9 each year. Hence, after 5 years, the value of the bike will be $1250 × 0.9^5 = $738.
- Ensure the pupils are familiar with the word *depreciation*.
- The students can now do Exercise 5D.

Plenary

- Talk about a tree that had been planted. Say that its height was expected to increase by 20% of its initial height every year, for the first ten years of growth. How long would it take to double in height?
- Students will probably want to know the initial height of the tree. Discuss whether this makes a difference to the problem. It does not, but they could use any suggested height to solve the problem.
- Ask, 'What do you need to know? You need to find the value of N for which 1.2^N is first equal to 2 or more.'
- Demonstrate how to find out by trial and improvement. Ask for suggestions, for example, try $N = 5$.
- $N = 5$ gives $1.2^5 = 2.49$, which is too large, so try a smaller number $N = 4$.
- $N = 4$ gives $1.2^4 = 2.07$, which is close to the right answer but is still too large, so try $N = 3$.
- $N = 3$ gives $1.2^3 = 1.7$, which is too small, so the best answer is after 4 years.

Homework and answers

Homework for this lesson is available in the homework folder. Consolidation and extension exercises are available in the further practice folder. Answers to the exercises can be found in the answers folder.

6.1 Inequalities

Syllabus references

Statement	Core
C1.5/E1.5	Order quantities by magnitude and demonstrate familiarity with the symbols $=, \neq, >, <, \geqslant, \leqslant$.

Collins references

- Student's Book pages 88–90

Learning aims

- Interpret the symbols $=, \neq, >, <, \geqslant, \leqslant$ to identify possible values.

Key words

- equals, greater than, less than

Prior knowledge

Students will need an understanding of the work covered in Lesson 4.3.

Common mistakes and remediation

When giving all the possible values, students may miss out zero altogether. Emphasise the importance of including it.

Useful tips

The 'less than' symbol (<) resembles a letter L and this is a way students can remember it. The 'greater than' symbol (>) is the opposite way around. Students may also remember it by the fact that the arrow tips in > and < points towards the smaller numbers.

"Greater than" and "more than" mean the same thing. Both terms are in common use. You can use either.

Starter

- Give out cards with a number on each. Include things such as decimals, fractions, squares and square roots. For students who need more **support**, you may want to reduce your range of numbers. Also give out cards with the symbols $=, <, >, \geqslant$ and $\leqslant$.
- Ask students to make correct statements using their cards,

  e.g. $2.5 = \frac{5}{2}$ $\quad +\sqrt{25} < +\sqrt{36}$ $\quad -4 < 1$ $\quad \frac{1}{4} < 0.5 < \frac{3}{4}$

Main lesson activity

- Explain the symbols as described in the Student's Book.
- Go through Examples 1 and 2. You may need to remind students about the meaning of the word 'integer'.
- Emphasise to students that they should consider carefully whether or not to include 'end values'.
- Students who need more **support** may need to draw number lines to help them.
- Students can now do Exercise 6A – in question 2 remind students that they will need to convert the fractions to decimals in order to answer this question.

Plenary

- Check students' understanding by writing the following inequalities on the board and asking them to find the integer values that satisfy both of them (i.e. 2, 3, 4).

  $2 < x < 7$ $\quad$ $-6 < x \leqslant 4$

Homework and answers

Homework for this lesson is available in the homework folder. Consolidation and extension exercises are available in the further practice folder. Answers to the exercises can be found in the answers folder.

6.2 Sets and Venn diagrams

Syllabus references

Statement	Core
C1.2/E1.2	Understand and use set language, notation and Venn diagrams to describe sets.

Collins references

- Student's Book pages 90–93

Learning aims

- To understand the notation of Venn diagrams.

Key words

- Venn diagram, set, element, universal set, union, intersection

Prior knowledge

Students need to understand, and be able to apply, the inequality symbols explained in Lesson 6.1.

Common mistakes and remediation

Students often confuse the notation union and intersection. Reminding them that union means altogether and that u looks like ∪, the symbol for union, should help.

Sometimes students write the elements in the intersection twice. That can be a problem if at some point they need to count the number of elements in the intersection. Ask them to check carefully that no element is written twice.

Useful tips

Advise students that when they are working with sets described in words or with inequality symbols, they should write down all the elements in the set before they attempt the question, irrespective of whether they are specifically asked to do so.

Starter

Revise work on types of numbers and use of inequality signs, as this will be required when working on the exercise in this chapter.

Ask students the following:

- What are the factors of 24? (1, 2, 3, 4, 6, 8, 12, 24)
- What are the prime numbers less than 20? (2, 3, 5, 7, 11, 13, 17, 19)
- What are the prime factors of 70? (2, 5, 7)
- What are the integer values that satisfy the following inequalities?

  a $2 \leqslant x \leqslant 8$ (2, 3, 4, 5, 6, 7, 8) **b** $2 < x \leqslant 5$ (3, 4, 5) **c** $\frac{1}{2} \leqslant x \frac{8}{3}$ (1, 2)

Main lesson activity

- Explain to students that this lesson is going to focus on the notation associated with Venn diagrams.
- Explain to students that a 'set' is a collection of items, which can be numbers, objects or symbols.
- Items are listed or described inside curly brackets, like this, { }, and separated by commas.
- We usually use capital letters to name sets.
- Explain to students that dots are used when we cannot list all the elements, e.g. the set of even numbers {2, 4, 6, 8...} and that we can describe sets using inequality symbols, e.g. {integer: $5 \leqslant x \leqslant 10$} = {5, 6, 7, 8, 9, 10}.
- Ask students to give you the set of the first six even numbers, and call this '*A*', and the set of the first three odd numbers, and call this '*B*'.
- So, we can say A = {2, 4, 6, 8, 10, 12} and B = {1, 3, 5}.
- The items in the set are called 'elements'.
- Explain that the number of elements in a set can be shown as n(set), which means n(A) = 6 and n(B) = 3
- Ensure that students understand how you arrive at the Venn diagram shown in the Student's Book and highlight the fact that the two circles are used to represent the sets T and F, and the rectangle that encloses them represents the universal set.
- Draw two copies of the Venn diagram shown in the Student's Book.
- Ask students the following questions:
  - 'Which elements are in both T and F?' (1, 5)
    Explain that this is called the 'intersection' of T and F. We write this as $T \cap F$. Shade $T \cap F$ on your first diagram.
  - 'Which elements are in T and F or both?' (1, 2, 3, 4, 5, 10, 15, 20)
    Explain that this is called the 'union' of T and F. We write this as $T \cup F$. Shade $T \cup F$ on your remaining diagram.
- Ask students to read through Example 3 in the Student's Book, to check their understanding. Students can now do Exercise 6B.
- Some students who need more **support** may find the intersection of three sets difficult so may need extra support on questions 6 onwards.

Plenary

Go through question 8 of Exercise 6B to confirm that students have understood this lesson.

Homework and answers

Homework for this lesson is available in the homework folder. Consolidation and extension exercises are available in the further practice folder. Answers to the exercises can be found in the answers folder.

6.3 More about Venn diagrams

Syllabus references

Statement	Extended
E1.2	Understand and use set language, notation and Venn diagrams to describe sets and represent relationships between sets.

Collins references

- Student's Book pages 93–99

Learning aims

- To understand and use language and notation associated with sets.
- To represent sets using Venn diagrams.

Key words

- Venn diagram, element, complement, empty set, universal set, subset, union, intersection

Prior knowledge

Students need to be able to recall the language and notation introduced in Lesson 6.2.

Common mistakes and remediation

Students sometimes confuse the notation for union and intersection. Suggest that the 'U' symbol looks like a 'u', which is the first letter of the word 'union'.

Useful tips

Advise students that when drawing Venn diagrams, they should draw them to be a reasonable size so that the information can be read easily.

Starter

- Ask students to list the elements of the following sets, given that x is a single-digit positive integer.
  $A = \{x: x \text{ is even}\}$
  $B = \{x: x < 5\}$
- What is n(A)? What is n(B)?

Main lesson activity

- Explain to students that this lesson is going to focus on using the notation of sets and subsets.
- The symbol $\notin$ is used to mean 'is not an element of', and so we can say $10 \notin B$.
- Now introduce the term 'subset', which means that all the elements of one set are also in another. Therefore, $C \subseteq D$.
- Define the 'empty set' as the set with no elements. The symbol $\varnothing$ is used to mean the empty set.
- Draw a set $M = \{1, 3, 5\}$ on the board. Ask students to give you all the subsets of M – they will probably give the sets {1}, {3}, {5}, {1, 3}, {1, 5}, {3, 5}. Explain to them that we also include the set itself and the empty set, so {1, 3, 5} and $\varnothing$ are also subsets, giving eight subsets altogether.
- Work through the example at the start of 6.3 in the Student's Book.
- Ensure that students understand how you arrive at the Venn diagram shown in the Student's Book and highlight the fact that the two circles are used to represent the sets A and B, and the rectangle that encloses them represents the universal set.
- Extend to consider Examples 4 and 5.
- Students can now do Exercise 6C.

Plenary

Discuss a selection of questions from Exercise 6C to confirm that students have understood this lesson.

Homework and answers

Homework for this lesson is available in the homework folder. Consolidation and extension exercises are available in the further practice folder. Answers to the exercises can be found in the answers folder.

7.1 Ratio

Syllabus references

Statement	Core
C1.11/E1.11	Give ratios in their simplest form. Divide a quantity in a given ratio. Use proportional reasoning and ratios in context.

Collins references

- Student's Book pages 102–108

Learning aims

- Simplify a ratio.
- Express a ratio as a fraction.
- Divide amounts in a given ratio.
- Use map scales.
- Complete calculations from a given ratio and partial information.

Key words

- ratio, cancel, simplest form, map scale

Prior knowledge

Students should know their multiplication tables up to 10 × 10, how to cancel fractions, how to find a fraction of a quantity and how to multiply and divide, both with and without a calculator.

Common mistakes and remediation

A common error is to forget to express all the quantities in the same units. Keep reminding students to sort out the units first. When expressing 3 : 5, for example, as portions of a whole quantity, students may write it as $\frac{3}{5}$. Always ask students for the number of parts in the ratio (8).

Useful tips

Avoid relating ratios to probability or chance as this will lead to errors later when students think a ratio of 1 : 3 is equal to $\frac{1}{3}$ instead of $\frac{1}{4}$ and $\frac{3}{4}$.

Starter

- Give students practice in converting between metric units by using quick-fire questions around the room.
- Ensure that all the conversions between units in Exercise 7A are covered: dollars, cents, minutes, hours, days, weeks, centimetres, millimetres, metres, kilograms, grams.
- Write down the conversions for students who need more **support**. Present more stretching conversions for students who need more of a **challenge**: litres and centilitres, tonnes and kilograms.
- Write pairs of fractions on the board with either the numerator or denominator from one fraction missing. Ask students for the missing number to make the fractions equivalent.
- **Challenge** students with improper fractions using large numbers, for example: $\frac{120}{40} = \frac{?}{5}$ or $\frac{720}{9} = \frac{?}{3} = \frac{?}{1}$

Main lesson activity

Common units and ratios as fractions

- Explain that a ratio is a way of comparing the sizes of two or more quantities, and any quantities should be given in the same units, e.g. compare centimetres to centimetres.
- A ratio can be expressed as a fraction of a quantity. The denominator of the fraction is obtained by adding the numbers in the ratio.
- Explain that if the units of the quantities are different, then one must be converted before the ratio can be formed and simplified. This is because a ratio does not have units. Ask students how many times bigger 1 kilometre is than 1 metre (1000). Write 1 metre to 1 kilometre = 1 m : 1 km = 1 : 1000.
- Write 1 : 2, 2 : 4, 3 : 6 on the board and explain that these are all equivalent.
- Now write 4 : 12 on the board and ask for some equivalent ratios (1 : 3, 2 : 6, etc.).
- Show students Examples 1 and 2 in the Student's Book, which combine the work on units with the work on ratios. Students can now do Exercise 7A.

Dividing amounts in a given ratio

- Ask students to tear a sheet of A4 paper into 16 pieces by tearing in half then half again and so on.
- Write the ratio 1 : 3 on the board. Tell the students to put the pieces into two piles, with three in the second pile for every one in the first pile. Explain that they have shared the 16 pieces in the ratio 1 : 3. Now ask what fraction of the 16 pieces are in the first pile ($\frac{1}{4}$) and the second pile ($\frac{3}{4}$). Reinforce this by stating that $\frac{1}{4}$ of 16 pieces = 4 pieces and $\frac{3}{4}$ of 16 pieces = 12 pieces. Then show students Example 3.
- Discuss map scales as in the Student's Book, and then work through Example 4. Students can now do Exercise 7B.

Calculating with ratios when only part of the information is known

- Explain that sometimes students will only be given the value of one part of a ratio. For example: I share some money between myself and my brother in the ratio 2 : 3; I get $2. How much does my brother get? ($3). Tell students that this question could be: divide $5 in the ratio 2 : 3.
- Give students who need more of a **challenge** other examples and ask them to rewrite the questions like this. For example the weight of a car and a lorry are in the ratio 1 : 20; the weight of the lorry is 40 tonnes, what is the weight of the car? (Divide 42 tonnes in the ratio 1 : 20.)
- For other students, emphasise that they must decide what information the question is giving. Talk through a few questions from Exercise 7C and discuss what information is given. Show students Examples 5 and 6. Allow them to choose their favoured method.
- Students can now do Exercise 7C.

Plenary

Show the students a spider diagram as on Resource 7.1. Ask students to complete it, giving the correct amount of money for each ratio. Increase the value to **challenge** students, e.g. $108.

Homework and answers

Homework for this lesson is available in the homework folder. Consolidation and extension exercises are available in the further practice folder. Answers to the exercises can be found in the answers folder.

7.2 Speed

Syllabus references

Statement	Core
C1.12/E1.12	Use common measures of rate. Apply other measures of rate. Solve problems involving average speed.

Collins references

- Student's Book pages 108–111

Learning aims

- Recognise the relationship between speed, distance and time.
- Calculate average speed from distance and time.
- Calculate distance travelled from the speed and the time taken.
- Calculate the time taken on a journey from the speed and the distance.

Key words

- average, speed, distance, time

Prior knowledge

Students need to know how to multiply and divide. Students should also be familiar with common units for speed.

Common mistakes and remediation

Students often fail to realise that, for example, 30 minutes is not 0.3 of an hour, or 15 minutes is not 0.15 of an hour. Pointing out these common errors before they are made can often prevent them.

Useful tips

Students do not relate units (km/h) to formulae. Explain that 'per' means 'over' in Latin and is related to division, so kilometres per hour means kilometres over hours, i.e. distance ÷ time.
Use this diagram to remind students of the relationships between distance (*D*), time (*T*) and speed (*S*).

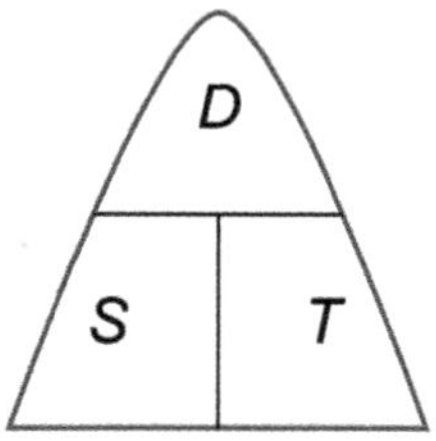

$D = S \times T \qquad S = \frac{D}{T} \qquad T = \frac{D}{S}$

Starter

- Give students simple quick-fire speed questions. Students could display answers on small whiteboards. For example:
  - Speed 30 km/h, how far in 30 minutes? (15 kilometres)
  - Speed 60 km/h, how far in 2 hours? (120 kilometres)
  - Speed 20 km/h, how far in 1 hour 30 minutes? (30 kilometres)
  - Distance 40 kilometres, time 1 hour, what is the average speed? (40 km/h)

- Distance 60 kilometres, time 2 hours, what is the average speed? (30 km/h)
- Distance 10 kilometres, time 30 minutes, what is the average speed? (20 km/h)
- Speed 50 km/h, distance 50 kilometres. How long to travel? (1 hour)
- Speed 60 km/h, distance 30 kilometres. How long to travel? (30 minutes)
- Speed 30 km/h, distance 90 kilometres. How long to travel? (3 hours)

Main lesson activity

- Remind students of a starter question using speed × time = distance and ask them how they worked out the answer, e.g. Speed 60 km/h, how far in 2 hours? Ask, 'How did you get 120 kilometres?' Prompt them to give you a formula, e.g. 60 kilometres in 1 hour, so in 2 hours $60 \times 2 = 120$.
- Repeat for questions using distance ÷ time = speed, e.g. Distance 60 kilometres, time 2 hours. Ask, 'How did you get 30 km/h?' Again, prompt them to give you the formula, e.g. 60 kilometres in 2 hours, $60 \div 2 = 30$ kilometres in 1 hour, so 30 km/h.
- Repeat for questions using distance ÷ speed = time, e.g. Speed 30 km/h, distance 90 kilometres – how long to travel? Again, prompt them to give you a formula, e.g. 30 km/h is 30 kilometres per hour so 30 kilometres in 1 hour, $90 \div 30 = 3$, so 3 hours.

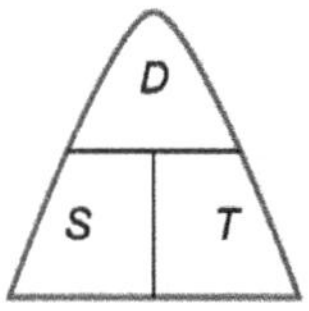

- Now draw the triangle to show the three formulae.
- Remind students how to convert hours and minutes into hours and vice versa, e.g. 20 minutes = $20 \div 60$ hour = $\frac{1}{3}$ hour; 0.8 of an hour = 0.8×60 minutes = 48 minutes.
- Show students Examples 7, 8 and 9.
- Students who need more **support** will need reminding to change hours and minutes into hours, so reinforce this point when looking at Examples 7, 8 and 9.
- Students can now do Exercise 7D. Tell students to note the hints provided with this exercise. Stress that they should relate the questions to real-life experience and decide whether their answers sound reasonable or sensible. For example, for question 11, they should ask themselves if the average speed for the whole journey is between the two speeds given. If not, this suggests they have made a mistake. Students who need more of a **challenge** should also realise that as the train is travelling for longer at the higher speed the answer will be closer to this speed.

Plenary

- Give students a two-part journey, e.g. 100 kilometres at an average speed of 40 km/h followed by 90 kilometres at an average speed of 60 km/h. Go through the process of calculating the average speed of the whole journey. (190 kilometres in 4 hours = 47.5 km/h)
- Discuss the checking methods outlined earlier to test whether answers are sensible, e.g. point out that 100 kilometres at 40 km/h followed by 100 kilometres at 60 km/h would give an average of 50 km/h. So the answer to the problem will be less than 50 km/h.

Homework and answers

Homework for this lesson is available in the homework folder. Consolidation and extension exercises are available in the further practice folder. Answers to the exercises can be found in the answers folder.

7.3 Rates

Syllabus references

Statement	Core
C1.12/E1.12	Use common measures of rate.

Collins references

- Student's Book pages 111–115

Learning aims

- Recognise that rate is the relationship between two sets of quantities.
- Work with density and pressure.

Key words

- rate, density, pressure

Prior knowledge

Students need to understand the work on speed in Lesson 7.2.

Common mistakes and remediation

Students often fail to realise that, for example, 30 minutes is not 0.3 of an hour, or 15 minutes is not 0.15 of an hour. Students also need to be careful when different units are involved, e.g. in question 5 of Exercise 7E, the rate is given as 50 cm per 1000 years but the question involves mm/year. Remind students that they need to change one of the units when dealing with problems such as this.

Useful tips

Encourage students to check that their answers are sensible when compared with the given information.

Starter

- Revisit Example 7 from Chapter 7.2 of the Student's Book:

  Paula drove a distance of 270 km in 5 hours. What is her average speed? (54 km/h)
- Explain that we can describe this as the rate at which Paula travels, i.e. 54 km in 1 hour.

Main lesson activity

- Explain to students that the concept of speed can be applied more generally to show the way quantities change.
- Consider Example 10 in the Student's Book:
  Using the formula you can substitute the values to find the density.
- Then there are basically two ways to proceed.
  The first method is to use the idea of proportion and note that since $\frac{1.6}{2.4} = \frac{2}{3}$, the volume is $\frac{2}{3}$ that of the first piece of gold.
  The second way is to use the formula.
  If students are familiar with the speed triangle, then you might want to build on that idea with the density triangle.
- If necessary, go over Example 11 and show the pressure triangle (like the speed triangle).
- To illustrate the effect of pressure, why not demonstrate with your weight and your shoes?
  You can measure the area your soles cover by drawing round them on cm^2 paper, and the force in Newtons will be your mass in kg multiplied by 9.8.
  Now if you wear stilettos you could show your students the force exerted by them on the ground is quite immense, which is why places of historic interest do not like people walking on their floors in stilettos.
- Students can now do Exercise 7E.

Plenary

Discuss students' answers to questions 8 and 9, to ensure that they understand the concepts involved.

Homework and answers

Homework for this lesson is available in the homework folder. Consolidation and extension exercises are available in the further practice folder. Answers to the exercises can be found in the answers folder.

7.4 Proportion

Syllabus references

Statement	Core
C1.11/E1.11	Use proportional reasoning and ratios in context.

Collins references

- Student's Book pages 115–117

Learning aims

- Recognise and solve problems using proportion.

Key words

- unitary method, proportion, single unit value

Prior knowledge

Students should know how to multiply and divide without using a calculator.

Common mistakes and remediation

Errors will occur when some information in the question leads to an answer that is not sensible – for example, when adapting a recipe, students will give answers of 1.5 eggs. (See main lesson activity.) Point out that in practical situations answers must always be sensible.

Useful tips

Encourage students to check that their answers are sensible when compared with the given information.

Starter

- Write the question below on the board and ask the students to give you an answer.

  6 cost 36 cents. How much do 5 cost?
- Students will have an intuitive idea of this and will answer 30 cents.
- Discuss the idea of finding the cost of one item and then scaling up to the required number.
- Repeat with other values, for example 8 cost 56 cents, how much will 6 cost?
- For students who need more **support**, use whole number answers.
- For students who need more of a **challenge**, use decimals, e.g. 7 cost $10.50, how much do 6 cost?

Main lesson activity

- Explain to students that the method they have used in the starter, i.e. finding the cost of one item and then scaling, is called the *unitary method.*
- Write this recipe on the board.

  For 4 people

  100 g butter

  300 g flour

  2 eggs

  1 cup of milk

- Ask students how they could change this for 3 people – don't worry about the eggs at this stage. (Divide by 4 and multiply by 3.)
- Now tell students that you only have 300 g of butter and 600 g of flour but plenty of eggs and milk, and ask how many you could serve and why (8, only enough flour for 8).
- Now work through Examples 13 and 14.
- Students can now do Exercise 7F. Students should be comfortable with the unitary method as a safe, reliable method that can always be used but students who need more of a **challenge** will be expected to adapt questions to use more efficient methods, e.g. in question 7b recognising that 150 kg is three-quarters of 200 kg and using mental methods to work out that three-quarters of 12 weeks = 9 weeks.

Plenary

- Write on the board a simple recipe for six people. Ask students how much of each ingredient would be needed for one person, two people, 12 people, and so on. Include an item, such as two eggs, which cannot be easily divided.
- Discuss the idea of 'best buys' by asking students which is better value, 3 chocolate bars for 90 cents or 5 chocolate bars for $1.50. (Neither, both work out at 30 cents for each bar.)
- Now discuss with students how they would compare values.

Homework and answers

Homework for this lesson is available in the homework folder. Consolidation and extension exercises are available in the further practice folder. Answers to the exercises can be found in the answers folder.

8.1 Rounding whole numbers

Syllabus references

Statement	Core
C1.9/E1.9	Make estimates for calculations involving numbers, quantities and measurements.

Collins references

- Student's Book pages 120–121

Learning aims

- Round a number.

Key words

- approximation, rounded up, rounded down

Prior knowledge

Students should know that numbers are sometimes approximated because it is not always sensible to use exact numbers.

Common mistakes and remediation

Students often have difficulty when rounding a number involving hundreds to the nearest 10 (e.g. rounding 197 to the nearest 10). Students will find extra practice on this type of rounding very useful. Counting in tens may also help. It is very important to identify the 'decider' digit and to ask 'Has this digit gone *at least half-way*?' If the answer is yes, the digit to the immediate left is rounded up.

Useful tips

Students may find it easier to turn a question back on itself so that it becomes a standard question. For example, question 4 states: There are 789 people living in Elsecar. To discover if this statement is correct, students may try to work out if 789 is in the range for 800 to the nearest 100. However, they may find it easier to write the number to the nearest 100. This will give them an answer of 800, showing that the statement could be true.

Starter

- Write multiples of 10 on the board: 10, 20, 30, 40, 50, 60.
- Read out numbers at random and ask students to pick the nearest value from the numbers on the board; for example, if you call out 43, they answer 40 and if you call out 65, they answer 70.
- Use this exercise to reinforce the idea that numbers ending with the digits 5, 6, 7, 8, 9 are rounded up and those ending with the digits 0, 1, 2, 3, 4 are rounded down (truncated).
- Write 100, 200, 300, 400, etc. on the board and repeat the exercise.

Main lesson activity

- Guide students to the idea that rounding gives an approximation and the accuracy of this approximation depends on the number of figures to which the number was rounded. Emphasise that even if the number is rounded, we can still work out what the maximum and minimum values are (as the bullet points in the Student's Book show).
- Discuss with students examples of where we use rounded information. These include numbers of people attending a concert or sports event, times taken to travel large distances, population, cooking times, etc. Discuss everyday examples of average amounts (such as the weight of the contents of a packet of cornflakes, the life of a light bulb, the number of sweets in a packet).
- Work through questions 1, 2 and 3 from Exercise 8A as a class, asking each student to answer one question. If the students need more **support**, allow them to work in pairs to work out the correct answer.
- If the students need more of a **challenge**, repeat the starter, writing multiples of 5 on the board (5, 10, 15, 20, 25, 30, etc.) and asking for values to the nearest 5. Then, progress on to picking a time (2:30:30, 4:15:00, 12:45:15) and asking for the time to the nearest 5 minutes, 30 seconds or 20 seconds.
- Students can now complete Exercise 8A.

Plenary

- Use real-life examples to practise rounding. Examples include the following:
  - The number of sweets in a tube is 30, to the nearest 10 – how many could there be in the tube?
  - There are 25 000 people attending a rugby match to the nearest 1000; what is the least possible attendance? What is the greatest possible attendance?

Homework and answers

Homework for this lesson is available in the homework folder. Consolidation and extension exercises are available in the further practice folder. Answers to the exercises can be found in the answers folder.

8.2 Rounding decimals

Syllabus references

Statement	Core
C1.9/E1.9	Round values to a specified degree of accuracy.

Collins references

- Student's Book pages 121–122

Learning aims

- Round a decimal number.

Key words

- round, digit, decimal place

Prior knowledge

Students need to be able to round to the nearest 10, 100 and 1000.

Common mistakes and remediation

Students often have difficulty rounding numbers such as 13.987 to 1 decimal place (14.0) or 169.995 to 2 decimal places (170.00). Students will find extra practice on this type of rounding very useful.

Useful tips

Students may find it easier to think about the values the given number lies between and 'fill in zeros', e.g. 13.987 lies between 13.900 and 14.000. The students can then clearly see that the number is closer to 14.000 or 14.0 to 1 decimal place.

Starter

- Revise the work from the previous lesson by going around the class asking students to round a given number to the nearest 10, 100, 1000. Examples:
  - to the nearest 10: 13 (10), 47 (50), 128 (130), 364 (360), 1058 (1060).
  - to the nearest 100: 85 (100), 469 (500), 732 (700), 2115 (2100), 8888 (8900).
  - to the nearest 1000: 657 (1000), 1983 (2000), 4318 (4000), 8563 (9000).

Main lesson activity

- Show students the place value table below. Then ask them to read out the number (8742.163).

Thousands	Hundreds	Tens	Units	·	Tenths	Hundredths	Thousandths
8	7	4	2	·	1	6	3

- Now ask students to round the number to the nearest thousand (9000), nearest hundred (8700), nearest ten (8740), nearest whole number (8742), nearest tenth (8742.2) nearest hundredth (8742.16).
- Explain that nearest tenth is the same as saying 'to 1 decimal place', nearest hundredth is the same as saying 'to 2 decimal places' and so on.
- Remind students that the rounding rule using the value of the next digit applies, i.e. less than 5, just remove the unwanted places, 5 or more, add 1 onto the digit in the last decimal place. Show students the examples in the Student's Book.
- Students can now do Exercise 8B. Tell students that question 5 of Exercise 8B shows a practical application of rounding numbers up to work out an estimate of the price of goods in a shopping basket.

Plenary

Use the diagram of an alien with six legs, provided as Resource 8.2. Ask for volunteers to round the number given, i.e. 63.598 346 to 5, 4, 3, 2, 1 and then no decimal places, filling in legs as they go. Repeat with another number on the second alien.

Homework and answers

Homework for this lesson is available in the homework folder. Consolidation and extension exercises are available in the further practice folder. Answers to the exercises can be found in the answers folder.

8.3 Significant figures

Syllabus references

Statement	Core
C1.9/E1.9	Round values to a specified degree of accuracy.

Collins references

- Student's Book pages 123–124

Learning aims

- Round to a given number of significant figures.

Key word

- significant figure

Prior knowledge

Being able to round to a given number of decimal places (10, 100, 1000) will help students learn how to approximate.

Common mistakes and remediation

These are best shown with two examples, both rounded to 2 significant figures (2 sf):

- 275 986 = 28, rather than 280 000 (the zeros are forgotten)
- 0.000 371 04 = 0.00 or 37, rather than 0.000 37 (2 dp instead of 2 sf, or zeros forgotten)

Students often have difficulty with the power of 10 in the answer. Sometimes students try to be too accurate in approximating calculations, rounding to more than 1 sf, or not rounding at all when using a calculator, assuming a totally correct answer is better than an approximation (when sometimes answers are better as 'roughly' or 'nearly').

Useful tips

Emphasise the value of estimating answers before carrying out calculations, or estimating to save time at every opportunity. Understanding the importance of this will enhance students' skills. Encourage students also to use rounding and estimation as a way of checking that their answers are sensible.

Starter

- Display: 0.04, 0.4, 4, 40, 400. Ask students what they have in common. (All have 4.) What else? (Zeros.)
- Now write: 0.065, 0.65, 6.5, 65, 650. Ask how this set differs from the first set. (Two non-zero digits.)
- Ask students for a set of numbers with three non-zero digits, e.g. 0.0123, 0.123, 1.23, 12.3, 123, 1230.
- Explain to students that you will be referring to these numbers in the main lesson activity.

Main lesson activity

- Refer to the numbers in the starter. The first set had one non-zero digit (4). Those in the second set had two (65) and those in the third, three (123). The non-zero digits are significant figures or sf. Zeros before or after these digits maintain the value of the number but are not sf. Highlight the number 0.7002 in the table in the Student's Book. Explain that this has 4 significant figures, since each 00 in the middle is holding a place.
- Ask students to explain why, in the table, 45 281 is 50 000 to one sf. (Only one non-zero digit, but place value must be correct.) Writing 45 281 to 1 sf requires the nearest number to 45 281 with only one non-zero digit, which is 50 000; to 2 sf it is 45 000, and so on. Ask quick-fire questions to reinforce the method.
- Now work backwards. If a number is 80 to 1 sf, what could the original number have been? (Any number from 75 to 84.999...)
- Students can now do Exercise 8C, in which question 6 is a problem-solving (PS) question.

Plenary

- Write on the board: 571.8 to 1 sf = 6, 0.003276 to 2 sf = 0.00, 25.37 to 1 sf = 30.00.
- Discuss with the students whether or not these statements are correct.

Homework and answers

Homework for this lesson is available in the homework folder. Consolidation and extension exercises are available in the further practice folder. Answers to the exercises can be found in the answers folder.

8.4 Estimating answers to calculations

Syllabus references

Statement	Core
C1.9/E1.9	Round answers to a reasonable degree of accuracy in the context of a given problem.

Collins references

- Student's Book pages 124-126

Learning aims

- Make estimates by rounding numbers to 1 significant figure

Key words

- Estimate, significant figure

Prior knowledge

Students should know how to round a number to 1 sf.

Common mistakes and remediation

The purpose of rounding is to give an easy calculation that you can do in your head. Encourage students not to use a calculator and to make the calculations easy ones.

Useful tips

The idea is to get an approximate answer. There can be more than one possible answer.

Starter

- Ask students to write down these numbers rounded to 1 significant figure:
  283 (300) 73199 (70000) 18.7654 (20) 3.14159 (3) 0.090909 (0.09)
- Now asks student to tell you a number that is 400 to 1 sf. Write down all the answers you are given. Any number in the interval $350 \leqslant 400 < 450$ is an acceptable answer.

Main lesson activity

- Draw a rectangle and label the sides as 63.9 cm and 19.2 cm. Ask students, without using a calculator, to give you an estimate of the answer. You might need to explain that an estimate is a number that you think is close to the correct answer.
- Take suggestions from a few students and ask them to explain how they worked out their estimate.
- Show how to find an estimate by rounding each length to 1 sf.
  So $63.9 \times 19.2 \approx 60 \times 20 = 1200$ cm^2 is an estimate of the area.
- Point out that by rounding to 1 sf you will get a calculation that you can do easily in your head without a calculator.

- Give students these to try:
  - 4.2913 × 4.827 (4 × 5 = 20)
  - 6.3585 × 327 (6 × 300 = 1800)
  - 8063 ÷ 18.05 (8000 ÷ 20 = 400)
  - 28.67 × 0.2155 (30 × 0.2 = 6)
  - 48.12 ÷ 1.903 (50 ÷ 2 = 1.903)
  - 1.837 ÷ 23.61 (2 ÷ 20 = 0.1)
- Now give them this more complicated example: $\frac{18.2 \times 9.41}{7.14^2}$
- Ask students to explain the steps they would take to find an estimate of this. You are looking for the following:
- $18.2 \times 9.41 \approx 20 \times 9 = 180$ $7.14^2 \approx 7^2 = 49$
- So you want $180 \div 49 \approx 200 \div 50 = 4$
- Explain that you cannot divide by 49 in your head so round 180 and 50 to 1 sf to make an easier calculation.
- Now look at Example 2 in section 8.4. Say that the final calculation of $35 \div 8$does not have to be given exactly. The answer is 4 with a remainder of 3 so just write it to 1 sf.
- It would also be acceptable to round the 35 to 40 and work out $40 \div 8 = 5$
- Both 4 and 5 are acceptable estimates of the exact answer. (The exact answer is 4.586...)
- Students can now do Exercise 8D.
- If the students need more of a **challenge**, you could ask them to explain why the following calculation gives an underestimate of the correct answer:
- $5.471 \times 8.235 \approx 5 \times 8 = 40$
- It is an underestimate because both 5 and 8 are less than the number so the answer will be less to. You could then challenge them to give you an example of a multiplication where the estimate will be more than the estimate.
- For students who need more **support** give more examples of simple calculations to make sure they understand the method.

Plenary

- Ask students to explain how to estimate the number of seconds in a year.
  Go through the following steps:
  - There are 60 seconds in a minute, 60 minutes in an hour, 24 hours in a day and 365 days in a year
  - The number of seconds in a year is 60 × 60 × 24 × 365
  - An estimate is 60 × 60 × 20 × 400
  - 60 × 60 = 3600 which is 4000 to 1 sf
  - 4000 × 24 is approximately 4000 × 20 = 80 000
  - 80 000 × 365 is approximately 80 000 × 400 = 32 000 000 seconds
  - Compare this to a calculator answer (31 536 000)

Homework and answers

Homework for this lesson is available in the homework folder. Consolidation and extension exercises are available in the further practice folder. Answers to the exercises can be found in the answers folder.

8.5 Upper and lower bounds

Syllabus references

Statement	Core
C1.10/E1.10	Give upper and lower bounds for data rounded to a specified accuracy.

Collins references

- Student's Book pages 126–128

Learning aims

- Find the limits of accuracy of numbers that have been rounded to different degrees of accuracy.

Key words

- upper bound, lower bound, limits of accuracy

Prior knowledge

Students should be familiar with all types of rounding: to the nearest 1, 10, 100, as well as to a given number of decimal places or significant figures (sf). It is particularly important that students understand the 'five rule' for rounding up. Students should also know the difference between discrete and continuous data.

Common mistakes and remediation

Students often give an incorrect upper bound, e.g. saying that 22 cm to the nearest centimetre has an upper bound of 22.49 cm or even 22.4 cm (rather than 22.49999… cm, or more usually, 22.5 cm). Pointing out this error to students frequently should help to minimise it.

Useful tips

Students who need more **support** should use number lines to help them understand the limits of accuracy. For example:

2 to the nearest whole number

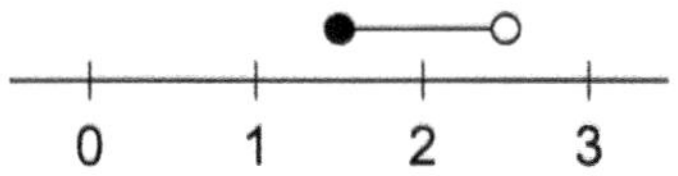

$1.5 \leqslant 2 < 2.5$

1.7 to 1 decimal place

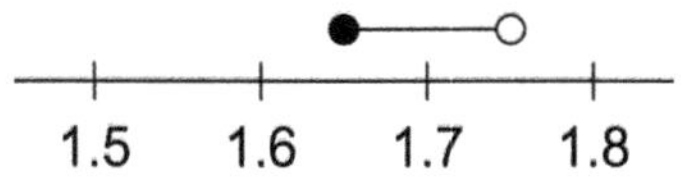

$1.65 \leqslant 1.7 < 1.75$

Students who need more **support** may need a brief reminder of the difference between decimal places and significant figures before doing the starter.

Starter

- Give students a number, e.g. 6513.9604, and ask them to round it to the nearest 1000, 100, 10, 1, 1 dp, 2 dp, 3 dp, 1 sf, 2 sf, 3 sf, 4 sf, as a check of their prior knowledge. (7000, 6500, 6510, 6514, 6514.0, 6513.96, 6513.960, 7000, 6500, 6510, 6514.)
- Repeat with other numbers, e.g. 9178.276 or 999.99.
- Now check that students know the difference between discrete and continuous data.

Main lesson activity

Discrete data

- Tell students that, to the nearest 10, there are 30 students in the next room. Ask them how many people could be in the room (25 to 34 inclusive). Discuss why it could not be 35. (This rounds to 40.) Remind students that the number of people is discrete data. Write 25 $\leqslant$ number of people $\leqslant$ 34 on the board to introduce the idea of using inequalities.
- For students who need more **support**, draw a number line to illustrate the answer. Point out that as the data is discrete only the whole numbers are used, i.e. you cannot have 25.3 people.

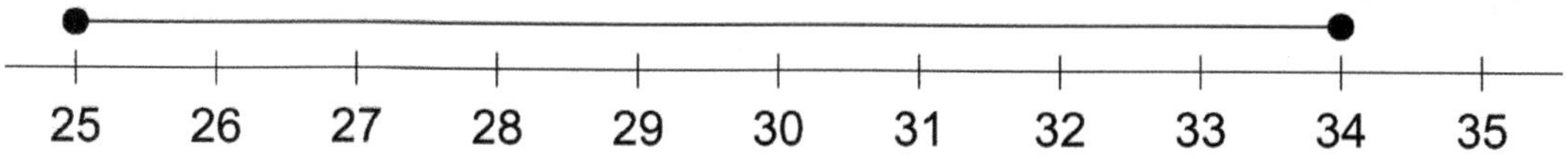

- Now discuss Example 2 and ask students to think of other quantities that are discrete data, e.g. money, number of letters in the post or shoe size. Follow this up with a money question, e.g. 'I have $5 to the nearest dollar, what is the most I could have?' ($5.49)

Continuous data

- Move on to continuous data. Ask students how far it is to a nearby town.
  Pick out an answer, e.g. 8 km, and ask if it is exactly 8 km.
  Tell them to assume that this is to the nearest kilometre.
  Ask them to suggest the minimum and maximum distance it could be.
- Show the information on a number line. Discuss the open (excluding) and closed (including) circles.

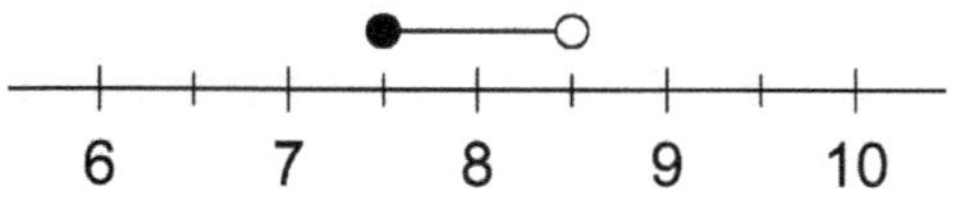

- Explain how the upper bound would actually round to the next value (in this case 9 km) but point out that they do not always have to use recurring decimals although these are correct (in this case 8.49999... km).
- Now work through Examples 3 and 4 before students complete Exercise 8E. Students who need more **support** can complete question 1, but may need to work in pairs to finish the exercise.

Plenary

- Prepare students for the next lesson by introducing the idea of multiple items, as in the last question of Exercise 8E. Discuss what would happen in reality with the marbles. (Some would weigh more than 65 grams and some would weigh less than 65 grams.)
- Follow this up by asking what the biggest difference between the masses of any two of the marbles could be. (65.5 grams – 64.5 grams = 1 gram)

Homework and answers

Homework for this lesson is available in the homework folder. Consolidation and extension exercises are available in the further practice folder. Answers to the exercises can be found in the answers folder.

8.6 Upper and lower bounds for calculations

Syllabus references

Statement	Extended
E1.10	Find upper and lower bounds of the results of calculations which have used data rounded to a specified accuracy.

Collins references

- Student's Book pages 128–131

Learning aims

- Combine limits of two or more variables together to solve problems.

Key words

- upper bound, lower bound

Prior knowledge

Students should be able to write down upper and lower bounds for values that are given to any degree of accuracy. They should know and be able to use basic area and volume formulae.

Common mistakes and remediation

Students sometimes use a maximum value rather than a minimum value (or vice versa) in problems involving subtraction, multiplication or division. If students are not sure which bounds to use they should work out the answers to all combinations and then select the correct answer.

Useful tips

Operation	Minimum	Maximum
Addition ($a + b$)	$a_{min} + b_{min}$	$a_{max} + b_{max}$
Subtraction ($a - b$)	$a_{min} - b_{max}$	$a_{max} - b_{min}$
Multiplication ($a \times b$)	$a_{min} \times b_{min}$	$a_{max} \times b_{max}$
Division ($a \div b$)	$a_{min} \div b_{max}$	$a_{max} \div b_{min}$

Students who need more **support** may have difficulty coping with all four rules or even two of the rules, so break down the exercise into smaller, more manageable parts. Rather than dealing with the rules first, simply ask students to work through each question, writing down the limits (upper and lower bounds) of each value.

You could do this as an oral starter, instead of the starter below.

Starter

- Give the class this problem. A square tile has a side length of 15 cm to the nearest cm (or 2 sf). What are the limits of accuracy? (14.5 cm $\leqslant$ length < 15.5 cm) What are the upper and lower bounds? (15.5 cm and 14.5 cm)

- Now give them this problem. 10 of these tiles are in a row. What is the longest possible length? (155 cm) What is the shortest possible length? (145 cm)
- Extend the question further. If the length to be tiled is 3 m, how many tiles are needed to cover one row? (21) If the length to be tiled is 12 m, how many tiles are needed to cover one row? (83) Ask students why the answer for 12 m is *not* four times the answer for 3 m. (300 ÷ 14.5 = 20.7 → 21 tiles, 1200 ÷ 14.5 = 82.8 → 83 tiles)

Main lesson activity

- Display two objects, e.g. books. State that each book has a mass of 1700 g to the nearest 100 g. Ask: 'What is the greatest possible total mass of the books?' (3500 g)
Now ask: 'What is the greatest difference between the masses?' (100 g)
- Now say that you have a pile of 30 books (to the nearest 10). Ask students how to calculate, without working out the answer, the maximum possible mass of these books. (34 × 1750 g)
Repeat for the minimum possible mass of these books. (25 × 1650 g)
- Now say that you are going to work the question backwards so that a pile of books has a total mass of 34 kg to the nearest kilogram and that each book weighs 1700 g to the nearest 100 grams. Ask for the calculation needed to work out the least possible number of books in the pile.
(33 500 ÷ 1750 (= 19.14 so 20))
- Discuss why it is: minimum mass of pile ÷ maximum mass of each book. If students are not convinced, ask them to use calculators to work out all other possible answers, using upper and lower bounds. Repeat for the calculation for the greatest possible number of books in the pile.
(34 500 ÷ 1650 (= 20.91, so 20))
- Write the answers to the calculations on the board; say that the number of books is discrete, as you cannot have 19.14 books; it has to be rounded up to 20. Similarly, you cannot have 20.91 so this must be rounded down. Repeat for a pile of books of mass 50 kg to show that the answers can vary (working in kilograms, minimum = 49.5 ÷ 1.75 = 28.3 → 29 books,
maximum = 50.5 ÷ 1.65 = 30.6 → 30 books).
- Work through Examples 5 and 6. Then students can do Exercise 8F. In question 2, students need to assume that the mass of each book is at the upper bound. You can extend this question by stating that the trolley safety limit is also given to the nearest kilogram, meaning that it is unsafe to exceed 24.5 kg.

Plenary

- Say these three amounts: 3000 g, 4000 g and 5000 g. Add that each amount is accurate to the nearest 100 g. Ask which of the following are possible total amounts:

  11 500 g   12 050 g   12 500 g   12 600 g   13 000 g (12 050 g)
- Now say that the amounts are accurate to 1 sf. Ask students which of them are possible total amounts for all three items (all are). Show students who need more **support** the calculations to demonstrate the limits of accuracy each time.

Homework and answers

Homework for this lesson is available in the homework folder. Consolidation and extension exercises are available in the further practice folder. Answers to the exercises can be found in the answers folder.

9.1 Standard form

Syllabus references

Statement	Core
C1.8/E1.8	Use the standard form $A \times 10^n$ when n is a positive or negative integer and $1 \leqslant A < 10$. Calculate with values in standard form.

Collins references

- Student's Book pages 134–136

Learning aims

- Change a number into standard form.

Key words

- standard form, index

Prior knowledge

Students should be able to multiply and divide by 10, 100 and 1000. They should know powers of 10, such as: $10^5 = 10 \times 10 \times 10 \times 10 \times 10 = 100\,000$, and be shown that: $10^{-3} = \frac{1}{10} \times \frac{1}{10} \times \frac{1}{10} = \frac{1}{1000} = 0.001$

Common mistakes and remediation

When inputting numbers on a calculator, students often insert an extra × 10; e.g. to key in 1.3×10^6 they key in [1] [.] [3] [×] [10] [×10ˣ] [6], giving an answer that is 10 times larger than required. Point out this error to help students avoid making it.

Useful tips

Suggest that students write these facts into their notes:

- When multiplying a number by a positive power of 10, digits move to the left.
- When multiplying a number by a negative power of 10, digits move to the right.

Starter

- Ask these quick-fire questions to familiarise students with multiplications by powers of 10:
  - What are 5×10, $5 \times 10 \times 10$, $5 \times 10 \times 10 \times 10$? (50, 500, 5000)
  - What are 10^1, 10^2, 10^3? (10, 100, 1000)
  - What are 5×10^1, 5×10^2, 5×10^3? (50, 500, 5000)
  - What are $5 \div 10$, $5 \div 10 \div 10$, $5 \div 10 \div 10 \div 10$? $\left(\frac{1}{2}, \frac{1}{20}, \frac{1}{200}\right)$
  - What are 10^{-1}, 10^{-2}, 10^{-3}? $\left(\frac{1}{10}, \frac{1}{100}, \frac{1}{1000}\right)$
  - What are 5×10^{-1}, 5×10^{-2}, 5×10^{-3}? $\left(\frac{1}{2}, \frac{1}{20}, \frac{1}{200}\right)$
- Now ask quick-fire questions using decimals, e.g. 3.6 instead of 5. Repeat using three digits, e.g. 9.27.

Main lesson activity

Arithmetic powers of 10

- Use the notes in the Student's Book to introduce this topic. Emphasise to students that the power of 10 is called the **index**. Ask students to draw a place value table as shown below.

thousands	hundreds	tens	units	.	tenths	hundredths	thousandths
			5				
		5	0				

- Then ask students to add the powers of 10 to the column headings, as shown below.

thousands	hundreds	tens	units	.	tenths	hundredths	thousandths
1000	100	10	1		$\frac{1}{10}$	$\frac{1}{100}$	$\frac{1}{1000}$

- Now ask students to add in the powers of 10 in their index form, as shown below.

thousands	hundreds	tens	units	.	tenths	hundredths	thousandths
1000	100	10	1		$\frac{1}{10}$	$\frac{1}{100}$	$\frac{1}{1000}$
10^3	10^2	10^1	10^0		10^{-1}	10^{-2}	10^{-3}

- Students should now be able to see that 5 can be written as 5×10^0, as the 5 is in the 10^0 column. Similarly, 50 can be written as 5×10^1.
- Explain to students that standard form is of the form $a \times 10^n$ where $1 \leqslant a < 10$ and n is a whole number. Deduce that $52 = 5.2 \times 10^1$, and $625 = 6.25 \times 10^2$.
- Now extend further to consider numbers with digits after the decimal point.
- Conclude that $0.4 = 4 \times 10^{-1}$, $0.007 = 7 \times 10^{-2}$, $0.123 = 1.23 \times 10^{-1}$.
- Refer students to the other examples in the Student's Book and check their understanding. Students can now do Exercise 9A.

Standard form on a calculator

- Ask students to use their calculators. Input 2 into the display and press the 'equals' (=) sign. Now press '× 2' and then repeatedly press the = sign. The answer will double each time, up to 8 589 934 592; the next time they press =, the display will change to standard form ($1.717\ 986\ 918 \times 10^{10}$). Discuss what has happened: once the number exceeds a certain size (number of digits) the calculator needs another way to display it.
- Repeat by inputting 2 again, but this time use '÷ 2' and repeatedly press =. Eventually $7.450\ 580\ 597 \times 10^{-9}$ will be displayed. Point out that the same has happened.
- Write 17 179 869 184 on the board. Say that this was the exact answer when $1.717\ 986\ 918 \times 10^{10}$ was displayed. Explain that some rounding has taken place. So $1.717\ 986\ 9184 \times 10^{10} =$ 17 179 869 184 or $1.717\ 986\ 918 \times 10^{10} =$ 17 179 869 180. Say that the calculator is displaying numbers in standard form. Show students the [×10^x] key and input [4][.][5][×][×10^x][6][=] and check the display (4 500 000). Repeat for other numbers.

Plenary

- Allow students to check answers to questions in Exercise 9A by using their calculators.
- Students who need more of a **challenge** can experiment with the 'scientific notation' button to convert from ordinary numbers to standard form. [SHIFT][SETUP][7 SCI].

Homework and answers

Homework for this lesson is available in the homework folder. Consolidation and extension exercises are available in the further practice folder. Answers to the exercises can be found in the answers folder.

9.2 Calculating with standard form

Syllabus references

Statement	Core
C1.8/E1.8	Use the standard form $A \times 10^n$ when n is a positive or negative integer and $1 \leqslant A < 10$. Convert numbers into and out of standard form.

Collins references

- Student's Book pages 136–139

Learning aims

- Calculate using numbers in standard form.

Prior knowledge

Students must understand how to write ordinary numbers in standard form and how to convert numbers in standard form to ordinary form.

Common mistakes and remediation

When carrying out calculations with standard form, students may leave answers that are not in standard form, e.g. 23×10^7. Remind students to check that the first number is between 1 and 10.

Useful tips

When multiplying and dividing numbers given in standard form, discourage students from writing the numbers as ordinary numbers and then doing the calculation. Although this is a valid method it can lead to errors with so many digits.

Starter

This starter aims to confirm that students understand the work covered in Lesson 9.1.

- Write the following on the board:

  a 55 **b** 16 000 **c** 23.4 **d** 0.0008

- Ask students to write these numbers in standard form (5.5×10^1, 16×10^3, 2.34×10^1, 8×10^{-4}).
- Now write the following on the board:

  a 2×10^4 **b** 3.85×10^5 **c** 2.62×10^{-5} **d** 12.4×10^{-1}

- Ask students to convert these numbers to ordinary form (20 000, 385 000, 0.000 026 2, 1.24).

Main lesson activity

- Write $2 \times 10^7 \times 6 \times 10^3$ on the board and discuss ways of working it out.
- One possible answer is to deal with the numbers 2 and 6 and then the powers of 10 giving 12×10^{10} and then convert to standard form, 1.2×10^{11}.
- A valid method is to first write the numbers as ordinary numbers before doing the calculation. However, this should be discouraged as it can lead to errors with so many digits.
- Repeat for division, e.g. $4 \times 10^3 \div 8 \times 10^5$ (5×10^{-3}). Work through Examples 1 and 2.
- Now students can do Exercise 9B.

Plenary

Discuss the answers that students have given for the problem-solving questions in Exercise 9B to confirm their understanding of the work covered in this chapter.

Homework and answers

Homework for this lesson is available in the homework folder. Consolidation and extension exercises are available in the further practice folder. Answers to the exercises can be found in the answers folder.

9.3 Standard form without a calculator

Syllabus references

Statement	Extended
E1.8	Calculate with values in standard form.

Collins references

- Student's Book pages 140-142

Learning aims

- Add, subtract, multiply and divide numbers in standard form without using a calculator.

Key words

- No new key words

Prior knowledge

Students should know how to write large and small numbers in standard form.

Students should be able to change numbers from standard form to usual notation.

Common mistakes and remediation

Students often have difficulty changing a number from an index notation into standard form. For example, writing 32.5×10^7 as 3.25×10^8. If students are unsure, write the number out in full (in this examples 325 000 000) as an interim step.

Useful tips

Writing numbers out in full is always helpful but it takes more time to do this. The aim is to be able to do it without that step.

Starter

- Say that throughout this lesson students should not use a calculator. The aim is to be able to work with standard form without a calculator.
- Ask students to write out in full:
- **a** 32×10^5 **b** 57×10^{-4} **c** 8.72×10^6 **d** 0.58×10^{-4} **e** 0.3×10^6
- Check the answers. (**a** 3 200 000 **b** 0.0057 **c** 8 720 000 **d** 0.000 058 **e** 300 000)
- Ask which numbers were originally written in standard form. (only **c**)
- Ask students to write the other 4 numbers in standard form.
  (**a** 3.2×10^6 **b** 5.7×10^{-3} **d** 5.8×10^{-5} **e** 3×10^5
- Ask students if having the number written out in full was helpful. In all cases the index changed by one. The hard part was deciding whether to increase it or decrease it.

Main lesson activity

- Write down these four numbers: A = 4.2×10^5 B = 8.3×10^4 C = 8.5×10^{-3} D = 6.8×10^{-2}
- Ask students to explain how to work out A + B and give the answer in standard form.
- There are three possible methods:
  - Write each number out in full:
    A + B = 420 000 + 83 000 = 503 000 = 5.03×10^5
  - Change A so that both indices are 4 and add:
    A + B = $42 \times 10^4 + 8.3 \times 10^4 = (42 + 8.3) \times 10^4 = 50.3 \times 10^4 = 5.03 \times 10^5$
  - Change A so that both indices are 5 and add:
    A + B = $4.2 \times 10^5 + 0.83 \times 10^5 = (4.2 + 0.83) \times 10^5 = 5.03 \times 10^5$
- Show students all three methods. They can use which ever method they want.
- Point out that it is **incorrect** to add 4.2 and 8.3 because the indices are different.
- Ask students to use a similar method to work out A – B.
- Check that all can get the correct answer of 3.37×10^5
- Now ask students to work out C + D and D – C, giving the answers in standard for.
- Check they get the answers 7.65×10^{-2} and 5.95×10^{-2}
- Students can now do Exercise 9C Questions 1 to 4.
- When students have time to finish that move on to multiplication and division. This is easier than addition and subtraction.
- Write down X = 4×10^5 Y = 8×10^4 Z = 5×10^{-3}
- Say that to multiply X and Y you **multiply** 4 and 8, and **add** the indices.
- $X \times Y = 4 \times 8 \times 10^{5+4} = 32 \times 10^9 = 3.2 \times 10^{10}$
- To work out $X \div Z$ you **divide** 4 by 5 and **subtract** the indices.
- $X \div Z = 4 \div 5 \times 10^{5--3} = 0.8 \times 10^8 = 8 \times 10^7$
- Ask students to work out $Y \times Z$ and $Y \div Z$ in a similar way.
- If they get the correct answers (4×10^2 and 1.6×10^7) they can do the rest of Exercise 9C.

Plenary

- Write down these three numbers: 4×10^4 4×10^5 4×10^6
- Ask students to find the sum and the product of these numbers. Give the answers in standard form.
- Check they get the correct answers: 4.44×10^6 and 6.4×10^{16}
- Now ask for the answers if 4×4^7 is added to the list.
- Check they get 4.444×10^7 and 2.56×10^{24}

Homework and answers

Homework for this lesson is available in the homework folder. Consolidation and extension exercises are available in the further practice folder. Answers to the exercises can be found in the answers folder.

9.4 Surds

Syllabus references

Statement	Extended
E1.18	Understand and use surds, including simplifying expressions.

Collins references

- Student's Book pages 142-145

Learning aims

- Simplify simple expressions involving surds

Key words

- surd

Prior knowledge

Students should know the meaning and notation of a square root

Common mistakes and remediation

A common error is to write, for example, $\left(\sqrt{3}\right)^2 = 9$ instead of 3. Emphasise the correct answer when this situation occurs.

Useful tips

Answers can be checked with a calculator but students should be able to manipulate expressions without one.

Starter

- Ask students to complete this inequality with two consecutive integers: $\ldots < \sqrt{60} < \ldots$
  The answer is 7 and 8. Students should use the facts that $7^2 = 49$ and $8^2 = 64$
- Ask students to do the same for: $\sqrt{31}$ (5 and 6); $\sqrt{150}$ (12 and 13); $\sqrt{210}$ (14 and 15)

Main lesson activity

- Ask students to use a calculator to find $\sqrt{2} \times \sqrt{3}$ and then $\sqrt{6}$. They will see that they are the same.
- Demonstrate why this is the case:
  $\left(\sqrt{2} \times \sqrt{3}\right)^2 = \sqrt{2} \times \sqrt{3} \times \sqrt{2} \times \sqrt{3} = \sqrt{2} \times \sqrt{2} \times \sqrt{3} \times \sqrt{3} = 2 \times 3 = 6$; Hence $\sqrt{2} \times \sqrt{3} = \sqrt{6}$
- This is an example of a general result: $\sqrt{a} \times \sqrt{b} = \sqrt{ab}$
- Ask this to check this with other pairs of numbers. For example, does $\sqrt{3} \times \sqrt{7} = \sqrt{21}$?
- Now ask students to find $\sqrt{2}$ and $\sqrt{8}$ on their calculators. What do they notice?
- They should see that one is twice the other.
- Write down: $\sqrt{8} = \sqrt{4 \times 2} = \sqrt{4} \times \sqrt{2} = 2\sqrt{2}$. This works because 4 is a square number.
- Ask learners to simplify in the same way: $\sqrt{18}$ and $\sqrt{500}$
- They should see that $\sqrt{18} = \sqrt{9 \times 2} = 3\sqrt{2}$ and $\sqrt{500} = \sqrt{100 \times 5} = 10\sqrt{5}$. In each case, look for a multiple of a square number.
- Say that expressions that contain a square root are called **surds**.
  Examples are $6\sqrt{7}$ or $5 + 3\sqrt{11}$.
  These are irrational numbers and cannot be written exactly as decimals.
- Look at this product: $\sqrt{2}\left(8 - 3\sqrt{2}\right)$. You can simplify it by multiplying out the bracket:
  $\sqrt{2}\left(8 - 3\sqrt{2}\right) = 8\sqrt{2} - 3\sqrt{2} \times \sqrt{2} = 8\sqrt{2} - 3 \times 2 = 8\sqrt{2} - 6$ or $4\left(2\sqrt{2} - 3\right)$
- Students can check with a calculator that these two expressions have the same value.
- Ask students to simplify in the same way: $5\sqrt{3}\left(\sqrt{3} + 4\right)$
- They should get $15 + 20\sqrt{3}$ or $5\left(3 + 4\sqrt{3}\right)$
- Students can now complete Exercise 9D.

Plenary

- $\sqrt{2} \times \sqrt{8} = \sqrt{16} = 4$ which is an integer.
- Ask students to try to find other example of two different surds whose product is an integer.
  After a few minutes ask them to share any examples.
- Some possible answers are $\sqrt{3} \times \sqrt{27} = 9$; $\sqrt{32} \times \sqrt{2} = 8$; $\sqrt{2} \times \sqrt{18} = 6$; $\sqrt{20} \times \sqrt{5} = 10$.
  The need to look for two numbers that multiply to make a square number.

Homework and answers

Homework for this lesson is available in the homework folder. Consolidation and extension exercises are available in the further practice folder. Answers to the exercises can be found in the answers folder.

9.5 Rationalising the denominator

Syllabus references

Statement	Extended
E1.18	Rationalise the denominator.

Collins references

- Student's Book pages 145-147

Learning aims

- Rationalise the denominator of a fraction containing surds

Key words

- rationalise

Prior knowledge

Students should be able to simplify simple expressions involving surds.

Common mistakes and remediation

Students can have difficulty choosing the correct multiplier for a denominator of the form $a+b\sqrt{c}$. Emphasise the fact that the only difference is the sign between the terms.

Useful tips

Students may find it easier at first to write out separately the multiplication for the numerator and the multiplication for the denominator.

Starter

- Write down $5\times10=50$
- We could write this as $5(4+6)=5\times4+5\times6=20+30=50$ which gives the same answer.
- We could also write $(3+2)(4+6)=3(4+6)+2(4+6)=12+18+8+12=50$
- Make sure students can see where the four terms come from.
- Another way is $(2+3)(1+9)=2(1+9)+3(1+9)=2+18+3+27=50$
- Or with a minus sign: $(7-2)(4+6)=7(4+6)-2(4+6)=28+42-8-12=50$
- The answer will always be 50. Ask each student to write one of their own. Ask for some examples and check the result is always 50.

Main lesson activity

- Look at this fraction $\frac{5}{\sqrt{2}}$. Multiply the numerator and the denominator by $\sqrt{2}$:

  $\frac{5}{\sqrt{2}} = \frac{5 \times \sqrt{2}}{\sqrt{2} \times \sqrt{2}} = \frac{5\sqrt{2}}{2}$ or $\frac{5}{2}\sqrt{2}$
- The denominator (2) is now a rational number. This process is called rationalising the denominator.
- Ask students to rationalise the denominators of these: $\frac{4}{\sqrt{3}}$; $\frac{10}{\sqrt{5}}$; $\frac{3}{\sqrt{6}}$; $\frac{\sqrt{8}}{\sqrt{2}}$

  Answers: $\frac{4}{3}\sqrt{3}$; $2\sqrt{5}$; $\frac{1}{2}\sqrt{6}$; 2
- Here is a more complicated example: $\frac{10}{3-\sqrt{2}}$
- In this case multiply the numerator and the denominator by $3+\sqrt{2}$:

  $\frac{10}{3-\sqrt{2}} = \frac{10(3+\sqrt{2})}{(3-\sqrt{2})(3+\sqrt{2})}$
- The denominator is $(3-\sqrt{2})(3+\sqrt{2}) = 3(3+\sqrt{2}) - \sqrt{2}(3+\sqrt{2}) = 9 + 3\sqrt{2} - 3\sqrt{2} - 2 = 7$
- Notice that the difference in signs means that the surds cancel out and the answer is an integer.
- So $\frac{10}{3-\sqrt{2}} = \frac{10(3+\sqrt{2})}{(3-\sqrt{2})(3+\sqrt{2})} = \frac{10(3+\sqrt{2})}{7}$ or $\frac{10}{7}(3+\sqrt{2})$
- Ask students to rationalise the denominator of $\frac{2}{\sqrt{10}+1}$

  Answer: $\frac{2}{\sqrt{10}+1} = \frac{2(\sqrt{10}-1)}{(\sqrt{10}+1)(\sqrt{10}-1)} = \frac{2}{9}(\sqrt{10}-1)$
- Now rationalise the denominator of $\frac{1+\sqrt{5}}{3-\sqrt{5}}$

  Answer: $\frac{(1+\sqrt{5})(3+\sqrt{5})}{(3-\sqrt{5})(3+\sqrt{5})} = \frac{3+\sqrt{5}+3\sqrt{5}+5}{9-5} = \frac{8+4\sqrt{5}}{4} = 2+\sqrt{5}$
- Make sure students understand how to simplify the numerator in this case.
- Students who need more of a **challenge** may be able to go directly from $(3+\sqrt{5})(3-\sqrt{5})$ to 9 – 5 but others will prefer to write out the four terms and simplify.
- Students can now complete Exercise 9E.

Plenary

- Ask students to rationalise the denominator of $\frac{\sqrt{3}+\sqrt{2}}{\sqrt{3}-\sqrt{2}}$
- They should find $\frac{\sqrt{3}+\sqrt{2}}{\sqrt{3}-\sqrt{2}} = \frac{(\sqrt{3}+\sqrt{2})(\sqrt{3}+\sqrt{2})}{(\sqrt{3}-\sqrt{2})(\sqrt{3}+\sqrt{2})} = \frac{3+\sqrt{6}+\sqrt{6}+2}{3-2} = 5+2\sqrt{6}$
- Check that students can multiply each pair of brackets correctly.

Homework and answers

Homework for this lesson is available in the homework folder. Consolidation and extension exercises are available in the further practice folder. Answers to the exercises can be found in the answers folder.

10.1 Units of measurement

Syllabus references

Statement	Core
C5.1/E5.1	Use metric units of mass, length, area, volume and capacity in practical situations and convert quantities into larger or smaller units.

Collins references

- Student's Book pages 150–151

Learning aims

- Decide which units to use when measuring length, weight and capacity.

Key words

- metric system, length, mass, volume, capacity

Prior knowledge

Students should be familiar with the basic metric units (millimetre, centimetre, metre, kilometre, tonne, millilitre, litre) and, more importantly, should have a rough idea, at least, of the size of each unit.

Common mistakes and remediation

There may be confusion about which units relate to what measures, e.g. whether centimetres measure length, volume or weight.

Useful tips

It might be helpful to display pictures illustrating the different types of measure, such as a jug labelled with litres, a bag of flour labelled with grams, a ruler labelled with centimetres, as a reminder. Also stress to students that we all use units every day and it is important for them to appreciate that we need to use the appropriate unit at the right time.

Starter

- Give students some practice in using the numbers they will need, to manipulate metric measures.
- Multiply various numbers by 10, 100, 1000. Divide various numbers by 10, 100, 1000.

Main lesson activity

- Ask the class to guess how tall you are. Make sure you know, in centimetres. Talk about height and say that the average height of an adult male is about 176 cm, and of an adult female, about 163 cm. Using your own height as a guide, start to estimate the heights of other members of the class.
- Ask for an estimate of the thickness of a pen or pencil. Make clear that millimetres are used for small measurements.
- Now discuss the height of the room, which is probably about 3 or 4 metres. Explain that we estimate bigger heights in metres, rather than centimetres. Ask for estimates of heights of things you can see from the window, such as trees or other buildings, noting that we measure or estimate these in metres.
- Ask how far it is from school to somewhere the students will know. Talk about using metres or kilometres. The discussion is still not about accuracy but choosing the right unit.
- Go through the units, from millimetres to metres, and summarise which to use, when. Check with the class the unit they would use for the thickness of a book (millimetres), the length of a baby (centimetres), the height of a building (metres), the distance between two towns (kilometres).
- Now talk about weight, discussing what unit they would use for a bag of potatoes (kilograms), the weight of a coin (grams), the weight of a load of bricks (tonnes).
- Ask how much petrol will fill up a petrol tank and lead into capacity being measured in litres.
- Ask how much medicine they might take, if they were ill. Expect answers such as one spoonful, but try to bring out or introduce a typical medicine spoon as being 5 millilitres, a very small metric measure (like millimetres).
- Talk about the units used for a glass in a restaurant – centilitres (like centimetres).
- Students who need more **support** will need a lot of practice with each type of unit. Stay with length at first, until they are confident with that, then move on to weight and finally cover liquid capacity.
- Students can now do Exercise 10A. There are lots of parts to questions 1 and 2, so students who need more of a **challenge** could try answering every other one, then move on to question 3. Students who need more **support** will need the practice. Question 3 also brings in estimation. Encourage students to remember certain key estimates, such as the height of an average male adult being about 176 cm.

Plenary

Go over the links between millimetre, centimetre, metre, kilometre, millilitre, litre, milligram, gram, kilogram and tonne. Try to link them to everyday objects, e.g. a bag of sugar weighs 1 kg.

Homework and answers

Homework for this lesson is available in the homework folder. Consolidation and extension exercises are available in the further practice folder. Answers to the exercises can be found in the answers folder.

10.2 Converting between metric units

Syllabus references

Statement	Core
C5.1/E5.1	Use metric units of mass, length, area, volume and capacity in practical situations and convert quantities into larger or smaller units.

Collins references

- Student's Book pages 151–152

Learning aims

- Convert from one metric unit to another.

Key words

- centimetre (cm), millimetre (mm), kilometre (km), gram (g), kilogram (kg), tonne (t), litre (l), millilitre (ml), centilitre (cl)

Prior knowledge

Students should know (or have written down) the links between the metric units for length, weight and capacity/volume.

Common mistakes and remediation

Mistakes occur when students multiply and divide by the incorrect number, for example, 1000 instead of 100.

Useful tips

Display on the wall a copy of this chart so that students become familiar with its contents.

Length

10 **millimetres**	=	1 **centimetre**
1000 millimetres	=	100 centimetres
	=	1 **metre**
1000 metres	=	1 **kilometre**

Mass

1000 **grams**	=	1 **kilogram**
1000 kilograms	=	1 **tonne**

Capacity

10 **millilitres**	=	1 **centilitre**
1000 millilitres	=	100 centilitres
	=	1 **litre**

Volume

1000 litres	=	1 metre^3
1 millilitre	=	1 centimetre^3

Starter

- Set students some mental divisions and multiplications by 10, 100 and 1000. Ensure they are familiar with such reminders as moving the decimal point as many places as the number of zeros.

Main lesson activity

- Ask 'How many metres are there in a kilometre?' (1000), 'How many grams are there in a kilogram?' (1000). Talk to the students about 'kilo-' meaning a thousand.
- Ask 'How many centimetres are there in a metre?' (100), 'How many centilitres are there in a litre?' (100). Talk about 'centi-' meaning a hundredth.
- Ask: 'How many millimetres are there in a metre?' (1000), 'How many millilitres are there in a litre?' (1000). Talk about 'milli-' meaning a thousandth.
- Tell the students who need more **support** that they need to remember these three prefixes, kilo-, centi- and milli-, as they will help them to find the correct unit.
- Introduce the fact that 10 millimetres = 1 centimetre (because 'centi-' is ten times bigger than 'milli-'). Ask the students who need more of a **challenge** to explain why. Repeat for 10 millilitres = 1 centilitre.
- Ask students about the connection between kilograms and tonnes, encouraging them to make guesses and reminding them to think metric (in powers of ten). Establish that 1000 kg = 1 tonne.
- Explain to students that 1000 litres = 1 cubic metre (m^3) and 1 millimetre = 1 cubic centimetre (cm^3).
- Ask students to look at the table in the Student's Book; they need to remember these relationships as they are expected to know them at this level.
- Go through Example 1 with the class, emphasising that changing from a small unit to a large unit gives a smaller number, so they need to divide; for example, 350 cm in metres will be 350 ÷ 100 = 3.5 m.
- Remind students who need more **support** of the rules for dividing by 10 and 100: count how many zeros and move the digits that many places.
- Similarly, go through Example 2, emphasising that changing from large units to small units means more, **so multiply**.
- Students can now do Exercise 10B. They can refer to the table of units. In this exercise, question 4 is the problem-solving (PS) question. Tell the students to note the hint alongside this question. There are lots of conversion questions at the beginning of the exercise, so students who need more of a **challenge** could try answering every other one, then move on to questions 3 and 4. Students who need more **support** will need the practice. Encourage the students who need more of a **challenge** to do this exercise without referring to the table, unless they really need to look something up. Question 3 is a 'real-life' example. In this question, encourage students to convert each unit to centimetres, for example, so that they can compare what Sarif wanted with what he could buy.

Plenary

- Play a metric quiz, perhaps the students in the left-hand half of the room playing those on the right.
- Ask each team, in turn, questions such as, 'How many metres are there in 238 cm?'
  Use all the links in the table.
- If a team answers straight away, then award 2 marks. If they need to confer, award 1 mark.
  Don't pass any questions over. Give the team time to consult the table.

Homework and answers

Homework for this lesson is available in the homework folder. Consolidation and extension exercises are available in the further practice folder. Answers to the exercises can be found in the answers folder.

10.3 Time

Syllabus references

Statement	Core
C1.15/E1.15	Calculate with time: seconds (s), minutes (min), hours (h), days, weeks, months, years, including the relationship between units. Calculate times in terms of the 24-hour and 12-hour clock. Read clocks and timetables.

Collins references

- Student's Book pages 153–155

Learning aims

- Use the 12- and 24-hour clocks.
- Solve problems involving time and timetables.

Key words

- 24-hour clock, 12-hour clock, timetable

Prior knowledge

Students need to know that there are 60 minutes in an hour.

Common mistakes and remediation

Students often make mistakes with time questions if they use a calculator because this works on a base of 10 while hours and minutes work on a base of 60. Advise them not to use calculators when working out time intervals.

Useful tips

Advise students to break journeys into sections when calculating time intervals, as in Example 3 of the Student's Book.

Starter

- Check students can convert between the 12- and 24-hour clocks, by considering the following examples.
  - Convert the following times to the 24-hour clock.

    a 2.30 pm **b** 5 past 5 in the morning **c** 10 past 8 in the evening **d** 11.38 pm
    (1430, 0505, 2010, 2338)

  - Convert the following times to the 12-hour clock.

    a 1712 **b** 0058 **c** 0322 **d** 1045
    (5.12 pm, 58 minutes past midnight, 3.22 am, 10.45 am)

Main lesson activity

- Ask students where we use the 24-hour clock. (E.g. transport timetables.)
- Ask why we use the 24-hour clock. (To avoid confusion between am and pm times.)
- Move on to calculating the time taken on journeys. Ask how long a journey takes if they set out at 0538 and arrive at 1115. Some may take one figure from the other and suggest an answer of 577 minutes. (The answer is 5 hours 37 minutes.)
- Emphasise that they cannot simply subtract the start from the end time using a calculator because there are 60 minutes in an hour, and the best way to calculate time taken is to break it into sections based on the hours in the journey. So in this example:
  - From 0538 to 0600 is 60 – 38 minutes = 22 minutes
  - From 0600 to 1100 is 11 – 6 hours = 5 hours
  - From 1100 to 1115 is 1115 – 1100 minutes = 15 minutes
- Now write the journey shown in Example 3 on the board and work it through with the students.
- Reinforce this by dividing students into pairs and asking them to set their own time problems for each other using times expressed in the 24-hour clock, perhaps based on real journeys they have made. They then work out the time taken in the journey described by their partners.
- At the end of the exercise ask the pairs to give you their problems and answers to write up on the board. Check their understanding.
- Students can now do Exercise 10C.
- Students who need more **support** may need additional support with questions 8 and 9 when local time zones also have to be taken into account. One effective way of tackling these problems is to work out the arrival time based on the time zone at the start point and then adjust the result by the time difference at the destination.

Plenary

Discuss the answers to question 11 of Exercise 10C. It is about time differences around the world.

Homework and answers

Homework for this lesson is available in the homework folder. Consolidation and extension exercises are available in the further practice folder. Answers to the exercises can be found in the answers folder.

10.4 Currency conversions

Syllabus references

Statement	Core
C1.16/E1.16	Calculate with money and convert from one currency to another.

Collins references

- Student's Book pages 156–157

Learning aims

- To convert from one currency to another.

Key words

- exchange rate

Prior knowledge

Students need to understand the work on rates in Lesson 7.4 and be able to round values to a given degree of accuracy.

Common mistakes and remediation

Students sometimes multiply when they should divide and vice versa. Plenty of worked examples and repeated practice should help to address this issue.

Useful tips

Encourage students to consider whether their answers are sensible when compared with the given information.

Starter

- Ask students to give examples of currencies used in different countries.
- Explain to students that every country has its own monetary system and there needs to be a way of converting the currency of one country to the currency of another if travel and trade are to happen. There must be a rate at which the money is converted, and this is called the exchange rate.
- Show the example below (taken from a week in 2010) and explain that the figure in the second column is the amount of the first currency you will get from one unit of the second, e.g. 1.62 US dollars for one UK pound, or 81.40 yen for one US dollar.

Currencies	Exchange rate
US dollar to UK pound	1.62
Euro to UK pound	1.16
US dollar to euro	1.40
Yen to US dollar	81.40

- Highlight to students that this is an example of a rate as described in Chapter 7.

Main lesson activity

- Explain to students that exchange rates vary all the time so conversion has to be carried out every time you change money depending on the exchange rate on that day.
- Consider how to convert 242 euros into pounds given that the rate of exchange is £1 = €1.15. Explain to students that there are two methods they can use:
  - Method 1 (unitary method)

    $€1.15 = £1$

    $€1 \quad = £\frac{1}{1.15}$

    $€242 \; = £\frac{1}{1.15} \times 242 = £210.43$

  - Method 2 (fractional method)

    $€242 = £\frac{242}{1.15} = £\ 210.43$

- Work through Example 4 with the class.
- Then use the table above to work through some practice examples with students – for example:
  - How many US dollars would you get for 200 yen? (2.46)
  - How many UK pounds would you get for 300 euros? (258.62)
- Ensure that students can also convert in the other direction – for example:
  - How many yen would you get for 200 US dollars? (16 280)
  - How many UK pounds would you get for 200 US dollars? (123.46)
- Students who need more of a **challenge** can be asked to do multiple conversions using the information from two rows in the table, e.g. convert yen into euros.
- Students can then do Exercise 10D. Students who need more **support** may need guidance in questions 7 and 8, which require comparison between the buying power of currencies and between currency exchange at different times.

Plenary

Discuss the answers to questions 7 and 8 of Exercise 10D, and ensure that students understand the steps they need to take to arrive at the answers.

Homework and answers

Homework for this lesson is available in the homework folder. Consolidation and extension exercises are available in the further practice folder. Answers to the exercises can be found in the answers folder.

10.5 Using a calculator efficiently

Syllabus references

Statement	Core
C1.14/E1.14	Use a calculator efficiently. Enter values appropriately on a calculator. Interpret the calculator display appropriately.

Collins references

- Student's Book page 157-158

Learning aims

- To use a calculator to work out calculations as efficiently as possible.

Prior knowledge

Students should understand how to use the brackets, square, square root, cube and cube root buttons on a scientific calculator. They should also understand the order of operations, i.e. BIDMAS/BODMAS, and be able to round answers to a given number of decimal places.

Common mistakes and remediation

Students often make mistakes with calculations such as $\frac{4.2 + 8.3}{0.12 + 0.15}$.

They type in 4.2 + 8.3 ÷ 0.12 + 0.15, which leads to an incorrect answer. Keep reminding them that they must use brackets if they wish to do this in one operation. Repeated practice of questions such as this should solve this problem.

Useful tips

Encourage students to always check whether their answer is reasonable. They should do this by approximating the numbers in the calculation, and where possible, have a calculation that they can easily do in their head.

Ensure that students understand that they may *not* use algebraic or graphical calculators in their examinations and that they should therefore not use them throughout the course. It is important they understand the theory and steps to calculations.

Very simple calculators may not have all the functions required – check that all students have calculators that can calculate roots and powers.

It is good practice to give calculations to 3 sf unless otherwise stated.

Starter

- Write the following calculations on the board:

  a 86 × 42 **b** 62 ÷ 16 **c** 4.2 × 9.6 **d** 152.74 ÷ 51.9 **e** $\frac{3.6 \times 4.9}{15}$

- Ask students to estimate the answers to these calculations by approximating the numbers. Stress that they must not use a calculator. (Answers in the region of: 3600, 4, 40, 3, 1.)

Main lesson activity

- Ensure students understand how to use the brackets, square, square root, cube and cube root keys on their calculators. These may be slightly different on different makes of calculator. Establish which makes are being used in the class and discuss where the main functions are on each one.

  Look at the example: $\dfrac{4.2 + 8.3}{0.12 + 0.15}$

  Explain how putting 4.2 + 8.3 ÷ 0.12 + 0.15 into the calculator will lead to an incorrect answer and that brackets must be used to do it correctly if the calculation is to be done in one operation.

  - 4.2 + 8.3 ÷ 0.12 + 0.15 = 104.317
  - (4.2 + 8.3) ÷ (0.12 + 0.15) = 46.296
- Emphasise that it is a good idea to check that the answer is reasonable by doing a quick approximation – here roughly 12 divided by 0.25, which would be 48.
- The students can now do Exercise 10E. Stress to students that you want them to carry out the calculations as efficiently as possible, without writing down any intermediate values.
- Also remind students that they need to be careful about the order of operations. Although many calculators will apply the rules of BIDMAS/BODMAS, students will need to apply these themselves when working out approximate answers.
- Students who need more **support** will need help with question 6.

Plenary

Discuss the answers to selected parts of question 6 of Exercise 10E.

Homework and answers

Homework for this lesson is available in the homework folder. Consolidation and extension exercises are available in the further practice folder. Answers to the exercises can be found in the answers folder.

11.1 The language of algebra

Syllabus references

Statement	Core
C2.1/E2.1	Know that letters can be used to represent generalised numbers.
C2.5/E2.5	Construct simple expressions, equations and formulas.

Collins references

- Student's Book pages 172–175

Learning aims

- Use letters, numbers and mathematical symbols to write algebraic expressions and formulae.

Key words

- expression, symbol, variable, formula, formulae, equation, term, solving

Prior knowledge

Students need to know the order of operations for arithmetic. This is known as the BIDMAS/BODMAS rule and should be used whenever different arithmetic operations occur together.

Common mistakes and remediation

Students will often write '7 less than x' as $7 - x$.

Students will often confuse x^2 and $2x$. Emphasise the difference so that students are clear.

Useful tips

The rules for algebra given in this lesson in the Student's Book give the best tips for dealing with basic algebraic questions.

Starter

- Throw three dice. Tell students the numbers, say 6, 3, 5, and ask them to make up five calculations, using mathematical signs, including brackets. For example:

  $6 + 3 + 5 = 14$ $(6 - 3) \times 5 = 15$ $6 \times 3 + 5 = 23$

  $6 \times 5 - 3 = 27$ $6 + 3 - 5 = 4$

- Collect in examples, particularly those using brackets incorrectly, and discuss with the class what the errors are. Recall the rules of BIDMAS/BODMAS. Repeat if there is time.

Main lesson activity

- Tell students you have a rule in your head, for example, 'add 2'. Ask them to give you the result for a number that you give them, e.g. if you say 7, they should say 9.
- Go around the class giving individuals a number so they can give you the result after applying the rule.
- Challenge students who need more of a **challenge** with larger numbers.
- Do this with numbers for about six students, then give a student x.
- This will probably cause some hesitation.
- It will help students who need more **support** if you write the operations on the board. For example, if you say 7, the result is 7 + 2 = 9; if you say 5, the result is 5 + 2 = 7. This will help when you explain that the result for x is $x + 2$.
- Explain that this is a general rule that describes the operation 'add 2' and that x represents any number that they care to substitute.
- Repeat with other rules to get the expressions $x - 4$, $3x$ and $x \div 5$ or $\frac{x}{5}$, for example. The last two will need to be explained as students will probably understand $3 \times x$ and will need to be told that there is an implied multiplication sign between the number (coefficient) and the letter. The notation $\frac{x}{5}$ is also a standard way of writing $x \div 5$.
- Go through the introductory text and Example 1.
- Students can now do Exercise 11A. This gives plenty of practice in manipulating and using basic algebra.
- Students who need more **support** may need some more reinforcement. Working through some of the questions in the exercise will help reinforce the key processes involved.

Plenary

Go through the seven important rules and check that students understand them.

- 3 more than $x = x + 3$
- 5 less than (or fewer than) $y = y - 5$ (not $5 - y$). (Make sure students understand the difference.)
- 4 multiplied by $z = 4z$
- b divided by $2 = \frac{b}{2}$ (not $\frac{2}{b}$). (Again, check to make sure students understand the difference.)
- $7a = 7 \times a$
- $1 \times c = c$ (not $1c$)
- $t \times t = t^2$ (Again, discuss the difference between this and $2t$.)

Homework and answers

Homework for this lesson is available in the homework folder. Consolidation and extension exercises are available in the further practice folder. Answers to the exercises can be found in the answers folder.

11.2 Substitution into formulas

Syllabus references

Statement	Core
C2.1/E2.1	Substitute numbers into expressions and formulas.

Collins references

- Student's Book pages 175–177

Learning aims

- Substitute numbers for letters in formulas and evaluate the resulting numerical expression.

Key words

- substitute

Prior knowledge

Students must have a working knowledge of BIDMAS/BODMAS in order to complete this lesson successfully. They must also be able to work with negative numbers.

Common mistakes and remediation

Common mistakes occur when operations are ordered incorrectly and when there is a negative before a bracketed expression. Examples used when teaching substitution should include and reinforce the correct method of working.

Useful tips

Encourage students to use brackets around numbers, especially negative numbers, when substituting values in a formula.

Starter

- Show or give students a copy of the puzzle supplied on Resource sheet 11.2. Ask them to work out values of A, B, C and D that make the puzzle work ($A = 4$, $B = 3$, $C = 1$, $D = 5$).
- Repeat with a puzzle of your own, or ask the students to make one up.

Main lesson activity

- Ask students if they know any examples of a mathematical formula. If they come up with any valid formulas, use these for the introduction. If not, suggest the area of a rectangle.
- Ask students to define their own variables. So, for example, they may establish a formula such as $A = l \times w$. Write this on the board.
- Ask students what the area would be for a rectangle of length 5 and width 3.
- They will probably do this mentally, but take the opportunity to remind students that when answering on paper it is good practice to write the values into the formula before working it out, i.e. $A = 5 \times 3 = 15$.
- Remind students to use brackets around the numbers, such as $A = (5) \times (3)$. This is particularly useful when dealing with negative numbers.
- Work through Example 2.
- Students can now do Exercise 11B. Tell them to note the many hints and tips that accompany this exercise. Question 9 requires students to find values that make the area of a rectangle and a triangle the same. Before jumping in and just trying out random values, students should think about any connections between the two shapes. A few moments' thought will – hopefully – lead them to the conclusion that a triangle is half of a rectangle.

Plenary

- Make sure students understand BIDMAS/BODMAS and the impact of the negative sign on an expression.

Homework and answers

Homework for this lesson is available in the homework folder. Consolidation and extension exercises are available in the further practice folder. Answers to the exercises can be found in the answers folder.

11.3 Rearranging formulas

Syllabus references

Statement	Core
C2.5/E2.5	Change the subject of simple formulas.

Collins references

- Student's Book pages 177–179

Learning aims

- Rearrange formulas by doing the same to both sides.

Key words

- rearrange, subject

Prior knowledge

Students need to understand the principle of inverse operations.

Common mistakes and remediation

The usual rearrangement errors will be common, so emphasise the importance of making the same change to both sides every time.

Useful tips

Remind students that, in algebra, the letters are just numbers with unknown values – there is no real difference between $2x - 4 = 6$ and $wx - y = z$.

Starter

Ask students:

- What is the area of a rectangle with width 4 cm and length 6 cm? ($6 \times 4 = 24$ cm^2)
- What is the length of a rectangle with width 9 cm and area 54 cm^2?
  (6 cm, as $6 \times 9 = 54$)
- What is the length of a rectangle with width 12 cm and area 96 cm^2?
  (8 cm, as $12 \times 8 = 96$)

Main lesson activity

- Tell students that you are now going to consider the examples used in the starter algebraically, using the formula for the area of a rectangle: $A = lw$. So, you want to find the rule for working out the length l when you know the area A and the width w.
- Explain that you are going to use the idea of 'doing the same to both sides'.

  If $A = l \times w$, then if we divide the right hand side by w, we get l. Stress that if this is to stay the same relationship then we have to do the same to the other side, in which case we get $\frac{A}{w}$.

  Hence $\frac{A}{w} = l$. Reversing the formula gives $l = \frac{A}{w}$; l is now the 'subject' of the formula.
- Repeat with the formula $y = x + 2$ and ask how to find x if we know y ($y = x - 2$ by subtracting 2 from both sides and reversing the formula).
- Go through the text and Examples 3 and 4 to give more practice.
- Students can now do Exercise 11C. Advise students who need more **support** to complete questions 1–10 first, then move on to questions 11–17, working with a partner.
- Students who need to more **support** find this topic difficult. Talking through the question 11, which involves several steps, will be useful for all students.

Plenary

- Ask students for the relationship between speed, distance and time.
- They should know that distance = speed × time.
- Show them the D, S, T road sign. Students may have seen this before.

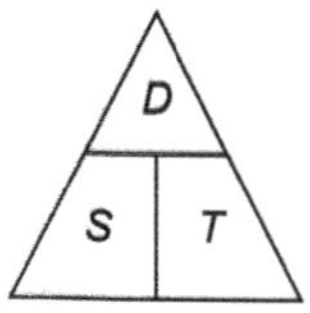

- Either ask or show how to obtain the formulas for speed and time from this.
- Point out that this format can be used for many other formulas such as $A = lw$, which links back to the main lesson activity.

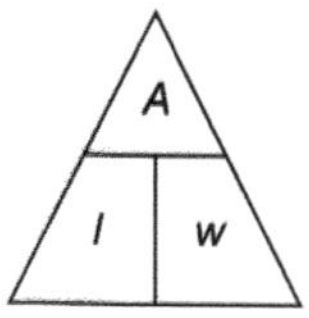

Homework and answers

Homework for this lesson is available in the homework folder. Consolidation and extension exercises are available in the further practice folder. Answers to the exercises can be found in the answers folder.

11.4 More complicated formulas

Syllabus references

Statement	Extended
E2.5	Change the subject of formulas. Construct expressions, equations and formulas.

Collins references

- Student's Book pages 179–181

Learning aims

- To rearrange more complicated formulas by doing the same to both sides.

Prior knowledge

Students need to be able to rearrange simple formulas as covered in Lesson 11.3.

Common mistakes and remediation

Students often make mistakes when squares and square roots are involved; for example, in question 12 of Exercise 11D students may establish correctly that $u^2 = v^2 - 2as$.

However, they may go on to give $u = \sqrt{v^2} - 2as$ rather than $u = \sqrt{(v^2 - 2as)}$.

In question 16, students may forget to square the 2 when squaring both sides. Highlighting and working through errors like this will help to address the problem.

Useful tips

Encourage students to write down what they do at each stage, as in the worked example, as it will help them identify any mistakes.

Starter

- Write the following on the board:

  a $y = \sqrt{x} + 1$ **b** $y = \sqrt{x+1}$

- Tell students that you want them to rearrange each of these formulas to make x the subject. Ask them to work in pairs to discuss these problems and then discuss the answers as a class. Students often give the same answer for both of these problems, i.e. $x = y^2 - 1$, but it is only correct for **b**.

Main lesson activity

- Explain to students that this lesson develops the ideas in Lesson 11.3 and looks at more complicated examples.
- Work through Example 6 with the students. They can then do Exercise 11D. Some of these examples are quite challenging and students who need more **support**, certainly, will need guidance.

Plenary

- Write the following on the board:

  Make v the subject of $\quad uv = fu + fv$

  Divide both sides by u gives $\quad v = \dfrac{fu + fv}{u}$

- Ask students to discuss this answer, in pairs. Highlight to students that this is an example of another common mistake, where students rearrange a formula so that the required subject is on the left-hand side but leave a term that includes the required subject on the right-hand side.

Homework and answers

Homework for this lesson is available in the homework folder. Consolidation and extension exercises are available in the further practice folder. Answers to the exercises can be found in the answers folder.

12.1 Simplifying expressions

Syllabus references

Statement	Core
C2.2/E2.2	Simplify expressions by collecting like terms.

Collins references

- Student's Book pages 184–188

Learning aims

- Simplify algebraic expressions by multiplying terms.
- Simplify algebraic expressions by collecting like terms.

Key words

- simplifying, coefficient, like terms

Prior knowledge

Students should know the basic rules of algebra.

Common mistakes and remediation

Students will multiply or divide indices instead of adding or subtracting, or multiply the bases together. Students will often write $3a + 7b$ as $10ab$. Allow plenty of practice so students are aware of potential errors.

Useful tips

- Remind students that 'simplifying', in this context, means writing the expression in as neat, or short, a form as possible: 'simpler means smaller'.

Starter

- This work involves lots of basic numeracy, so run through some basic multiplication tables and number bonds. For example, ask students 4 × 5, 7 + 8, 2 – 5.
- Students can write their answers on whiteboards or this activity can be done as a 20-question mental test.

Main lesson activity

- Tell students you are going to give them two numbers and you want them to multiply them together. For example, 2 and 3 gives the result 6. Give individuals two numbers so they can give the result.
- After giving numbers to about six students, ask for '3 and x'. Building on the previous lesson, students should be able to say $3x$.
- Now give a and b. Once again, students should be able to say ab.
- Now give '$2x$ and y'. This may cause some hesitation, but write '$2x \times y$' on the board, then students should accept that the answer is $2xy$.
- Repeat with similar examples such as 2 and $3m$, 4 and $5p$, $2m$ and m. The latter is important as it leads to the results $2m^2$ and students will need to be familiar with powers.
- Go through Examples 1 and 2 for extra reinforcement.
- Students can now do Exercise 12A.
- Now repeat the process above, but this time students will be adding the two numbers they are given.
- Start with numbers and work up to a and a. From previous work, students should be able to say $2a$. Increase the complexity to expressions such as $7x^2$ and $2x^2$.
- Finally, write the two expressions $2a + 3b$ and $4a + 7b$ on the board. Ask students what the sum of these would be. There may be some confusion, and answers such as $16ab$ will be common. Explain that only like terms can be added or 'collected'. The answer is $6a + 10b$. Repeat with similar examples, including some that involve negative signs.
- Go through Example 3, which shows a method of rearranging the terms. This method can be useful for students who need more **support**.
- Students can now do Exercise 12B. Tell the students to note the hints provided.
- Show students the algebra pyramids (Resource sheet 12.1a). Ask them to complete each pyramid by first adding and then multiplying the expressions in adjacent boxes, writing the answer in the appropriate box in the row above. The answers are: (adding) $6 + 2m$ and $6 + 4u$; (multiplying) $9m^2$ and $12u^3$.
- Show students the array of algebraic expressions (Resource sheet 12.1b). Ask them to match up the expressions, for example linking $4a^2$ with $(2a)^2$. They can do this individually, in groups or as a class.

Plenary

- Check students' understanding by asking them questions such as, 'Give me two expressions that could add to $3x$.' 'Give me two expressions that multiply to give $4n^3$.'

Homework and answers

Homework for this lesson is available in the homework folder. Consolidation and extension exercises are available in the further practice folder. Answers to the exercises can be found in the answers folder.

12.2 Expanding brackets

Syllabus references

Statement	Core
C2.2/E2.2	Expand products of algebraic expressions.

Collins references

- Student's Book pages 188–191

Learning aims

- Expand brackets and simplify the resulting algebraic expressions.

Key words

- expand, simplification

Prior knowledge

Students need to be able to multiply algebraic terms together.

Common mistakes and remediation

Students may multiply the first term properly but then just write down the subsequent terms without multiplying them; for example, $5(3a + 2b) = 15a + 2b$ (it should be $15a + 10b$).

Useful tips

Encourage students to work out the number calculations and each different letter calculation separately. Emphasise that, when they are multiplying one term outside the brackets by one or more terms inside the brackets, students must multiply everything inside the brackets by the number or algebraic term outside the brackets.

Starter

- Ask a variety of multiplication questions. For example:

  5×2, $5 \times 2a$, $5 \times 2a^2$, $5a \times 2a$, $5a \times 2b$, $5a \times 2a^2$, $5a^2 \times 2a$, $5a \times 2a^3$

Main lesson activity

- Write on the board:

  $3 \times 4 + 3 \times 5$ and $3 \times (4 + 5)$

  Ask students to work out these calculations. They should get the same answer.

  Write on the board:

  $3 \times (4 + 5) = 3 \times 4 + 3 \times 5$.

- Now write $2 \times (x + 5)$ and ask students, using the previous example as a clue, to write this without brackets.
- Look for the answer $2 \times x + 2 \times 5$. Simplify this to $2x + 10$.
- Repeat with $3(y - 6)$ to get the answer $3y - 18$.

Remind students that need more **support** that there is an invisible multiplication sign between the number outside the brackets and the opening bracket. Show this as 3 × $(y - 6)$ and the answer as 3 × y – 3 × 6 if necessary.

Emphasise that students must evaluate the multiplications and simplify expressions.

- Go through this section in the Student's Book and familiarise the students with the term 'expand'.
- Students who need more **support** may need to write in the multiplication signs before the opening brackets, and expand the terms, again with multiplication signs, before simplifying them. They could draw arcs as they multiply the terms, to keep track. For example:

$$5(3a + 2a) = 15a + 10b)$$

- Students who need more **support** may also find the 'minus sign outside the bracket' a difficult concept, but it is essential that students who need more of a **challenge** understand this.
- Students can now do Exercise 12C. Tell the students to note the hint provided with question 3. Students who need more **support** should work carefully through question 1, to gain practice, before moving on to questions 2–4. If it will help these students, allow them to work in pairs.
- Remind students of the meaning of 'simplification', 'like terms' and 'coefficient'.
- Go through Examples 4 and 5.
- Students can then do Exercise 12D, noting the hints provided.

Plenary

- Write on the board: 3 × 4 + 3 × 5 = 3 × (4 + 5). Students will have seen this from the main lesson activity.
- Now write $2x + 10 = 2(x + 5)$. This was also part of the main lesson activity.
- Explain that this is the reverse process of expanding.
- Now ask the students if they can 'reverse' $3x - 18$, $4x - 20$, $4y + 6x$.
- Tell them that they will come across this later when they learn about factorisation.

Homework and answers

Homework for this lesson is available in the homework folder. Consolidation and extension exercises are available in the further practice folder. Answers to the exercises can be found in the answers folder.

12.3 Factorisation

Syllabus references

Statement	Core
C2.2/E2.2	Factorise by extracting common factors.

Collins references

- Student's Book pages 192–194

Learning aims

- Factorise an algebraic expression.

Key words

- common factors, factorise, factorisation

Prior knowledge

Being able to expand brackets is the first step to being able to factorise, because one is the inverse of the other.

Students should be able to find the highest common factor (HCF) of two or more numbers.

Common mistakes and remediation

Students may just find a factor, and not the *highest* factor.

In a question such as $4xy + 2x$, students may write $2x(2y + 0)$ as they think that there is nothing left when $2x$ is taken from the second term. Emphasise that they are dividing, not subtracting.

Useful tips

Advise students that they can always check that an expression has been factorised correctly by expanding the answer. When they are factorising, encourage students to look for the *HCF* and not just *any* factor.

Starter

- Ask students to give the HCF of:
  - 6 and 9 (3)
  - $6a$ and 9 (3)
  - $6a$ and $9a$ ($3a$)
  - a^2 and $9a$ (a)
  - $6a^2$ and $9a^2$ ($3a^2$)
- Students who need more **support** may find it easier to see the common factors if the expressions are broken down into products, e.g. 6 and 9 become 3 × 2 and 3 × 3, $6a$ and 9 become 3 × $2a$ and 3 × 3, $6a$ and
  $9a$ become $3a$ × 2 and $3a$ × 3, and so on.

Main lesson activity

- Write some expansions on the board, such as:
  - $2(x + 5) = 2x + 10$
  - $3(x - 2) = 3x - 6$
  - $5(x + 3) = 5x + 15$
- Now write $2x + 14 = \ldots$ and ask students if they can write down the expansion that it comes from. Most students should be able to work out that the answer is $2(x + 7)$.
- Repeat with other examples, such as $5x - 15$, $6x + 18$.
- Explain that what they have been doing is called factorisation and it is the reverse of expanding.
- Now ask students to factorise $4x + 12$. Many of them may say $2(2x + 6)$. Explain that this is a partial factorisation; the correct answer is $4(x + 3)$.
- Explain that they must take the HCF out of each term.
- Now ask for the factorisation of $x^2 + 3x$. This may need a further explanation, such as: $x \times x + 3 \times x = x(x + 3)$.
- Work through Example 6, which includes a variety of different factorisations.
- Emphasise to the students that they can multiply out the factored expression to check that the answer is the same as the original expression.
- All but the students who really need more of a **challenge** will be likely to find this topic difficult. Initially, restrict students who need more **support** to examples with positive values and those that do not require a letter to be taken out as a factor. Alternatively, encourage these students to write terms as products, as in the starter.
- Students can now do Exercise 12E. Encourage students who need more support to complete question 1 as far as they can on their own, but be prepared to offer support for questions 2–6.

Plenary

- Ask students to supply expressions with terms that have a HCF of 6, $6a$, $6a^2$, $6ab$, $6ab^2$, etc. Ask for several expressions for each HCF. (This is the reverse of the starter.)

Homework and answers

Homework for this lesson is available in the homework folder. Consolidation and extension exercises are available in the further practice folder. Answers to the exercises can be found in the answers folder.

12.4 Multiplying two brackets: 1

Syllabus references

Statement	Core
C2.2/E2.2	Expand products of algebraic expressions.

Collins references

- Student's Book pages 195–197

Learning aims

- Expand two linear brackets to obtain a quadratic expression.

Key words

- quadratic expression, linear, quadratic expansion

Prior knowledge

Students should be able to multiply out a single set of brackets.

Common mistakes and remediation

Students make errors with signs, particularly when multiplying two negatives (– × –). A reminder of the conventions for multiplying like and unlike signs would help.

Students may make mistakes by not writing down the brackets twice when expanding the square of an expression in brackets. Emphasise to students that they must remember to do this.

Useful tips

When expanding two linear brackets, there will always be four terms, two of which usually combine to make one term. Encourage students to always write down the four terms in their working.

Starter

- Revise multiplying out a single set of brackets, e.g. $5(2x + 3)$, $x(2x + 3)$, $5x(2x + 3)$.

Main lesson activity

Quadratic expansion

- Write on the board: $(8 + 2)(7 - 3)$. Ask students to work it out.
- They should work out the brackets and calculate $10 \times 4 = 40$.

  Write down: $8 \times 7 + 2 \times 7 - 8 \times 3 - 2 \times 3$. Then ask students to work out:
  $56 + 14 - 24 - 6 = 40$.

- Ask students for the connections between the two calculations (apart from the obvious one that the answer is the same). Then write $(x + 2)(x + 3)$ on the board and ask students if they can expand this in the same way as the numerical problem. Following the same pattern this gives $x \times x + 2 \times x + x \times 3 + 2 \times 3$. Simplify and evaluate this to $x^2 + 2x + 3x + 6 = x^2 + 5x + 6$. Then show three methods for doing problems like this.
- **Method 1**: Split the first set of brackets.
    - $(x - 3)(x + 4) = x(x + 4) - 3(x + 4) = x^2 + 4x - 3x - 12 = x^2 + x - 12$
- **Method 2**: FOIL
    - Multiply the **F**irst terms: $x \times x = x^2$.
    - Multiply the **O**uter terms: $x \times -2 = -2x$.
    - Multiply the **I**nner terms: $-3 \times x = -3x$.
    - Multiply the **L**ast terms: $-3 \times -2 = 6$.
    - Combine the terms to give $x^2 - 5x + 6$.

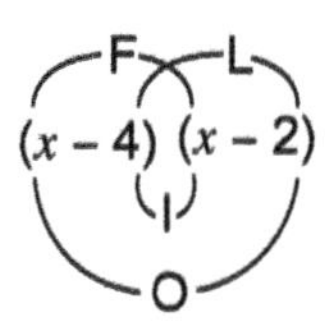

- **Method 3**: Box method.
    - $(x - 5)(x + 4)$
    - Collect the x terms 'diagonally': $x^2 - x - 20$.

x	x	-5
x	x^2	$-5x$
$+4$	$4x$	-20

- Go through Examples 7, 8 and 9 for extra practice. This topic is usually accessible to all students. Students who need more of a **challenge** can move on to the next section fairly quickly. The main problems will be errors with arithmetic and signs. Students can now do Exercise 12F. Tell them to note the hint provided.

Plenary

- Ask students if they can work out the brackets that lead to $x^2 + 12x + 27$ ($(x + 3)(x + 9)$).
- Repeat with $x^2 - 7x - 30$ ($(x - 10)(x + 3)$).

Homework and answers

Homework for this lesson is available in the homework folder. Consolidation and extension exercises are available in the further practice folder. Answers to the exercises can be found in the answers folder.

12.5 Multiplying two brackets: 2

Syllabus references

Statement	Core
C2.2/E2.2	Expand products of algebraic expressions.

Collins references

- Student's Book pages 197–200

Learning aims

- Expand two linear brackets to obtain a quadratic expression with non-unit square coefficient.

Key words

- quadratic expression, linear, quadratic expansion, non-unit, coefficient

Prior knowledge

Students should be able to multiply out two linear brackets with unit coefficients.

Common mistakes and remediation

Students make errors with signs, particularly when multiplying two negatives (– × –). A reminder of the conventions for multiplying like and unlike signs would help.

Students may make mistakes by not writing down the brackets twice when expanding the square of an expression in brackets. Emphasise to students that they must remember to do this. This should enable them to not make the mistake of writing $(x + a)^2 = x^2 + a^2$.

Useful tips

When expanding two linear brackets, there will always be four terms, two of which usually combine to make one term. Encourage students to always write down the four terms in their working.

Starter

- Revise multiplying out two sets of brackets, e.g. $(x + 4)(x + 3)$, $(x - 4)(x + 3)$.

Main lesson activity

Quadratic expansion with non-unit coefficients

- Now students have mastered Exercise 12F, work through Examples 10 and 11, which use the box method and the split brackets method. However, if FOIL is the preferred method, use this. Students can now do Exercises 12G and 12H. Students who need more of a **challenge** may not need to do all of 12G, but 12H requires them to spot a pattern; they should work through each question until they spot the pattern, then they can just write down the answers.

Expanding squares

- Go through this section, which deals with a very important fact about expanding squares: always write down the brackets twice. Then work through Examples 12 and 13.
- Students can now do Exercise 12I.

Plenary

- Ask students if they can work out the brackets that lead to $2x^2 + 7x + 3$ ($(2x + 1)(x + 3)$).

Homework and answers

Homework for this lesson is available in the homework folder. Consolidation and extension exercises are available in the further practice folder. Answers to the exercises can be found in the answers folder.

12.6 Expanding three brackets

Syllabus references

Statement	Extended
E2.3	Manipulate algebraic fractions.

Collins references

- Student's Book pages 200–202

Learning aims

- Expand three-bracket expressions.

Key words

- product, brackets

Prior knowledge

Students should be confident in expanding and simplifying two linear brackets with non-unit coefficients.

Common mistakes and remediation

Students may make simple errors when multiplying by, or adding, negative numbers.

Useful tips

- Remind students always to check their expansions.

Starter

- Ask the students to expand these expressions:
  - $(x + 3)(x - 2)$ $(x^2 + x - 6)$
  - $(x - 3)(2x + 1)$ $(2x^2 - 5x - 3)$
  - $(3x - 1)^2$ $(9x^2 - 6x + 1)$

Main lesson activity

- Tell students you are going to help them expand an expression containing three brackets.
- They need to be aware that they will be building on what they have already covered and they should always write down all the terms to avoid missing any, or making mistakes. This is especially important in tracking down slips when they do not get the correct answer.
- Work through the example given in the Student's Book: $(x + 2)(x - 4)(2x + 3)$. Let the students know it does not matter about the order in which the brackets are multiplied (multiplication is commutative), but they are less likely to make a mistake if they first find the product of brackets with unit coefficients.
- The FOIL method is rather cumbersome when one of the brackets contains more than two terms, so the box method is to be encouraged (especially if finding the product of two brackets each containing three terms).
- Students can now do Exercise 12J. Students who need more **support** will have difficulty, especially when the original expression contains minus signs. They should concentrate on getting the first four questions correct. Students who need more of a **challenge** may move through this exercise quickly as they become fluent in expanding two brackets: if you notice they are getting the questions correct, encourage them to write down the product of the first two brackets without showing any working.

Plenary

- Ask students to fully factorise $x^4 - 16$. $(x - 2)(x + 2)(x^2 + 16)$
- Ask, 'Do you think the difference of two identical powers can always be factorised?' (Yes, $(x - y)$ is always a factor of $x^n - y^n$. They will not meet a proof of this below IGCSE level however.)

Homework and answers

Homework for this lesson is available in the homework folder. Consolidation and extension exercises are available in the further practice folder. Answers to the exercises can be found in the answers folder.

12.7 Quadratic factorisation

Syllabus references

Statement	Extended
E2.2	Factorise expressions in the form $ax + bx + kay + kby$, $a^2x^2 - b^2y^2$, $a^2 + 2ab + b^2$, $ax^2 + bx + c$, $ax^3 + bx^2 + cx$. Complete the square for expressions in the form $ax^2 + bx + c$.

Collins references

- Student's Book pages 202–209

Learning aims

- Factorise a quadratic expression into two linear brackets.

Key words

- difference of two squares, brackets

Prior knowledge

Students should be confident in factorising simple expressions, e.g. $5x + 10$, $5x^2 + 10x$, $5xy - 20x$. They should also be able to expand and simplify two linear brackets.

Common mistakes and remediation

Students may use two negatives in the difference of two squares. They may ensure that the 'x^2' part and the '+ number' part are correct, but not check that the 'x' part is correct.

Useful tips

- Remind students always to check their expansions and factorisations by doing the reverse each time, to reach the original expression.

Starter

- Ask the students to factorise these expressions:
  - $8r + 2t$
  - $7acd - 5abc$
  - $12s + 6t$
  - $3t^2 + 6t$
  - $4a - 12ab$
  - $12ab^2 + 24b$
  - $16ab + 12b$
  - $9x^3 - 27x$
- Ask students to expand expressions such as $(2x)^2$, $(4y)^2$, $(3p)^2$.

Main lesson activity

- Refer back to the plenary of the lessons covering sections 12.4 and 12.5. Explain that this lesson is about reversing the process where two brackets are expanded. Start with $x^2 + 5x + 6$.
- Try to extract from students some of the rules by asking them to look back at the expansions from Exercise 12F. For example, they should see that this will have two sets of brackets, each of which will start with $(x \ldots)(x \ldots)$. They should also spot that the two numbers in brackets multiply to give the constant term. Students who need more of a **challenge** may spot that the sum of the numbers is the coefficient of x, and the rules about signs.
- Write down the answer to the factorisation: $(x + 2)(x + 3)$. Using the rules already given, go to the Student's Book, which covers the rules of expansion in more detail, followed by Examples 14 and 15.

- Students can now do Exercise 12K. Students who need more **support** will have difficulty, especially when the original expression contains minus signs. Giving one of the brackets may help, but this takes away somewhat from the point of the process. Students who need more of a **challenge** may move through this exercise and move on to the difference of two squares.

Difference of two squares

- Ask students to look back at their answers for Exercise 12H. Ask what rule they spotted. They should say $(x + y)(x - y) = x^2 - y^2$ or similar. Clearly this can be reversed, so the factorisation of $x^2 - y^2 = (x + y)(x - y)$. Explain that this is called the difference of two squares.
- Go through this topic in the Student's Book, followed by Examples 16 and 17.
- Students can now do Exercise 12L. Students who need more **support** may find the square roots of terms such as $4x^2$ difficult to recognise, so provide practice on these, as in the starter.

Factorising $ax^2 + bx + c$

- Ask students if they can factorise $2x^2 - x - 1$ ($(2x + 1)(x - 1)$). The process of factorising equations with a non-unit coefficient of x^2 is not as easy as with a unit coefficient of x^2. Essentially, it depends on intelligent guess work and experience. Work through Examples 18 and 19.
- Students can now do Exercise 12M questions 1 to 12. Students who need more **support** will find it difficult to match the factors of the coefficient of x^2 and the constant term, particularly if there is a minus sign involved. There is not much help that can be given except practice. Students must learn simple quadratic factorising and the method of the difference of two squares.

Completing the square

- Write the expression $x^2 + 12x + 29$ and then ask students to expand and simplify $(x + 6)^2$.
- They should find $(x + 6)^2 = x^2 + 12x + 36$
- Comparing this with the first expression we can see that $x^2 + 12x + 29 = (x + 6)^2 - 7$
- Rearranging a quadratic expression in this way is called **completing the square**.
- Point out that the number in the bracket is half the coefficient of x.
- Write down the expression $x^2 - 4x - 3$ and ask what we should square to complete the square.
- The answer is $(x - 2)^2 = x^2 - 4x + 4$
- It follows that $x^2 - 4x - 3 = (x - 2)^2 - 7$ in completed square form.
- Example 20 gives further practice if you need it.
- Students can now complete Exercise 12M from question 13 onwards. **Note:** The quadratic expressions here cannot be factorised simply. Completing the square is a technique that can be used to solve quadratic expressions, as will be shown in Chapter 13.

Plenary

- Ask students to factorise $x^2 + 2x$.
- Ask, 'Is this is a quadratic?' (Yes, but it is a special case that will be covered later.)

Homework and answers

Homework for this lesson is available in the homework folder. Consolidation and extension exercises are available in the further practice folder. Answers to the exercises can be found in the answers folder.

12.8 Algebraic fractions

Syllabus references

Statement	Extended
E2.3	Manipulate algebraic fractions. Factorise and simplify expressions.

Collins references

- Student's Book pages 209–213

Learning aims

- Simplify algebraic fractions.

Key words

- cancel, single fractions

Prior knowledge

Students should have a good understanding of adding, subtracting, multiplying and dividing normal fractions. They should also be confident in factorising, multiplying and simplifying algebraic expressions.

Common mistakes and remediation

Students often fail to cancel common factors, and make errors when multiplying two minus values, thus giving the answer as a negative. Remind students to watch out for these things.

Useful tips

When adding and subtracting algebraic fractions, students must find a common denominator, just as if they were adding and subtracting normal fractions. They should remember that they are only changing the appearance of the fraction and not its actual value – in effect multiplying each fraction by 1.
For example:

$$\frac{x}{3}+\frac{x}{2}=\frac{2}{2}\times\frac{x}{3}+\frac{3}{3}\times\frac{x}{2}=\frac{2x}{6}+\frac{3x}{6}=\frac{5x}{6}$$

$\frac{2}{2}=1$ $\frac{3}{3}=1$

Same value as before, they just look different.

Starter

- Ask students to use their calculators to work out:
  $\frac{1}{3}+\frac{2}{5}$, $\frac{3}{5}-\frac{1}{8}$, $\frac{3}{4}\times\frac{1}{5}$, $\frac{3}{5}\div\frac{2}{3}$
- They should write their answers in full, e.g. $\frac{1}{3}+\frac{2}{5}=\frac{11}{15}$

Main lesson activity

- Refer students to the four rules of fractions, as outlined in the Student's Book. Explain that these rules always work for any pair of fractions. Also explain that a, b, c and d can be any values, numbers or expressions.
- Now revisit the four calculations from the starter and ask individuals to come to the board and use the rules to work out the answers to the fraction calculations.
- Students' answers should agree with those from the calculators ($\frac{11}{15}, \frac{19}{40}, \frac{3}{20}, \frac{9}{10}$).
- Ask half the class to do the following calculations on their calculators and the other half to use the rules:
  $\frac{5}{8}+\frac{1}{6}$ $\quad\quad$ $\frac{3}{4}-\frac{3}{10}$
- Now swap groups so the previous half using calculators are now using the rules and vice versa. Ask students to work out the following.
  $\frac{5}{9}\times\frac{3}{10}$ $\quad\quad$ $\frac{8}{9}\div\frac{2}{3}$
- Compare answers, which should not agree, although some students may have cancelled their answers.
- Explain that the rules always work, but that they may not give the answer in its simplest form.
- Write $x^2 - 9$ on the board and ask students if they recognise it. They may need prompting, but this is the difference of two squares. Remind students that this can be factorised to $(x - 3)(x + 3)$. Repeat with $4a^2 - 25$ and other examples if necessary.
- Exercise 12N builds up the algebraic skills, so after each example ask students to attempt the questions as asked. Both students who need more support and those who need more of a challenge will benefit from doing this.
  - Work through Example 21 and ask students to complete questions 1 and 2 of Exercise 12N.
  - Then work through Example 22 and ask students to complete questions 3 and 4.
  - Work through Example 23 and ask students to complete questions 5 and 8.
  - Work through Example 24 and ask students to complete questions 6, 7 and 9. Question 7 requires students to reverse the process. The clue to getting started is the missing factor at the bottom. Students should see that this is the difference of two squares. If not, remind them of this.

Plenary

- Ask students to expand $x^2 + (x + 2)^2 = 16$ and then rearrange this into the general quadratic form of $ax^2 + bx + c = 0$. This should give $2x^2 + 4x - 12 = 0$. When students reach this stage, say that this is not simplified fully. Press them to cancel the common factor of 2.
- Repeat with $x^2 + (2x + 5)^2 = 10$, which finally cancels to $x^2 + 4x + 3 = 0$. This equation can be solved to give $x = -3$ and -1.

Homework and answers

Homework for this lesson is available in the homework folder. Consolidation and extension exercises are available in the further practice folder. Answers to the exercises can be found in the answers folder.

13.1 Solving linear equations

Syllabus references

Statement	Core
C2.5/E2.5	Solve linear equations in one unknown.

Collins references

- Student's Book pages 216–221

Learning aims

- Solve linear equations, including those containing brackets, where the variable only appears on one side.
- Solve linear equations, including those containing brackets, where the variable appears on both sides.

Key words

- equation, variable, solution, brackets, solve

Prior knowledge

Students should be familiar with the basic language of algebra and know how to collect like terms. They should also understand that addition is the opposite (inverse) operation to subtraction (and vice versa), and that multiplication is the opposite (inverse) operation to division (and vice versa). Students should be familiar with inverse operations, setting up equations using flow diagrams, and inverse flow diagrams.

Common mistakes and remediation

Students often use the same operation for the inverse operation. Errors with minus signs are common. An equation such as $\frac{x}{2} = 4$ will often be solved as $x = 2$. Checking answers in the original equation is the only way students can eliminate these errors.

Useful tips

It may be helpful for students who need more **support** to think of an equation as being like a set of balance scales. To keep them balanced, they must always make sure they do the same thing on both sides. Advise students to pay particular attention to minus signs.

Starter

- Play 'Think of a number'. Say, 'I am thinking of a number. I multiply it by 6 and subtract 2. My answer is 46. What was my number?' (8). Repeat with different numbers and operations.
- Use the algebra grid, Resource sheet 13.1, to practise simplifying expressions by adding the expressions in the columns and the rows. Make up some new grids.

Main lesson activity

- Ask students each to think of a number, double it and add 3. Let some students give their result and ask other students to say what the original number was. Do this for about five or six results.
- Now ask students if they can say how they worked out the original number. Students who need more of a **challenge** should see that they first subtracted 3 from the result and then divided by 2.

- Show the original problem as an equation: $2x + 3 =$ result. Tell the students that x is the variable and takes whatever value students choose. To find the value of x, they need to 'solve the equation'. Tell students that the instruction 'solve the equation' means they need to find the value of the letter that makes it true.

Doing the same to both sides

- Use the text and diagrams in this section to demonstrate the process. Set up an equation such as $3x + 6 = 12$ and work through the solution progressively, doing the same operation on both sides.

$$3x + 6 - 6 = 12 - 6$$
$$3x = 6$$
$$3x \div 3 = 6 \div 3$$
$$x = 2$$

- Students may find this tedious but it is worth doing, to get them into good habits.
- Work through Example 1. Remind students about multiplying and dividing with negative numbers.
- Students can now do Exercise 13A. Students who need more **support** may find questions involving division harder to solve and may benefit from a group discussion of one or two questions.
- Now consider equations with brackets by working through Examples 2 and 3. Students can then do Exercise 13B, noting the hints provided.

Equations with variables both sides

- Say that you are thinking of a number. You multiply it by 2 then add 5. The final result is 8 more than the original number you thought of. Ask students if they know what the number you started with was.
- This will not be as easy to solve as previous similar questions. If no student comes up with the correct answer of 3, then tell them the answer and work through the problem, i.e. $2 \times 3 + 5 = 3 + 8$.
- Now set this up as an equation: $2x + 5 = x + 8$. Ask students why this is different from previous problems. Students should see that the variable is on both sides.
- Work through the method of solution: $2x - x = 8 - 5$

$$x = 3$$

- Emphasise that the variable terms need to be collected on one side (normally the left-hand side, but this is not always the case), and the numbers on the other side.
- Students who need more of a **challenge** can work through Examples 4, 5 and 6. Students who need more **support** may find Example 6 difficult and could instead focus on Examples 4 and 5. Students can then do Exercise 13C.

Plenary

- Give students a 'think of a number' problem, e.g. 'I am thinking of a number. I times it by 3 and subtract 5. The result is 5 more than the number I originally thought of.' (5)
- Whether they solve it mentally or not, set up the equation $3x - 5 = x + 5$ and solve it.
- Repeat with similar examples.

Homework and answers

Homework for this lesson is available in the homework folder. Consolidation and extension exercises are available in the further practice folder. Answers to the exercises can be found in the answers folder.

13.2 Setting up equations

Syllabus references

Statement	Core
C2.5/E2.5	Solve linear equations in one unknown.

Collins references

- Student's Book pages 222–224

Learning aims

- Set up equations from given information, and then solve them.

Prior knowledge

Students must be able to solve simple equations.

Common mistakes and remediation

Students often find that the most difficult aspect in accessing questions is in setting up the equations. Encourage them to follow the steps as outlined. This should limit the errors to those for solving any equations (mistakes with rearranging and incorrect calculations) covered in the previous lesson.

Useful tips

Tell students that this lesson is about setting up equations from real-life or practical situations, and that they should define the 'unknown' as a variable, say x, then set up an equation using x.

Starter

- Ask students to formulate equations for simple word problems such as: If you cut r cm from 2 metres of ribbon how much is left? ($(200 - r)$ cm) The side of a square is c metres, so what is its area? ($c^2\,\mathrm{m}^2$)
- Ask students to name any simple formulae they know. They may provide these in words that, as a class, you can convert into algebra.

Main lesson activity

- Explain to students that they will look at problems from real-life or practical situations and solve them by using a variable such as x to represent the unknown. While the unknown is often defined in the question, sometimes it needs to be identified from the information provided.
- Point out that they should read the question thoroughly and identify key words.
- Give students the following problem: My sister is five years younger than I am. The sum of our ages is 41. How old am I?

- Ask students to identify the unknown. Clearly the unknown is 'my age'. Calling this x (or A), ask students to identify the necessary information and translate it into mathematical symbolism. The sister's age is $x - 5$; sum of ages $x + x - 5 = 41$; $2x - 5 = 41$, $2x = 46$, $x = 23$.
- Having solved the problem (my age is 23), explain that students should check that the solution fits the original problem. In doing so, students should be aware of the process skills:
  - Define the unknown if necessary.
  - Put given information in terms of the unknown.
  - Set up an equation.
  - Collect terms if necessary and solve the equation.
  - Check that the answer fits the original problem.
- Students who need more **support** may need help in accessing the questions and deciding on the variable. They may also need help in breaking down the problems.
- Work through Example 7.
- Now students can complete Exercise 13D. Tell them to note the many hints provided. Students who need more **support** may find some questions challenging. If they are having difficulty suggest they work in pairs, possibly with a student who needs more of a **challenge**.

Plenary

- Ask students to make up simple problems such as: Suni goes to the shop with a \$10 note and spends \$$x$. She gets \$3 change. How much did she spend?
- Although this is a trivial question, set it up as an equation, e.g. $10 - x = 3$.
- Ask for similar problems that involve multiplication and division.

Homework and answers

Homework for this lesson is available in the homework folder. Consolidation and extension exercises are available in the further practice folder. Answers to the exercises can be found in the answers folder.

13.3 Solving quadratic equations by factorisation

Syllabus references

Statement	Extended
E2.5	Solve quadratic equations by factorisation.

Collins references

- Student's Book pages 224–229

Learning aims

- Solve a quadratic equation by factorisation.

Key words

- linear equation, quadratic expressions, quadratic equation, factors, difference of two squares

Prior knowledge

Students need to be able to factorise the quadratic expressions in Exercises 13E and 13F.

Common mistakes and remediation

Before factorising, ensure the equation ends '= 0', and is cancelled as much as possible. Once it has been factorised, write out the options, e.g. for $(x + 2)(x - 3) = 0$, write $x + 2 = 0$ or $x - 3 = 0$ before solving to get $x = -2$ or $x = 3$. This should prevent students from giving incorrect answers of $x = 2$ and $x = -3$.

Students often get the signs wrong when solving linear equations. Go through more worked examples to help students if this persists.

Useful tips

Tell those students who simply cannot learn to factorise that there are other ways to solve quadratics, which will be covered later. Work out the number calculations and each different letter calculation separately.

Starter

- Write on the board: $a \times b = 0$. Ask students for pairs of values of a and b that make it true. List these as coordinate pairs, e.g. (0, 7), (–6, 0), (0, 0), (2.3, 0).
- Now write on the board: $x(x - 1) = 0$. Ask for values of x that make it true.
- The students may feel that a coordinate pair of answers is necessary, but in fact there are only two possible answers: 0 and 1.
- Demonstrate this if necessary: $0 \times (0 - 1) = 0 \times -1 = 0$, or $1 \times (1 - 1) = 1 \times 0 = 0$
- Now ask what values make $(x + 3)(x - 2) = 0$ true (–3 or 2).
- Repeat with similar examples if necessary.

Main lesson activity

Solving the quadratic equation $x^2 + ax + b = 0$

- If students have mastered factorisation, then solving quadratics is relatively easy.
- Work through Examples 8, 9 and 10.
- Students can now do Exercise 13E. Tell them to note the hints provided. Students who need more **support** may find some of the questions challenging. If they are having difficulty, suggest that they work in pairs.

Solving the general quadratic equation by factorisation, special cases

- Providing students can factorise expressions, then even students who need more **support** will be able to solve quadratics. The main problem will be with minus signs in solving the linear equations.
- Work through Examples 11 and 12, noting the special cases where b or c is zero.
- Students can now do Exercise 13F. Students who need more **support** will need guidance for this exercise.

Plenary

- Ask students to solve the equation $(x + 2)^2 - 9 = 0$ and work through the process:
  $(x + 2)^2 = 9$, $x + 2 = 3$ or $x + 2 = -3$, $x = 1$ or $x = -5$.
- Now ask students to solve $x^2 + 4x - 5 = 0$: $(x - 1)(x + 5) = 0$, $x = 1$ or -5.
- Write down $(x + 2)^2 - 9 = x^2 + 4x - 5$.

Homework and answers

Homework for this lesson is available in the homework folder. Consolidation and extension exercises are available in the further practice folder. Answers to the exercises can be found in the answers folder.

13.4 Solving quadratic equations by the quadratic formula

Syllabus references

Statement	Extended
E2.5	Solve quadratic equations by use of the quadratic formula.

Collins references

- Student's Book pages 229–231

Learning aims

- Solve a quadratic equation by using the quadratic formula.

Key words

- quadratic formula, coefficients, constant, soluble

Prior knowledge

Students need to be able to substitute into formulae, understand what the implications of the ± sign are, and have good calculator skills.

Common mistakes and remediation

The diagram below shows common mistakes students make when using the quadratic formula. Plenty of practice should help to eliminate them.

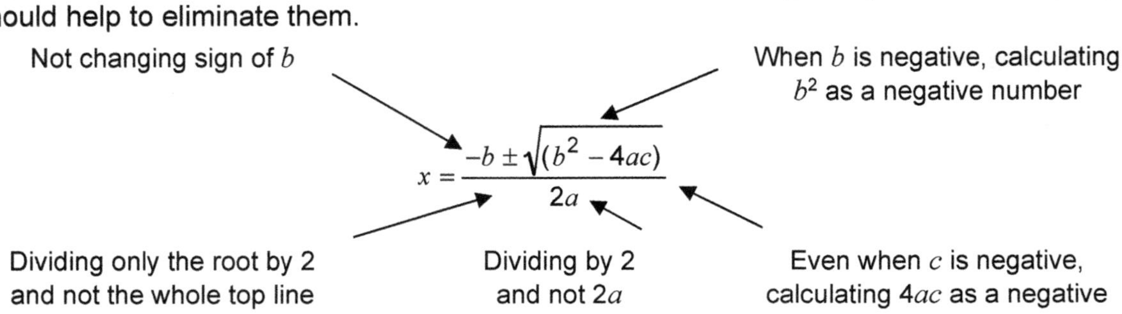

Students also make mistakes when using calculators. Encourage them to work in stages or to use brackets.

Useful tips

Make sure that students are aware of the need to apply the quadratic formula accurately.

The solution to the general quadratic equation $ax^2 + bx + c = 0$, $(a \neq 0)$ is given by:

$$x = \frac{-b \pm \sqrt{(b^2 - 4ac)}}{2a}$$

Tell students that they will know when the quadratic formula is needed as the question will say: Give your answer to 2 decimal places. Or: Give your answer to 3 significant figures.

Starter

- Ask students to use their calculators to work out these problems one at a time. If students write down answers, ask for at least 4 dp. Check the answers at each stage.

  $\frac{-3+2.5}{4}(-0.125)$, $\frac{-3-2.5}{4}(-1.375)$, $\frac{-3+\sqrt{65}}{4}(1.2655...)$, $\frac{-3-\sqrt{65}}{4}(-2.7655...)$

- It is unlikely that all students will complete these correctly so it will be worth spending time explaining brackets. As there are many different types of calculators, it may help to pair students with the same calculators to work together. It will help students who need more **support** if they are paired with students who need more of a **challenge**.

Main lesson activity

- Ask students to copy the diagram under 'Common mistakes and remediation' into their books. Say that it points out the main errors, and that if they give an incorrect answer they should use it to find out at which stage they went wrong.
- Write the following equation on the board: $x^2 + 5x + 6 = 0$.
- Identify a (1) as the coefficient of x^2, b (5) as the coefficient of x and c (6) as the constant term.

  Substitute these values into the formula using brackets around each term: $x = \frac{-(5) \pm \sqrt{(5)^2 - 4(1)(6)}}{2(1)}$

  Point out that this is an important step for students to include in their working.
- Now work out the square root and tidy up the terms and brackets to get: $x = \frac{-(5) \pm \sqrt{1}}{2}$
- Evaluate both answers, making sure students know that in the two solutions: $x = -2$ or -3.
- Compare this to the answer obtained from factorisation $(x + 2)(x + 3) = 0$. Obviously, they are the same and this would not be a typical quadratic formula example.
- Now write $2x^2 - 7x - 3 = 0$ and repeat the process to get: $x = \frac{-(-7) \pm \sqrt{(-7)^2 - 4(2)(-3)}}{2(2)}$
- Evaluate this as $x = \frac{7 \pm \sqrt{73}}{4}$, giving $x = 3.89$ or -0.39.
- Point out the very important fact that the denominator divides the whole top line and not just the square root.
- You may wish to explain that modern calculators will evaluate this without the intermediate step. Students who need more of a **challenge** should be able to use this method. However, encourage students who need more **support** to do the calculation in steps and to be careful with minus signs. Work through Example 13 for more practice.
- Now students can complete Exercise 13G. Students who need more **support** can complete the questions in pairs, if necessary. Provide support as necessary.

Plenary

- Solve $x^2 + 2x + 1 = 0$ using the formula. (A square root of 0 and an answer of $x = -1$.)
- Ask students if they can explain the fact that there is only one solution.
- Compare with the factorisation, which is $(x + 1)^2 = 0$, so there is only one answer.
- Point out that the quadratic formula always works (there is an exception in the next section).

Homework and answers

Homework for this lesson is available in the homework folder. Consolidation and extension exercises are available in the further practice folder. Answers to the exercises can be found in the answers folder.

13.5 Solving quadratic equations by completing the square

Syllabus references

Statement	Extended
E2.5	Solve quadratic equations by completing the square.

Collins references

- Student's Book pages 231–234

Learning aims

- Solve a quadratic equation by completing the square.

Key words

- completing the square

Prior knowledge

Being able to recognise and factorise, using the difference of two squares, is very useful.

Common mistakes and remediation

Students may forget to subtract a^2; they may forget to take the negative square root. Practice and more worked examples should help to remediate these problems.

Useful tips

If you have to complete the square on $x^2 + ?x$ …, always halve the ? to get the number at the end of the term in brackets, that is:

$$\left(x+\frac{?}{2}\right)^2$$

If you have to complete the square on $2x^2$..., always take out the factor 2 first, to give $2(x^2\ldots)$ then carry on to get:

$$2\left[\left(x+\frac{?}{2}\right)^2\ldots\right]$$

If you have to complete the square on $3x^2$…, always take out the factor 3 first, to give $3(x^2\ldots)$ then carry on to get:

$$3\left[\left(x+\frac{?}{2}\right)^2\ldots\right]$$ etc.

When solving, always remember that $\sqrt{?}$ = has a positive and a negative (±) value.

Starter

Ask these questions:

- Factorise $x^2 - 4$ $((x - 2)(x + 2))$
- Expand $(x - 2)^2$ $(x^2 - 4x + 4)$
- Expand and simplify these:
  - $(x + 2)^2 - 2^2$ $(x^2 + 4x)$
  - $(x + 3)^2 - 3^2$ $(x^2 + 6x)$
  - Guess $(x + 5)^2 - 5^2$ $(x^2 + 10x)$

Main lesson activity

- The main problem is that students do not link the x^2 and x terms with the $(x + a)^2 - b$, which is the starting point of the process.
- For a group who needs more **support**, this is worth exploring in depth.
  $(x + a)^2 = x^2 + 2ax + a^2$, which gives $x^2 + 2ax = (x + a)^2 - a^2$
  This is a result that students need to be very aware of.
- If necessary, work through many examples of the type $x^2 + 6x = \ldots$
- Students who need more **support** may find it helpful to complete question 1 of Exercise 13H before moving on.
- Now add a constant term.
  $x^2 + 2ax - b = (x + a)^2 - a^2 - b$
- Work through this section in the Student's Book to teach the main principles; then work through Examples 14, 15 and 16.
- Emphasise the steps of the process.
- Students can now do Exercise 13H. Students who need more **support** may work through questions 1 and 2 but will need guidance to complete the remaining questions in the exercise.

Plenary

- Review the steps involved in solving a quadratic by completing the square.

Homework and answers

Homework for this lesson is available in the homework folder. Consolidation and extension exercises are available in the further practice folder. Answers to the exercises can be found in the answers folder.

13.6 Fractional equations

Syllabus references

Statement	Extended
E2.5	Solve fractional equations with numerical and linear algebraic denominators.

Collins references

- Student's Book pages 234–236

Learning aims

- Solve fractional equations with the variable in the denominator.

Key words

- No new key words

Prior knowledge

Students should know how to solve linear equations with the variable on both sides and simple quadratic equations

Common mistakes and remediation

Students can make mistakes in rearranging an equation containing fractions. Emphasise the fact that this is done by multiplying both sides of the equation by a particular number or expressions.

Useful tips

Write each step of a solution clearly with each step on a separate line.

Starter

- Write down three equations: $12x = 5$; $\frac{x}{12} = 5$; $\frac{12}{x} = 5$, ask students to solve each one and to students to explain how they found each solution.
- For $12x = 5$, divide each side of the equation by 12: $x = \frac{5}{12}$; for $\frac{x}{12} = 5$, multiply each side of the equation by 12: $x = 5 \times 12 = 60$; for $\frac{12}{x} = 5$, multiply each side by x then divide both sides by 15: $x = \frac{12}{5} = 2.4$
- Point out that, in all three, you multiply both sides of the equation by the denominator of the fraction.

Main lesson activity

- Write down $\frac{4}{2x-3} = 5$ and ask students for the first step in solving this; they should say multiply both sides by $2x-3$ to get $4 = 5(2x-3)$.
- Ask students to complete the solution: multiply out the bracket: $4 = 10x - 15$; add 15 to both sides: $4 + 15 = 10x$; divide both sides by 10: $x = \frac{19}{10} = 1.9$

- Here is a check: if $x = 1.9$ then $\frac{4}{2x-3} = \frac{4}{2\times1.9-3} = \frac{4}{0.8} = 5$, which is the correct value.
- Next, ask students to solve this equation $\frac{x+6}{x-6} = 9$
- Here is a solution: multiply both sides by $x-6$: $x+6 = 9(x-6)$; multiply out the bracket: $x+6 = 9x-54$; subtract x from both sides: $6 = 8x-54$; add 54 to both sides: $60 = 8x$; divide by 8: $x = 7.5$
- Ensure that students can explain each step.
- Students can now do Exercise 13I questions 1 to 4 (they may need a hint with question 1 – for part **a** they need to multiply by a multiple of 3 and 5; 15 is the LCM).
- Next, ask students to look at this equation: $\frac{15}{x+2} = \frac{6}{x-1}$ and ssk for suggestions about the first step.
- Multiply by the product of the two denominators, $(x+2)(x-1)$; this gives $15(x-1) = 6(x+2)$.
  Solve this by multiplying out the brackets: $15x-15 = 6x+12 \Rightarrow 9x = 27 \Rightarrow x = 3$
- Now give this equation: $\frac{10}{x-3} + \frac{12}{x+1} = 7$
  Multiply by the product of the denominators $(x-3)(x+1)$; $10(x+1)+12(x-3) = (x-3)(x+1)$
- Ask the students how to continue.
  Multiply out the brackets: $10x+10+12x-36 = 7(x^2-2x-3)$; rearrange: $22x-26 = 7x^2-14x-21$;
  $7x^2-36x+5 = 0$; factorise: $(x-5)(7x-1) = 0$; So, $x = 5 \text{ or } \frac{1}{7}$
- There are two possible solutions. Students can check that each of these give the value 7 when substituted into the expression on the left of the equation.
- Students can now complete Exercise 13I.

Plenary

- Ask Students to solve the equation $x + \frac{12}{x} = 7$
- They should get $x^2 - 7x + 12 = 0$
- This factorises as $(x-3)(x-4) = 0$ and so $x = 3 \text{ or } 4$
- Now ask for the solution of $y+1+\frac{12}{x+1} = 7$
- The easy way to do this is to see that $y = x+1$ so the solution is $y = 4 \text{ or } 5$

Homework and answers

Homework for this lesson is available in the homework folder. Consolidation and extension exercises are available in the further practice folder. Answers to the exercises can be found in the answers folder.

13.7 Simultaneous equations

Syllabus references

Statement	Core
C2.5/E2.5	Solve simultaneous linear equations in two unknowns.

Collins references

- Student's Book pages 236–244

Learning aims

- Derive and solve simultaneous linear equations in two unknowns.

Key words

- simultaneous linear equations, derive, eliminate, substitute

Prior knowledge

Students must be able to rearrange and solve basic equations such as $4x - 6 = 2$.

Common mistakes and remediation

When deriving equations students often struggle to translate the written problem into symbols. This often happens with age problems. For example, if they let Tariq's age be x years and Meera's age be y years and are told 'Tariq is twice as old as Meera' they often write $2x = y$, rather than $x = 2y$. Ask them to check whether that works by letting x be some simple number. Students may also become confused with negatives, e.g. when subtracting the two equations and one of the numbers is already a negative.

Not labelling equations and explaining steps can lead to confusion and errors; encourage students to label equations and give a full explanation of the steps they take, as good practice.

Useful tips

Advise students always to try to use the elimination method in favour of the substitution method.

Starter

- With their books closed, ask students, 'If $x + y = 10$, what is the value of x?' Assuming someone ventures a number, say 5, ask, 'What is the value of y when $x = 5$?' Write down some of the pairs of solutions; make sure $x = 4$ and $y = 6$ are included. Get students to realise that there are many (indeed, infinitely many) values for both x and y.
- Ask the same of $2x + y = 14$. Write down possible values of x and y; make sure $x = 4$ and $y = 6$ are included. Point out that $x = 4$ and $y = 6$ satisfy both of the equations at the same time, that is, *simultaneously*.
- Some students might appreciate being shown a quick sketch of the two graphs, showing that they intersect at the point (4, 6).
- Now set up a simple problem like the example, but using two people they might know, such as yourself and someone a lot older than you in your establishment. Tell them that that person is twice as old as you and that when you add their age to yours you get 75 (or any multiple of 3). Then ask them to work out how old you are. Expect them to do quite a bit of trial and improvement techniques until they obtain the answer. Then go over how to derive the simultaneous equations, making sure that they define the variables before doing anything else.

Main lesson activity

Note: This is a very long section and, if your lesson is short, you could split the section into two lessons, the first dealing with Exercises 13J and 13K, the next with Exercises 13L and 13M.The first exercise deals with the methods of solving simple simultaneous equations and deriving them. These start with pairs of equations that do not need to have coefficients balanced. The next exercise (13K) moves on to pairs of equations where only one equation needs to be balanced, and finally to pairs of equations where both pairs need balancing. Depending on the ability of students, some of the questions could be worked through quickly or missed altogether. Students who need more **support** will need to work their way through the process slowly, assimilating each step before moving on to the next.

Elimination method, Substitution method

- Explain that there are six steps to consider when solving simultaneous equations.
  - Step 1: Make the coefficients of one of the variables the same.
  - Step 2: Depending on the signs of the balanced variables, add or subtract the equations to eliminate this variable.
  - Step 3: This will leave a linear equation in the other variable; solve this equation, using the methods in previous sections.
  - Step 4: Having found the value of one variable, substitute it back into one of the original equations to set up a linear equation in the other variable.
  - Step 5: Solve this linear equation, using the methods of the previous sections.
  - Step 6: Substitute both values into *both* of the *original* equations to check that they are correct.
- It is essential to get students into good habits in solving these equations, as there are so many ways to make errors.
- Emphasise the need to label equations and write down the operations performed on each one as part of their working. This helps students to keep track of their work and identify any mistakes.
- Work through the introduction and the section on the elimination method, including Examples 18 and 19.
- Next, work through the solution of simultaneous equations by the substitution method, and Example 20. Students who need more **support** can leave this out but it is essential for students who need more of a **challenge**, as they will need the skill later when solving pairs of simultaneous equations where one is linear and one is non-linear.
- Students can now work through Exercise 13K.

Balancing coefficients, Balancing coefficients in both equations

- Continue with the methods of balancing coefficients, working through Examples 22 and 23 before students do Exercise 13L.
- Go through Example 24, then students can do Exercise 13M.

Plenary

- Work through some of the more complex questions in Exercises 13K (question 2), 13L (question 2) and 13M (questions 2, 3 and 4).

Homework and answers

Homework for this lesson is available in the homework folder. Consolidation and extension exercises are available in the further practice folder. Answers to the exercises can be found in the answers folder.

13.8 Linear and non-linear simultaneous equations

Syllabus references

Statement	Extended
E2.5	Solve simultaneous equations, involving one linear and one non-linear.

Collins references

- Student's Book pages 244–246

Learning aims

- Derive and solve simultaneous equations, involving one linear and one quadratic.

Key words

- linear, non-linear

Prior knowledge

Students should be able to solve linear simultaneous equations, using the substitution method. They should also be able to factorise and solve quadratic equations.

Common mistakes and remediation

Students sometimes find the values of one unknown and forget to substitute this back into the linear equation to find the other unknown value. They should be reminded there are generally two pairs of solutions and it can help them to write these as co-ordinates.

Useful tips

Most, but not all, questions that the students meet will have integer value solutions, so if they end up with awkward values they will probably have made an error in their working. This is a good indication that they need to check their solution.

Students can substitute for either unknown, so, if one of the unknowns in the linear equation has a unit coefficient, it is easier if that is the unknown that is substituted as it eliminates the need for fractions.

Starter

- Write on the board the following pair of simultaneous equations: $xy = 10$ and $x + y = 7$. Then ask: How does this pair of simultaneous equations differ from what you have met before?
- Many will say that they have not met the product xy before.
- Now ask students how they think they should proceed.
- They might suggest substituting for y in one of the equations, and use their suggestion, which might be to substitute $y = 10/x$ in $x + y = 7$, or substitute $x + y = 7 - x$ in $xy = 10$.
- Either way, show that the resulting equation is $x^2 - 7x + 10 = 0$, and solve it to get $x = 2$ or $x = 5$.
- Then substitute back into either equation to find the corresponding values of y.
- Finally show the answer as both $x = 2$, $y = 5$ or $x = 5$, $y = 2$ and (2, 5) or (5, 2), mentioning that either format is acceptable.

Main lesson activity

- Work through Example 26 as shown in the Student's Book as this deals with a slightly different pair of equations to the starter activity.
- Students can now tackle Exercise 13N. You might want to start them on question 2, as question 1 deals with pairs of linear simultaneous equations and is there in case you wish your students to be reminded of them.
- Questions 6, 7 and 8 deal with deriving the pair of simultaneous equations before solving them.

Plenary

- Go over question 8 with your students, pointing out the subtle difference between the 'sum of the squares' and the 'square of the sum'.

Homework and answers

Homework for this lesson is available in the homework folder. Consolidation and extension exercises are available in the further practice folder. Answers to the exercises can be found in the answers folder.

13.9 Representing inequalities

Syllabus references

Statement	Core
C2.6/E2.6	Represent and interpret inequalities, including on a number line.

Collins references

- Student's Book pages 247–248

Learning aims

- Show an inequality on a number line.

Key words

- No new key words

Prior knowledge

Students should know the four different inequality signs and be able to distinguish between them.

Common mistakes and remediation

Students should remember that if the numbers in an inequality are reversed then the inequality is changed. For example, $x < 4$ is equivalent to $4 > x$

When the opportunity arises, reinforce this point by writing the inequality both ways.

Useful tips

Always write a double inequality with the smaller number on the left and the larger on the right.

Starter

- Write down $x \geqslant 5$ and ask for possible values of x. Ask for the smallest possible value. (5)
- Repeat with $x \leqslant 1$ and ask for possible values of x. Here there is a largest possible value. (1)
- Now say that n is an integer $n < -3$. What is the largest possible value of n ? (−4)
- Emphasise the difference between < and $\leqslant$.

Main lesson activity

- Draw a number line to show how to represent $x \geqslant -2$ like this:

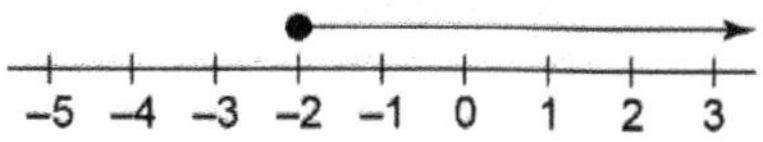

- The black dot shows that –2 is included.
- Now draw a line to represent $x < 1$ like this:

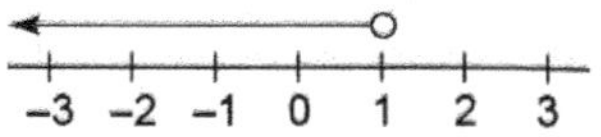

- The open dot shows that 1 is not included.
- This inequality $-2 \leqslant x \leqslant 3$means that x is between –2 and 3. Show it like this:

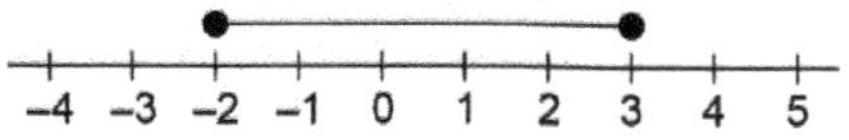

- Point out the black dots to show –2 and 3 are included.
- Ask students to draw number lines to show $2 < x < 5$and $-4 \leqslant x < 1$
- They should look like this:

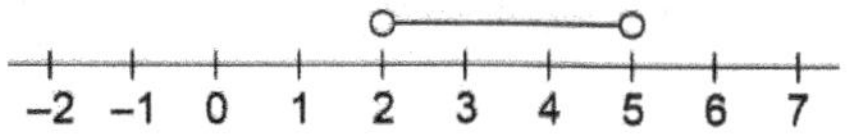

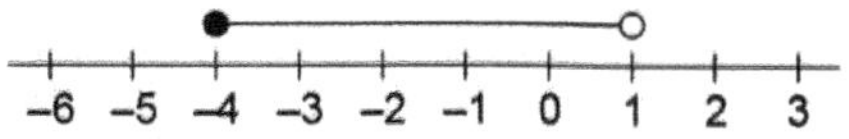

- Make sure the end points are correctly marked.
- Now say that n is an integer and $-3 < n \leqslant 3$. Ask for possible values of n. (–2, –1, 0, 1, 2, 3)
- Students can now complete Exercise 13O.

Plenary

- Ask students to draw a coordinate grid with positive axes only.
- Ask them to find all the points with coordinates (x, y) that satisfy the inequalities $4 < x < 7$and $2 < y < 5$
- They should find a square with vertices at (4, 2), (4, 5), (7, 5) and (7, 2)
- All the points inside the square are the ones required but not the points on the boundary.

Homework and answers

Homework for this lesson is available in the homework folder. Consolidation and extension exercises are available in the further practice folder. Answers to the exercises can be found in the answers folder.

13.10 Solving inequalities

Syllabus references

Statement	Extended
E2.6	Construct, solve and interpret linear inequalities.

Collins references

- Student's Book pages 249–253

Learning aims

- Derive and solve a simple linear inequality.
- Represent and interpret inequalities on a number line.

Key words

- inequality, number line

Prior knowledge

Students should be able to solve linear equations, including those with the variable occurring on both sides of the equation.

Common mistakes and remediation

Students often solve these as equations and do not put the inequality back in for the answer. Encourage students to keep the inequality in the solution at each step. Remind students that failure to recover the inequality in the final answer would result in zero marks, as the level of difficulty of the question has been changed.

Useful tips

Solve inequalities in the same way you solve equations, i.e. by rearrangement, but do not replace the inequality with an equals sign.

Starter

- Write on the board: $x < 5$. Ask students to show this on a number line. Make sure there is an open circle above 5 and the arrow points to the left.
- Then ask: What can you say about $2x$?
- They should say that $2x < 10$
- Ask how it is different if we write $x \leqslant 5$. Students should say that in this case the circle on the diagram is filled in. They should also say that $2x \leqslant 10$

Main lesson activity

- Ask students if they can give the values of x that are true for $x + 3 > 7$.
- They should be able to see that x can be any number bigger than 4
- Now show students how to solve this inequality:

  $x + 3 > 7$

  $x > 7 - 3$

  $x > 4$
- Students should recognise this as being the same basic method as solving an equation. Say that the methods are the same, but that the equals sign is replaced with an inequality sign.
- Work through Examples 27 and 28.
- Now students can do Exercise 13P. Students who need more **support** should be able to complete questions 1–5 on their own. They may need support to finish the exercise.

Plenary

- Write $2x + y \leqslant 7$ on the board and say that x and y are non-negative integers. Ask for possible values of x and y that make it true. For example, $x = 3$, $y = 1$ or $x = 0$, $y = 4$.
- Use these values to plot points on a coordinate grid with positive axes.
- They will find that all their points are inside the triangle with vertices at (0, 0), (0, 7) and (3.5, 7) or on the perimeter of the triangle. There are 20 points altogether.

Homework and answers

Homework for this lesson is available in the homework folder. Consolidation and extension exercises are available in the further practice folder. Answers to the exercises can be found in the answers folder.

14.1 Conversion graphs

Syllabus references

Statement	Core
C2.9/E2.9	Use and interpret graphs in practical situations including travel graphs and conversion graphs.

Collins references

- Student's Book pages 256–259

Learning aims

- Convert from one unit to another unit by using a graph.

Key words

- scale, estimate, conversion graph

Prior knowledge

Students should know how to plot coordinates.

Common mistakes and remediation

Students may misread scales, particularly when there are no divisions marked. Provide some examples to help to avoid this. Students may omit units when giving answers. Emphasise that without units the numbers are meaningless.

Useful tips

Remind students that they need to read conversion graphs accurately in order to get accurate answers.

Starter

- Draw these two graphs on the board.

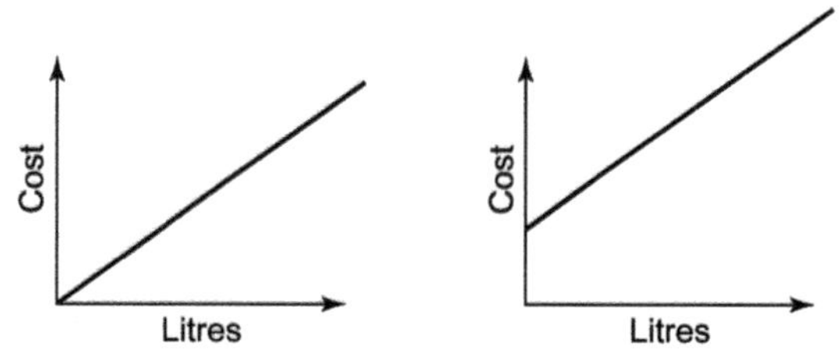

- Explain that the first graph shows the relationship between cost of car fuel and number of litres, and that the second graph shows the cost of delivering litres of heating oil.
- Discuss what each graph shows. For example, the first graph shows that 0 litres costs $0. Why does the second graph start some way up the cost axis?
- Establish the idea of a standing charge and a linear relationship between amounts and costs.

Main lesson activity

- Following on from the starter, ask students if they can think of any quantities that might need a conversion graph.
- Students can be prompted by suggestions about mass, volume, currency, gas bills, etc.

- Look at Example 1. Ask students if they know the approximate conversion factor between gallons and litres. (4.55 litres to the gallon.) Show them how they can work this out using the graph.
- Ask students why the graph goes through the origin. (If there are 0 litres, there must be 0 gallons.)
- Go through the two parts of Example 1.
- Extend this example by asking students how they could use the graph to convert 40 litres into gallons. (≈ 8.8 litres)
- Look at Example 2. Ask students why this graph does not go through the origin (reinforce the idea of a 'standing charge').
- Talk students through the two parts of Example 2.
- Ask if they can work out the cost per unit of electricity. (7 cents, because 500 units costs $35 more than the basic charge.)
- Students can now do Exercise 14A.
- All students will find this exercise accessible. Students who need more of a **challenge** could answer the first seven questions orally.
- Students who need more of a **challenge** could use the graph in question 7 to work out the temperature in °C when it is 0 °F and what the temperature in °F is when it is 100 °C (the boiling point of water). Or, they could use the internet to find out some significant temperatures (in °F and °C), such as the normal body temperature (37 °C, 98.6 °F) and the temperature when paper catches fire
(232 °C, 451 °F).

Plenary

- Tell students that a gas bill for 100 units is $20. Can they use this information to work out the cost of 1 unit of gas?
- They may say 20 cents, but others should say, rightly, that there could be a fixed charge.
- Now tell them that a different bill for 80 units was $17. Can students work out the cost per unit now?
- Sketch a graph of the given information, e.g. as shown below.

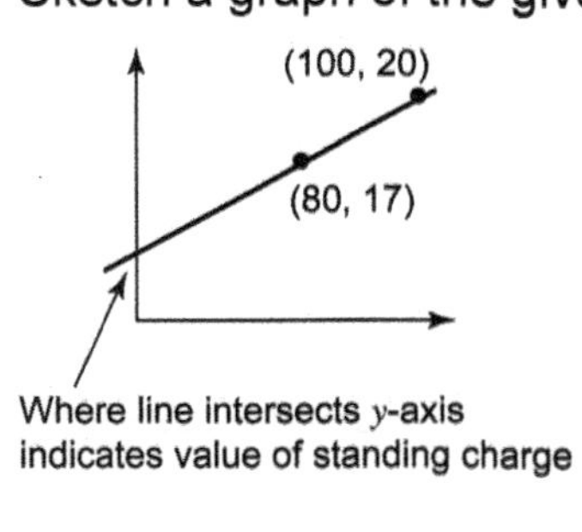

- They can work out that a unit costs 15 cents and the fixed charge is $5.

Homework and answers

Homework for this lesson is available in the homework folder. Consolidation and extension exercises are available in the further practice folder. Answers to the exercises can be found in the answers folder.

14.2 Travel graphs

Syllabus references

Statement	Core
C2.9/E2.9	Use and interpret graphs in practical situations, including travel graphs. Draw graphs from given data.

Collins references

- Student's Book pages 260–264

Learning aims

- Interpret distance–time graphs.

Key words

- average speed, distance–time graph

Prior knowledge

Students need to be confident in plotting points on coordinate axes, and they should be able to read scales on axes. Students should also know that:

distance = speed × time
time = distance ÷ speed
speed = distance ÷ time

Common mistakes and remediation

Students often interpret the minutes as a decimal, e.g. entering 2 hours 30 minutes as 2.3 on a calculator instead of 2.5. If this problem persists, provide some common conversions on the board or on an information sheet. Students often misread scales and thus make mistakes, so remind them to pay special attention to this and give them extra practice in reading scales of different types, if necessary.

Useful tips

Ask students to remember the following information.

- Time is always plotted as the horizontal axis.
- Parts of journeys are always shown as straight lines, showing a steady average speed. However, in reality, speed varies due to traffic conditions.

Starter

- Ask students to convert the following hours, minutes and decimals.
  - From minutes to hours and minutes: 230 mins (3 h 50), 165 mins (2 h 45), 84 mins (1 h 24)
  - From hours and minutes, to minutes only: 1 h 55 (115 mins), 2 h 15 (135 mins), 3 h 16 (196 mins)
  - From hours and minutes, to decimals: 1 h 30 (1.5 h), 2 h 45 (2.75 h), 3 h 20 (3.3333… h)
  - From decimals in hours, to hours and minutes: 2.666… h (2 h 40), 3.25 h (3 h 15), 1.1 h (1 h 6)
- Make sure students are familiar with the decimal equivalents of 15 minutes, 30 minutes, etc.

Main lesson activity

- On the board, sketch a graph as shown. Mark the axes as time and distance.

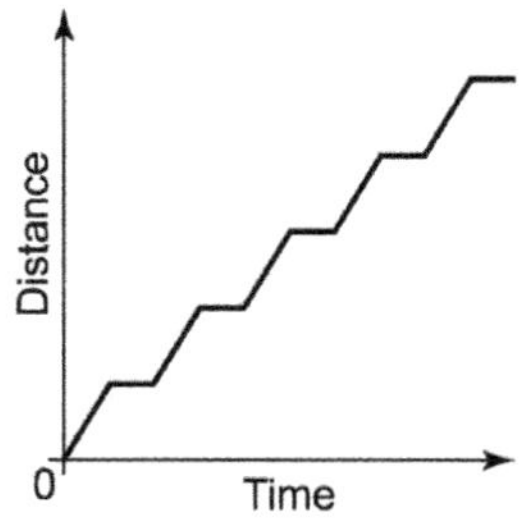

- Ask what sort of journey this could represent, e.g. a form of public transport. (Students should recognise this as being a bus journey.) Ask, 'Why are the sloping lines straight?' Explain that the graph models the real situation; the sloping lines show average or steady speed. In reality, speed varies and the line would not be straight. Ask what the horizontal lines represent. Make sure they understand that horizontal lines represent zero speed.
- Now sketch a graph as shown.

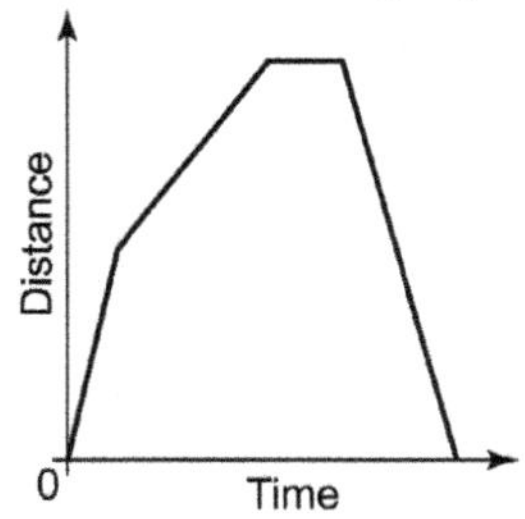

- Discuss the differences, compared to the previous graph. (It returns to the original point, so is an out-and-back graph.) Discuss the slopes of the lines. Ask which part shows the fastest part of the journey. Make sure they understand that the steeper the line, the higher the speed. Then ask which is the second-fastest part of the graph (the return journey). Make it clear that the slope direction is not relevant.
- Ask students if they can recall the connection between speed, distance and time. Then draw the speed–distance–time 'road sign'.

- Work through Example 3. Students who need more **support** may have difficulty in working out average speed. Make sure they understand that it is the total distance travelled divided by the total time taken.
- Students can now do Exercise 14B, noting the hint that accompanies question 6.

Plenary

- Ask some quick-fire questions about calculating speed: How fast would you be travelling if you covered 60 miles in 2 hours? 3 kilometres in 30 minutes? 10 kilometres in 15 minutes? 210 miles in 3 hours? (30 mph, 6 km/h, 40 km/h, 70 mph)
- Ensure students can convert to distance per hour.

Homework and answers

Homework for this lesson is available in the homework folder. Consolidation and extension exercises are available in the further practice folder. Answers to the exercises can be found in the answers folder.

14.3 Speed–time graphs

Syllabus references

Statement	Extended
E2.9	Apply the idea of rate of change to simple kinematics involving: distance–time and speed–time graphs, acceleration and deceleration. Calculate distance travelled as area under a linear speed–time graph.

Collins references

- Student's Book pages 264–268

Learning aims

- Interpret speed–time graphs.

Key words

- speed–time, constant speed, distance travelled, rate of increase, acceleration, deceleration

Prior knowledge

Students need to understand the work done in Lesson 14.2. They also need to know how to find the area of a rectangle and the area of a triangle.

Common mistakes and remediation

Students sometimes confuse speed with average speed. When we talk of speed, we mean how fast something is travelling at any given point on a journey. Average speed describes the speed over a period of time, during which the speed may be changing constantly.

Useful tips

Remind students that $\text{average speed} = \dfrac{\text{total distance travelled}}{\text{total time taken}}$

Starter

- Ask students to consider the following graphs, which show a car journey.
- Emphasise the fact that the first is a distance–time graph and the second a speed–time graph. Warn them that the graphs may look the same but each tells a very different story.

a The car is initially travelling at a constant speed, then it slows down until it stops for a period of time. Then the car returns to where it started from at a constant speed.	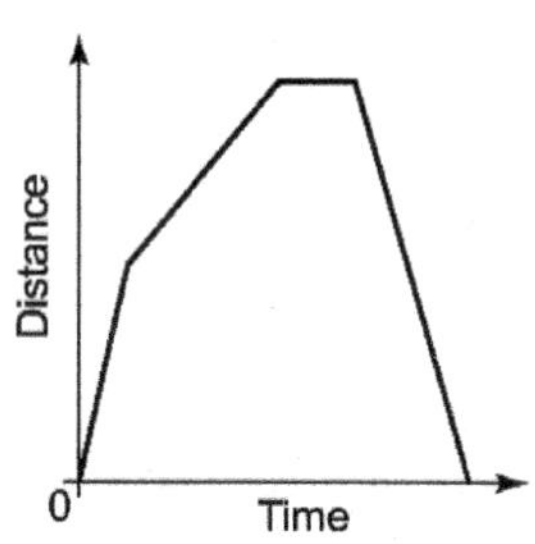
b The car is initially accelerating, then its speed still increases but at a slower rate. The car then travels at a constant speed for a period of time. The car then slows down, i.e. decelerates, until it comes to a stop.	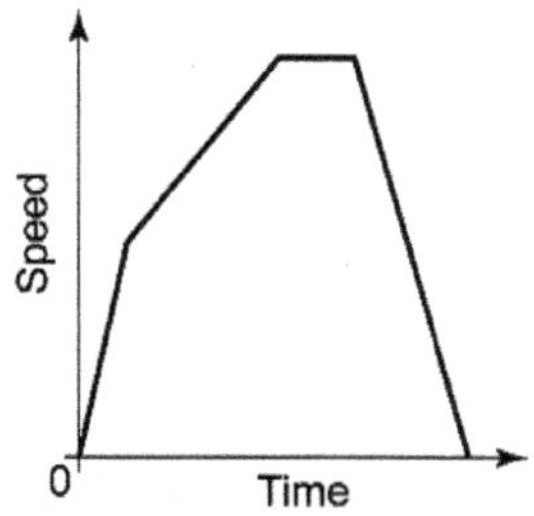

Main lesson activity

- Tell the students that you are now going to think about what else we can find out from a speed–time graph. Work through the text to establish that the area under a speed–time graph tells us the total distance travelled.
- Ensure that students understand that acceleration is the rate of increase of speed, and similarly, deceleration is the rate of decrease of speed.
- Work through Example 4 with the students. Highlight the fact that the process will be quicker if they can apply the formula for the area of a trapezium.
- Students can now do Exercise 14C. Students who need more **support** may require some guidance.

Plenary

- Discuss with the students their solutions to questions 7, 8 and 9 of Exercise 14C.

Homework and answers

Homework for this lesson is available in the homework folder. Consolidation and extension exercises are available in the further practice folder. Answers to the exercises can be found in the answers folder.

14.4 Curved graphs

Syllabus references

Statement	Extended
E2.9	Apply the idea of rate of change to simple kinematics involving distance–time and speed–time graphs, acceleration and deceleration.

Collins references

- Student's Book pages 268–271

Learning aims

- Estimate and interpret the gradient of a tangent at a point on distance–time and speed–time graphs.

Key words

- distance–time, speed–time, acceleration, deceleration, gradient

Prior knowledge

Students need to understand the work done in Lessons 14.2 and 14.3.

Common mistakes and remediation

Students sometimes confuse speed with average speed. When we talk of speed, we mean how fast something is travelling at any given point on a journey. Average speed describes the speed over a period of time, during which the speed may be changing constantly.

The same is true of acceleration.

Useful tips

Remind students that the gradient of a line is given by $\frac{\text{vertical change}}{\text{horizontal change}}$.

However, when dealing with speed, which is a scalar quantity, they should ignore any negative signs and describe the speed appropriately. In Example 5 the gradient of the tangent is negative for $x > 5$, but the runner has a positive speed directed towards the start.

When dealing with acceleration, a negative acceleration is often referred to as a deceleration. However, this can cause problems in students' understanding at a later stage beyond IGCSE as a deceleration implies a slowing down. A negative acceleration can mean that a body is in fact speeding up, but in the opposite direction to travel.

It is useful to have copies of the first graph (and that of Exercise 14D question 1) available so that students can draw in the tangents themselves at appropriate points.

Starter

- Ask students to look at the first graph showing the distance travelled by a runner in a 10-second interval and describe the runner's motion. They should also estimate the distance travelled under the curve (which is clearly 50 metres since the runner goes 25 metres and back!).
- If it is easy to go outside you might want to have a volunteer try and run to emulate the runner in the example.

Main lesson activity

- Give the students copies of the first graph and ask them to draw in a tangent at (3.4, 20).
- When drawing the triangle underneath the tangent, suggest that the horizontal distance they use is a whole number, say between 2 and 5 seconds: that makes the division fairly easy.
- They should now calculate the gradient (actually 5.74, but accept anything between 5.5 and 6).
- Work through Example 5 with the students if you feel it necessary.
- Students can now do Exercise 14D.

Plenary

- Discuss with the students their solutions to questions 4 and 5 of Exercise 14D.

Homework and answers

Homework for this lesson is available in the homework folder. Consolidation and extension exercises are available in the further practice folder. Answers to the exercises can be found in the answers folder.

15.1 Drawing straight-line graphs

Syllabus references

Statement	Core
C3.1/E3.1	Use and interpret Cartesian coordinates in two dimensions.
C3.2/E3.2	Draw straight-line graphs for linear equations.

Collins references

- Student's Book pages 274–277

Learning aims

- Plot negative coordinates.
- Draw graphs of linear functions.

Key words

- quadrant, coordinates, negative coordinates, Cartesian coordinates, equation of line, Cartesian plane, straight-line graph

Prior knowledge

Students need to know how to plot coordinates in all four quadrants and how to calculate coordinates. They should also know how to substitute numbers into formulae.

Common mistakes and remediation

Students sometimes plot the x and y coordinates the wrong way round. Provide plenty of examples to help them get used to the correct method.

Students sometimes make mistakes when substituting and so the points they plot are incorrect. Choosing to plot three points rather than two can often highlight this error and so prompt students to review their answers.

Useful tips

Encourage students to think about whether their plotted points are sensible, e.g. if either the x or y coordinate is negative, then the point cannot be in the first quadrant.

Remind students to draw graphs with a sharp pencil, and use a ruler to ensure the lines are straight.

Starter

- Ensure students know how to plot coordinates in the first quadrant, by displaying a set of coordinate axes on the board, and inviting them to plot a number of different points, all of which are in the first quadrant.
- Now consider points in all four quadrants. Once they are confident plotting both positive and negative coordinates, ask them to plot the following sets of points, with each set on a different grid.

  (2, –2); (2, 0); (2, 1); (2, 2); (2, 5)

  (–3, –1); (–1, –1); (0, –1); (3, –1); (5, –1)

  (0, –5); (0, –3); (0, 1); (0, 5); (0, 6)

  (–4, 0); (–2, 0); (0, 0); (5, 0); (8, 0)

  All the graphs should be straight lines.

Main lesson activity

- Write (x, y) on the board. Ask students for a number. Take this as x. Apply a rule to it to get y, for example $y = 2x - 1$, without telling the students what the rule is. Underneath the (x, y) write the coordinate pair resulting from the students' number. Repeat for further suggested starting numbers.
- Ask students to give the rule as a statement connecting x and y. So, for the example above, you may get (5, 9), (3, 5), (10, 19). Encourage students to be methodical so that they give you starting numbers of 1, 2, 3, etc. and make the rule easier to see.
- Repeat with other numbers. Include some divisions but, if dividing by 3, for example, insist that the students give you multiples of 3.
- Students can describe these lines ($x = 2$; $y = -1$; $x = 0$; $y = 0$). Highlight the fact that '$x =$' lines are parallel to the y-axis, and '$y =$' lines are parallel to the x-axis.
- Deduce with them that the equation of the x-axis is $y = 0$, and the equation of the y-axis is $x = 0$.
- Introduce the idea of straight-line graphs, which are usually referred to as linear graphs. Explain that any equation of the form $y = mx + c$, where m and c are numbers, will give a straight-line graph.
- Make sure students know the following:
  - Two points are the minimum needed to draw a straight line, but three are better as one can be used as a check.
  - Use a sharp pencil. Thick lines make values harder to read accurately.
  - Make sure points and lines are drawn accurately.
- Now work through Example 1 with the class. Students can now do Exercise 15A.
- Students who need more **support** may benefit from having the sizes of the axes that need to be drawn for each question.

Plenary

- Ask students to sketch the following graphs without plotting any points:

  $x = 5$ (straight line parallel to the y-axis, crossing the x-axis at (5, 0))

  $y = -0.5$ (straight line parallel to the x-axis, crossing the y-axis at (0, –0.5))

  $x = -7$ (straight line parallel to the y-axis, crossing the x-axis at (–7, 0))
- Ask students to look at the graphs they have drawn in the lesson and their equations (their graphs should be labelled). Can they see any connections?

Homework and answers

Homework for this lesson is available in the homework folder. Consolidation and extension exercises are available in the further practice folder. Answers to the exercises can be found in the answers folder.

15.2 The equation $y = mx + c$

Syllabus references

Statement	Core
C3.3/E3.3	Find the gradient of a straight line.
C3.5/E3.5	Interpret and obtain the equation of a straight-line graph in the form $y = mx + c$.

Collins references

- Student's Book pages 277–281

Learning aims

- Find the equation of a line, using its gradient and intercept.

Key words

- slope, gradient, coefficient, intercept

Prior knowledge

Students should know how to draw a straight-line graph.

Common mistakes and remediation

Students often fail to recognise a negative gradient. Remind them (if necessary, every time they begin to answer questions about a graph) that lines that slope from top left to bottom right always have a negative gradient: forward slope = positive, backward slope = negative.

Useful tips

Students often get the x- and y- distances the wrong way round when calculating gradients. Provide examples to compare with to help them get used to the correct method.

Starter

- Ask students to draw the following straight-line graphs:

  a $y = x$ **b** $y = x + 1$ **c** $y = 2x + 4$

  d $y = x - 7$ **e** $y = -4x + 3$ **f** $y = -0.5x - 1$

Main lesson activity

Gradient

- Introduce the concept of a *gradient*, which is the *slope* of a line. Emphasise to students that the steeper the line, the larger the value of the gradient.
- Work through the text in the Student's Book and Example 2. Make sure students know the following:
  - When drawing a triangle to measure the gradient, students should use grid points so that values are integers.
  - Always divide the distance in the y-direction by the distance in the x-direction.

- Lines that slope from top left to bottom right have a negative gradient. Justify this by considering the movement from one grid point to another. If you have to move down and across to get from one point to the other then this vertical move can be described as negative. Hence you are dividing a negative value by a positive one, which gives a negative answer.
- Define the y-intercept as the point at which the line crosses the y-axis.

The equation $y = mx + c$

- Use the graphs drawn in the starter to complete this table. Students should copy it into their books.

Equation	Gradient	Intercept
$y = x$		
$y = x + 1$		
$y = 2x + 4$		
$y = x - 7$		
$y = -4x + 3$		
$y = -0.5x - 1$		

- The completed table should look like this.

Equation	Gradient (m)	Intercept (c)
$y = x$	1	0
$y = x + 1$	1	1
$y = 2x + 4$	2	4
$y = x - 7$	1	–7
$y = -4x + 3$	–4	3
$y = -0.5x - 1$	–0.5	–1

- Students should link the gradient to m and the intercept to c. Explain that this means that if they know the gradient and the intercept then they can write down the equation of the line immediately.
- Work through the text and Example 3 to consolidate this. Ensure that students understand the term *coefficient*.
- Now students can complete Exercise 15B. Students who need more **support** may need guidance. Provide them with copies of the grids so that they can draw 'triangles' to find the gradients.

Plenary

- Show a sketch of a 'gas bill' graph, which consists of a basic charge and a price per unit. Do not put any values on the graph.
- Ask students to say how they can find the basic charge from the graph (the intercept). Then ask how they can find the price per unit (the gradient).

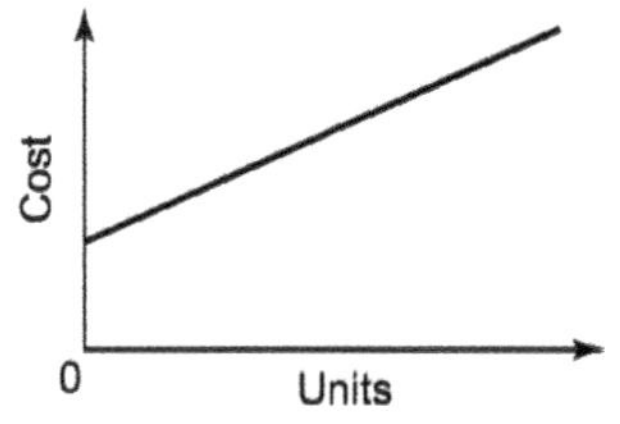

Homework and answers

Homework for this lesson is available in the homework folder. Consolidation and extension exercises are available in the further practice folder. Answers to the exercises can be found in the answers folder.

15.3 More about straight-line graphs

Syllabus references

Statement	Extended
E3.3	Calculate the gradient of a straight line from the coordinates of two points on it.

Collins references

- Student's Book pages 281–283

Learning aims

- Express the equation of a straight line in different forms other than $y = mx + c$.

Key words

- gradient, intercept

Prior knowledge

Students need to be able to draw straight-line graphs and be able to write down the equation of a straight line, given its gradient and intercept. They should also be able to re-arrange simple linear equations.

Common mistakes and remediation

Students often make sign and arithmetic errors because of poor algebraic skills. Extra practice and worked examples should help to remediate these problems.

Useful tips

Remind students that linear graphs are not always given in the form $y = mx + c$.

Starter

- Write the equation $y = 3x - 5$ on the board and ask students to give you an equivalent equation without a subtraction. ($y + 5 = 3x$)
- Now ask for an equivalent equation that has = 0 in it. ($y - 5 + 3x = 0$)
- Now ask for an equivalent equation that has = 5 in it. ($5 = 3x - y$)

Main lesson activity

- Students should be familiar with working with straight-line graphs now. Work through Example 4 to develop the method of rearranging the formula into the $y = mx + c$ form.
- Now students can complete Exercise 15C. Working at this level, students are expected to use algebraic methods to find the gradient and intercepts.

Plenary

Discuss the solutions to questions 7 and 8 of Exercise 15C.

Homework and answers

Homework for this lesson is available in the homework folder. Consolidation and extension exercises are available in the further practice folder. Answers to the exercises can be found in the answers folder.

15.4 Solving equations graphically

Syllabus references

Statement	Core
C3.5/E3.5	Interpret and obtain the equation of a straight-line graph in the form $y = mx + c$.

Collins references

- Student's Book pages 283–286

Learning aims

- Solve equations using straight-line graphs.

Key words

- coordinate, x-axis, y-axis

Prior knowledge

Students need to be able to recognise straight-line graphs.

Common mistakes and remediation

Students often read off the wrong axis. Stress the fact that the x-axis is the line $y = 0$ and the y-axis is the line $y = 0$.

Useful tips

Tell students to draw in the necessary lines as that helps them locate the correct answer.

All students should be provided with copies of the graphs so they can draw on them and not mark the books.

Starter

- Draw the equation $y = 2x + 1$ on the board with axes having integer values and x from −2 to 5 and ask students if they can explain how to use the graph to find the value of x when $y = 7$.
- Now repeat for $y = 4$ – this may cause a problem as there is no integer value, but students should be able to estimate the answer as $x = 1.5$.
- Explain that the graph could be used to find values of x for any value of y, though it might need to be extended.

Main lesson activity

- Students should be familiar with graphs by now.
- Explain that they now need to be able to read more detailed scales. Give them all a copy of the graph so they can work through Example 5 with you.
- Work through Example 5 to develop the method of drawing on the graph the appropriate lines. Pay particular attention to part **c** which has to be re-arranged to give $0.73x - 1.15 = 1.5$.
- Extend this example to show how to solve $7.3x - 11.5 = 10$, by dividing throughout by 10, to obtain $0.73x - 1.15 = 1$. ($x = 2.9$)
- Now students can complete Exercise 15D. Students should be given a sheet with all the graphs on them so that no time is wasted in them copying the graphs or making mistakes.

Plenary

Discuss the solutions to questions 5 and 6 of Exercise 15D

Homework and answers

Homework for this lesson is available in the homework folder. Consolidation and extension exercises are available in the further practice folder. Answers to the exercises can be found in the answers folder.

15.5 Parallel lines

Syllabus references

Statement	Core
C3.6/E3.6	Find the gradient and equation of a straight line parallel to a given line.

Collins references

- Student's Book pages 286–288

Learning aims

- Find the equation of a line that is parallel to another line of which the equation is given.

Key words

- parallel

Prior knowledge

Students need to be able to draw straight-line graphs and be able to write down the equation of a straight line, given its gradient and intercept.

Common mistakes and remediation

Students often make sign and arithmetic errors because of poor algebraic skills. Extra practice and worked examples should help to remediate these problems.

Useful tips

Tell students to remember that parallel lines are lines with the same gradient (m) but different intercepts (c).

Starter

- Write the equation $y = 3x - 5$ on the board and ask students to give you the equation of a line parallel to this line. They should provide equations such as $y = 3x + c$.
- Now ask for a line parallel to $3x + 2y = 6$.
- This may cause some pause and an attempt to rearrange into the form $y = \ldots$, but in fact any equation of the form $3x + 2y = c$ is parallel to the example given.

Main lesson activity

- Students should be familiar with drawing graphs, measuring gradients and $y = mx + c$, and after the starter they should be aware of the conditions for parallel lines.
- Explain that they now need to be able to find the equations of lines that are parallel to given lines. Work through Example 6 to develop the method.
- Now students can complete Exercise 15E. Working at this level, students are expected to use algebraic methods to calculate the equations of lines that are parallel to given lines.
- Students who need more of a **challenge** will cope with the questions in Exercise 15E but you could help students who need more **support** by drawing the required lines on grids.

Plenary

- Put a simple linear equation on the board. Ask students to give you an equation of a line parallel to it.
- After a few correct answers, put a more awkward equation on the board and again ask for the equation of a line parallel to it.
- Choose equations, such as those below, with increasing difficulty.

  a $y = 6x + 1$

  b $y = x + 1$

  c $y = 6x - 9$

  d $y = -6x + 5$

  e $6x + 4y = 15$

  f $9x - 1.5y - 0.5 = 0$

Homework and answers

Homework for this lesson is available in the homework folder. Consolidation and extension exercises are available in the further practice folder. Answers to the exercises can be found in the answers folder.

15.6 Points and lines

Syllabus references

Statement	Extended
E3.3 E3.4	Calculate the gradient of a straight line from the coordinates of two points on it. Calculate the length of a line segment. Find the coordinates of the midpoint of a line segment.

Collins references

- Student's Book pages 288–290

Learning aims

- Find the gradient of a line passing through two given points.
- Find the equation of a line joining a pair of points.
- Find the midpoint of a straight-line segment, given its end points.
- Find the distance between two points.

Key words

- midpoint, line segment

Prior knowledge

Students need to know how to find the gradient of a straight line and be able to establish the equation of a straight line using the gradient-intercept method. Students should also be able to apply Pythagoras' theorem.

Common mistakes and remediation

Students often make sign and arithmetic errors because of poor algebraic skills. Extra practice and worked examples should help to remediate these problems.

Useful tips

Tell students that when they are asked to find the gradient of a line drawn through a pair of points, it will help if they plot them on a grid.

Starter

- Recap with students how to find the gradient of a straight line. Develop this to establish the fact that if we know two points on the line, then we can find the gradient by using

  $$\text{gradient} = \frac{\text{difference between } y\text{-coordinates}}{\text{difference between } x\text{-coordinates}}$$

- Work through the example in the Student's Book to illustrate this. Students can then go on to complete question 1 of Exercise 15F.

Main lesson activity

The equation of a line

- Building on the starter activity, ask the students to discuss, in pairs, how they can go on to find the equation of the line that joins these two points. If necessary, lead them through the text in the Student's Book.
- Students can then work through question 2 of Exercise 15F.

Midpoints

- Make sure students understand the terms 'midpoint' and 'line segment'.
- Ask students to draw a grid and on it plot the points (–2, 2) and (4, 5). Tell them to join their points with a straight line. Ask them if they can establish the midpoint of the line segment they have drawn and deduce the rule outlined in the Student's Book. Students can now do questions 3 and 4 of Exercise 15F.

The distance between two points

- Ask students to refer back to their previous diagram and complete a triangle with their line segment as its sloping side (hypotenuse). Ask students if they know how to calculate the length of the line segment. Hopefully they will suggest using Pythagoras' theorem and be able to conclude that the length of the line is 6.71 to 2 decimal places. Students who need more **support** may need guidance with this.
- Students can then complete Exercise 15F.

Plenary

- Discuss the solutions to questions 6 and 7 of Exercise 15F.

Homework and answers

Homework for this lesson is available in the homework folder. Consolidation and extension exercises are available in the further practice folder. Answers to the exercises can be found in the answers folder.

15.7 Perpendicular lines

Syllabus references

Statement	Extended
E3.7	Find the gradient and equation of a straight line perpendicular to a given line.

Collins references

- Student's Book pages 290–293

Learning aims

- To understand the relationship between the gradients of perpendicular lines.

Key word

- perpendicular

Prior knowledge

Students will need to know how to find the gradient of a straight line.

Students need to be able to write down the equation of a line with a given gradient passing through a given point.

Common mistakes and remediation

Students will sometimes forget the negative aspect of the relationship and only use the positive reciprocal.

Useful tips

Ensure that students recognise the relationship between perpendicular gradients and that they learn it: 'If you know the gradient of a line, then to find the gradient of the line perpendicular to it, find the reciprocal and change the sign.'

Starter

- Put on the board a grid with a pair of perpendicular lines for which the gradients can easily be calculated.

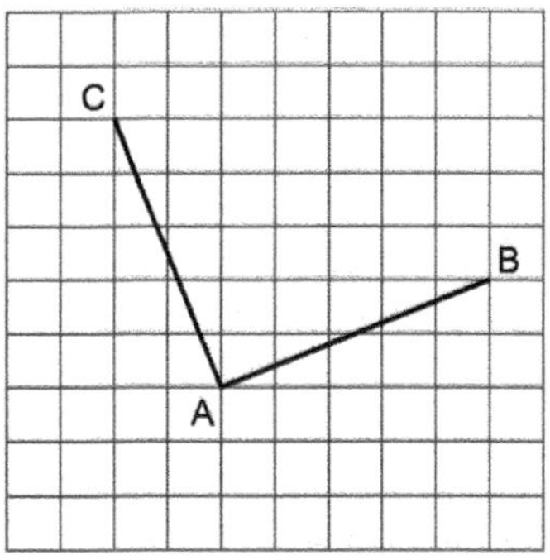

- Ask students how they know the lines are perpendicular. (They are at right angles. Rotating line *AB* through 90° gives line *AC*.)
- Ask the students for the gradient of line *AB*. Use the grid to show that it is $\frac{2}{5}$.
- Ask the students for the gradient of line *AC*. Use the grid to show that it is $-\frac{5}{2}$.

  Students who need more **support** may need reminding about negative gradients.
- Ask students what they notice about the gradients. (Use the same numbers, one is upside down, one is positive, the other negative.) Encourage them to use the correct term (inverse) and say, 'One is the negative inverse of the other.'

- Ask students to multiply the gradients together. Make sure that they get the result of –1.
- Ask if they think this will always happen.

Main lesson activity

- Ask students each to draw a pair of perpendicular lines in a grid, making sure that they can easily work out the gradients. Then ask them to investigate the gradients of their perpendicular lines.
- Once students have completed this, confirm that it is always true that the product of the gradients is –1, except where one line is horizontal.
- Put on the board the equation $y = 2x + 4$ and sketch a line to represent it. Say that the line is the graph of the equation.
- Ask students to suggest coordinates of points on this line, for example, (0, 4), (1, 6), (2, 8).
- Then ask students for the gradient of this line. (2)
- Now ask how to find the equation of the line perpendicular to this line, at one of the points previously listed, such as (2, 8).
- Inverting 2 and changing the sign, the gradient of the perpendicular will be $-\frac{1}{2}$.
- The equation of the line is $y = -\frac{1}{2}x + c$.
- Now ask students to calculate c by substituting the known values (2, 8), to give $8 = -\frac{1}{2} \times 2 + c$

  $\rightarrow 8 = -1 + c \rightarrow c = 8 + 1 = 9$. Then the equation is $y = -\frac{1}{2}x + 9$.
- Tell students who need more of a **challenge** that the gradient of a line is $\frac{a}{b}$ and ask them to give the

  gradient of the perpendicular to it. Make sure that they correctly state the negative inverse, $-\frac{b}{a}$.
- Talk students through Example 7 in the Student's Book.
- The students can now do Exercise 15G.

Plenary

- Ask students how they could check if the lines representing two equations were perpendicular. (Find their gradients and see if one is the negative reciprocal of the other.)
- Put on the board these equations and ask which lines are perpendicular to each other:
  a $y = 2x + 3$ **b** $y = 1 - 2x$ **c** $2y = 4 - x$ **d** $2y = x - 6$
- Encourage a class discussion and, if appropriate, take a vote. Lines **a** and **d** are perpendicular, as are lines **b** and **c**.

Homework and answers

Homework for this lesson is available in the homework folder. Consolidation and extension exercises are available in the further practice folder. Answers to the exercises can be found in the answers folder.

16.1 Quadratic graphs

Syllabus references

Statement	Core
C2.10/E2.11	Construct tables of values, and draw, recognise and interpret graphs for functions.
C2.11/E2.11	Recognise, sketch and interpret graphs of the following functions: linear, quadratic.

Collins references

- Student's Book pages 296–300

Learning aims

- Draw a quadratic graph, given its equation.
- Find equivalent values, using a quadratic graph.

Key words

- quadratic graph, quadratic equation, parabola, roots

Prior knowledge

Students should know how to plot coordinate points in all four quadrants, be able to substitute numbers into a formula and know how to draw linear graphs.

Common mistakes and remediation

Some students may assume that the square of a negative number is itself negative.

Students may fail to produce a continuous smooth curve. The most basic mistake is to fail to use a sharp pencil.

When a graph has its turning point at a point halfway between two plotted values, the bottom of the graph is often drawn flat.

Useful tips

Encourage students to note that quadratic graphs are always symmetrical. Once they have identified the turning point (which is not necessarily where $x = 0$) they should match the curve on both sides.

Starter

- Use quick-fire questions to check that students know the squares of numbers from 0 to 10. Then repeat for negative numbers from –10 to 0.

Main lesson activity

- Explain that a quadratic equation is one that includes a term in x^2. The most basic quadratic equation is $y = x^2$. The graph of this equation is a smooth curve, called a parabola. All quadratic graphs are based on this equation and they all have the same basic shape.
- Work through Example 1, to demonstrate drawing a quadratic graph. Make sure that students have centimetre-squared paper, so that they can draw the graph.
- Make the following basic points about answering questions on quadratic graphs.
  - There will always be a range of values of x.
  - There is usually a table giving the x-values, with space for the corresponding y-values.
  - Intermediate lines to help calculation of the values are not usually given.
  - Most values will already be filled in and there will only be two or three to calculate.
  - Coordinate axes are always provided.
- On the board, draw up a table of values and complete it for $y = x^2$. Students should copy the table into their notes.
- Demonstrate how to plot the points on a coordinate grid.
- Students who need more **support** may have difficulty in reading the scales. Check that students plot the points correctly.
- Demonstrate how to join the points with a smooth curve. Let students read through Example 1, and the notes that follow it about drawing graphs. Then let them draw their curves through the points they have plotted.
- Make sure that students label the graphs.
- Work through Example 2, which includes intermediate lines in the table for calculating values for the different parts of the equation.
- Make sure that students know how to use the graph to find corresponding values. Starting from an x value, move vertically to the curve, then horizontally to the y value. Practise several times.
- Now start from a y-value and move across to the curve, then down to find the x-values. Stress that there will generally be two x-values for every one y-value, except at the turning point of the curve. Mention that these two x-values are called the 'roots' of the equation.
- Students can now complete Exercise 16A. The questions include a wide range of quadratic graphs. They also use some halfway-values, so the points plotted may be more accurate.

Plenary

- Ask students to look at the tables they have filled in. Ask them if they can spot anything about the y-values. They should see that the y-values have symmetry about a point.
- Ask them to look at the graph and see if this can be related to the symmetry. They should see that the x-value about which the table is symmetrical is also the line about which the graph is symmetrical.

Homework and answers

Homework for this lesson is available in the homework folder. Consolidation and extension exercises are available in the further practice folder. Answers to the exercises can be found in the answers folder.

16.2 Solving equations graphically

Syllabus references

Statement	Core
C2.10/E2.10	Construct tables of values, and draw, recognise and interpret graphs for functions. Solve associated equations graphically.

Collins references

- Student's Book pages 301–302

Learning aims

- To use graphs of functions to solve equations.

Key words

- No new key words

Prior knowledge

Students should know how to complete a table of values of a function and use it to draw a graph.

Common mistakes and remediation

Students often make mistakes in plotting points. Always look and see whether the points seem reasonable. If it is difficult to join the points with a smooth curve there is probably an error.

Useful tips

If there are positive and negative values of x, work out the values of the function for the positive values first. The arithmetic will be easier.

Starter

- Ask students to draw a graph of the line $y = 7 - x$
- Then, on the same axes, draw the line $y = 0.5x - 2$
- It should look like the graph, right.
- Ask where the lines cross (at (6, 1)).
- The x coordinate of 6 gives the same value for $7 - x$ and $0.5x - 2$ (The value is 1, of course.)
- So, the solution of the equation $7 - x = 0.5x - 2$ is $x = 6$
- The important point: drawing two graphs gives the solution of the equation where the graphs cross.
- An optional extra for students who need more of a **challenge** is to say that the equation can be solved by rearranging the equation and you could invite students to do this. Rearranging gives $1.5x = 9$, and so $x = \frac{9}{1.5} = 6$.

  However, the method can still be used when rearranging is not so easy.

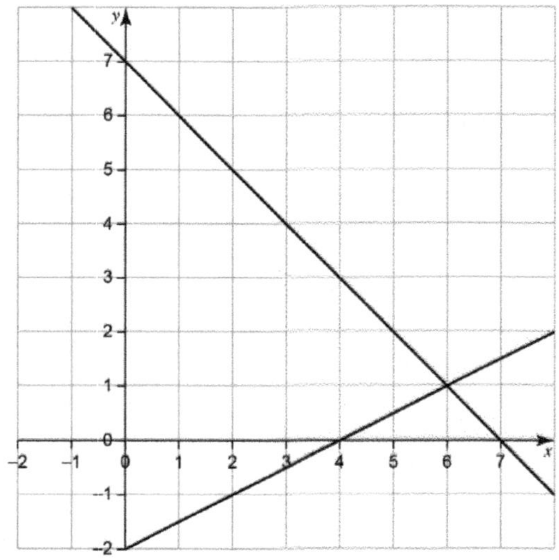

Main lesson activity

- Here is an equation: $x^2 = 2x + 6$
- Say that we can solve this by drawing graphs of $y = x^2$ and $y = 2x + 6$
- Ask the students what each graph will look like; then draw the graphs.
- They should recognise $y = x^2$ as a symmetrical curve with a lowest point at (0, 0)
- They should also say that $y = 2x + 6$ is a straight line with a gradient of 2 that crosses the y-axis at 6
- Ask student to look at the start of section 6.2 where they will see he two lines drawn on the same grid. Notice that the axes are not the same.
- The curve and the straight line cross at two points so the equation has two solutions. Ask the students to read the two solutions from the diagram. They should see that x = 3.6 or −1.6
- You can check these answers. For example, if $x = 3.6$, then $x^2 = 3.6^2 = 12.96$ and $2x + 6 = 2 \times 3.6 + 6 = 13.2$. These are not exactly the same because the answers are estimated to 1 decimal place. Finding solutions from a graph will not usually give exact solutions.
- Similarly, if $x = -1.6$ then $x^2 = (-1.6)^2 = 2.56$ and $2x + 6 = 2 \times (-1.6) + 6 = 2.8$
- Students following the Extended syllabus who need more of a **challenge** can be asked to solve the equation exactly. For example, the could do it by completing the square:

$$x^2 - 2x = 6$$
$$(x-1)^2 - 1 = 6$$
$$(x-1)^2 = 7$$
$$x = 1 \pm \sqrt{7} = 3.646 \text{ or } -1.646 \text{ to 3 dp}$$

- Students can now do Exercise 16B

Plenary

- Earlier, the equation $x^2 = 2x + 6$ was solved by drawing two graphs.
- An alternative method is to rewrite it as $x^2 - 2x = 6$ and draw a graph of $y = x^2 - 2x$
- Ask students to tell you the missing numbers in this table:

x	-2	-1	0	1	2	3	4
$x^2 - 2x$	8	3					

  The missing values are 0, −1, 0, 3, 8
- Show students a graph of this function or ask them to draw it.
- They can solve the equation by seeing where the line $y = 6$ crosses the curve.
- They should get the same values of x

Homework and answers

Homework for this lesson is available in the homework folder. Consolidation and extension exercises are available in the further practice folder. Answers to the exercises can be found in the answers folder.

16.3 Turning points on a quadratic graph

Syllabus references

Statement	Extended
E2.10	Solve associated equations graphically, including finding and interpreting roots by graphical methods.

Collins references

- Student's Book pages 303–304

Learning aims

- Recognise, sketch and interpret quadratic graphs.
- Identify turning points.

Key words

- completing the square, turning points

Prior knowledge

Students need to be able to draw quadratic graphs.

They should also be able to complete the square (see Section 13.5).

Common mistakes and remediation

Making an error in the amount needed to compensate for the squared constant term when completing the square. Students should always check that they return to the original equation after completing the square.

Another common mistake is to give the x-coordinate of the turning point with the wrong sign: if the equation of the graph is $y = (x + a)^2 + b$, then the coordinates of the turning point are $(-a, b)$.

Useful tips

Remind students that when <u>sketching</u> graphs (as opposed to <u>drawing</u> them, when more points are required), all they need to do is plot the turning point, the intercept on the y-axis and the corresponding symmetrical point. The curve can then be sketched using those points.

Starter

- Ensure students know how to complete the square by asking them to put the following equations into the completed square format:
  $y = x^2 + 4x + 6$ $\quad y = (x + 2)^2 + 2$
  $y = x^2 - 8x + 5$ $\quad y = (x - 4)^2 - 11$
  $y = x^2 - 5x - 4$ $\quad y = (x - 2.5)^2 - 10.25$
- Now ask them what the minimum value of each of these equations are: they may need reminding that any square is non-negative, i.e. greater than or equal to zero.
  The corresponding minimum values are 2, –11 and –10.25 and the turning points are (−2, 2), (4, −11) and (2.5, −10.25).

Main lesson activity

- Work through Example 3 on the board.
- Discuss with students that it is not necessary to plot more than the three points when sketching the curve, as they should be able to use the symmetry of the graph.
- Students can now do Exercise 16C.

Plenary

- Ask students to consider how the turning point would be different if the coefficient of x^2 was negative.
- They can then complete the square and find the turning points of the following:
- $y = -x^2 + 6x + 6$ $y = -(x - 3)^2 + 15$
  $y = -x^2 - 8x + 1$ $y = -(x + 4)^2 + 17$
  $y = -x^2 - 3x - 4$ $y = -(x + 1.5)^2 - 1.75$
- Now ask them what the maximum value of each of these equations is.
  The corresponding maximum values are 15, 17 and –1.75 and the turning points are (3, 15), (–4, 17) and (–1.5, –1.75).

Homework and answers

Homework for this lesson is available in the homework folder. Consolidation and extension exercises are available in the further practice folder. Answers to the exercises can be found in the answers folder.

16.4 Reciprocal graphs

Syllabus references

Statement	Core
C2.10/E2.11	Construct tables of values, and draw, recognise and interpret graphs for functions.
C2.11/E2.11	Recognise, sketch and interpret graphs of the following functions: linear, quadratic.

Collins references

- Student's Book pages 304–306

Learning aims

- Recognise and plot reciprocal graphs.

Key words

- reciprocal, asymptote

Prior knowledge

Students need good calculator skills to complete the table, especially in some graphs. Drawing these graphs requires the same skills as for drawing quadratic graphs.

Common mistakes and remediation

Miscalculating reciprocals is a common mistake here, but more practice and worked examples may help to reduce the frequency of this happening. When drawing reciprocal graphs, students often draw them touching the axes. Keep reminding them that although the graph gets closer and closer to the axes, it never actually touches them.

Useful tips

Remind students that when drawing graphs of reciprocal functions, they should use the same methods as for quadratic graphs – complete the table, plot the points, draw the curve. These graphs also tend to require more calculator use when completing the table.

Asymptotes are not specifically mentioned in the syllabus for Core, but since reciprocal graphs are included it is worth mentioning them in this lesson.

Starter

- Ensure students know how to calculate reciprocals by asking them to copy and complete the following table, rounding their answers to 2 decimal places.

x	–4	–3	–2	–1	0	1	2	3	4
$y = \frac{1}{x}$	–0.25								0.25

- The completed table is given in Example 4 in the Student's Book.

Main lesson activity

- Refer back to the table of results used in the starter, and hence plot the graph of the equation $y = \frac{1}{x}$.

  Discuss with students that this isn't a particularly helpful graph and we need to see what happens between $x = 0$ and $x = 1$, and $x = 0$ and $x = -1$.
- Draw another table with values from –0.8 to 0.8 in steps of 0.2, again giving answers to 2 decimal places. Plot these values on your graph.
- The resulting graph is shown in the Student's Book. Highlight the fact that reciprocal graphs consist of two separate parts.
- Ask students to consider their graphs carefully and discuss, in pairs, the special features of the graph. They should be able to spot the symmetry of the graph, but may need help in describing these lines of symmetry. Point out to the students that these lines of symmetry that are not crossed by the graph are called asymptotes. This and the other properties of reciprocal graphs are described in the Student's Book.
- Students can now do Exercise 16D.

Plenary

- Ask students to consider the graphs they have drawn in Exercise 16D. What do they notice? (They should see that graphs of positive reciprocal functions are drawn in the first and third quadrants, whereas graphs of negative reciprocal functions are drawn in the second and fourth quadrants.)

Homework and answers

Homework for this lesson is available in the homework folder. Consolidation and extension exercises are available in the further practice folder. Answers to the exercises can be found in the answers folder.

16.5 More graphs

Syllabus references

Statement	Extended
E2.10	Construct tables of values, and draw, recognise and interpret graphs for functions Solve associated equations graphically, including finding and interpreting roots by graphical methods.
E2.11	Recognise, sketch and interpret graphs of functions.

Collins references

- Student's Book pages 306–309

Learning aims

- Recognise and plot cubic and other graphs.
- Understand turning points and asymptotes.

Key words

- cubic, turning point, asymptote

Prior knowledge

Students need good calculator skills to complete the table, especially in some graphs. Drawing these graphs requires the same skills as for drawing quadratic graphs – one smooth curve passing through all points plotted. Students should also be able to use the power button on their calculators.

They should be able to draw lines on the graphs to find solutions of relevant equations.

Knowledge of turning points and asymptotes is required.

Common mistakes and remediation

Miscalculating powers is a common mistake here, but more practice and worked examples may help to reduce the frequency of this happening.

Useful tips

Remind students that in the questions on graphs of cubics and sums of simple powers, they should use the same methods as for quadratic graphs – complete the table, plot the points, draw the curve. These graphs also tend to require more calculator use when completing the table. When calculating values near asymptotes, the numbers often get very big (positive or negative) very quickly.

Starter

- Ask students to use their calculators to find values of $1/x^2$ when x gets close to 0 from above (i.e. small positive values): what do they notice?
- Repeat for values of $1/x^2$ when x gets close to 0 from below (i.e. small negative values): what do they notice?
- Explain that the values of y get larger and larger and the graph therefore gets closer and closer to $x = 0$. We call this line an asymptote.
- Repeat for values of $1/x^2$ when x gets very large and positive: what do they notice this time? Again, repeat for values of $1/x^2$ when x gets very large and negative.
- Explain that the values of y get closer and closer to 0 and the graph therefore gets closer and closer to $y = 0$. This line is also an asymptote.

Main lesson activity

- Work through the text on cubic graphs.
- Students can then do questions 1–4 of Exercise 16E.
- Now introduce the other graphs by working through Example 5.
- Notice that this time there is an asymptote at $y = 2$.
- Now students can now complete Exercise 16E.

Plenary

- Draw together Lessons 16.1, 16.2, 16.3 and 16.4, and ask students to summarise what the graphs of quadratic, reciprocal and cubic functions look like. Ask them what they notice about graphs that have asymptotes. You may want to give them a set of graphs and a set of equations, and ask them to match one with the other.

Homework and answers

Homework for this lesson is available in the homework folder. Consolidation and extension exercises are available in the further practice folder. Answers to the exercises can be found in the answers folder.

16.6 Square root graphs

Syllabus references

Statement	Extended
E2.10	Construct tables of values, and draw, recognise and interpret graphs for functions. Solve associated equations graphically, including finding and interpreting roots by graphical methods.

Collins references

- Student's Book pages 310–312

Learning aims

- Draw and interpret graphs of functions involving square roots.

Key words

- No new key words

Prior knowledge

Students should know how to complete a table of values and use it to draw a graph.

Common mistakes and remediation

Reinforce the fact that a negative number does not have a square root.

Useful tips

The scale on the two axes do not need to be the same. A larger graph will give a more accurate solution to an equation.

Starter

- Write down $\sqrt{1}=1$ and $\sqrt{4}=2$
- Ask students to tell you the positive square roots of some other numbers without using a calculator.
- They should be able to tell you the square roots of square numbers such as $\sqrt{49}=7$ or $\sqrt{121}=11$
- Do not forget $\sqrt{0}=0$ and remind students that negative numbers do not have a square root.

Main lesson activity

- Here is a table of square roots:

x	0	1	2	3	4	5	6	7	8	9
$\sqrt{x}$	0	1	1.414	1.732	2	2.236	2.449	2.646	2.828	3

- The values are given to 3 decimal places. Fill in a couple of numbers in the second row and ask students to complete the table, using a calculator.
- Ask students to draw a graph of $y=\sqrt{x}$ on graph paper. Use a scale of 1 cm to 1 unit on each axis.

- It should look like this:

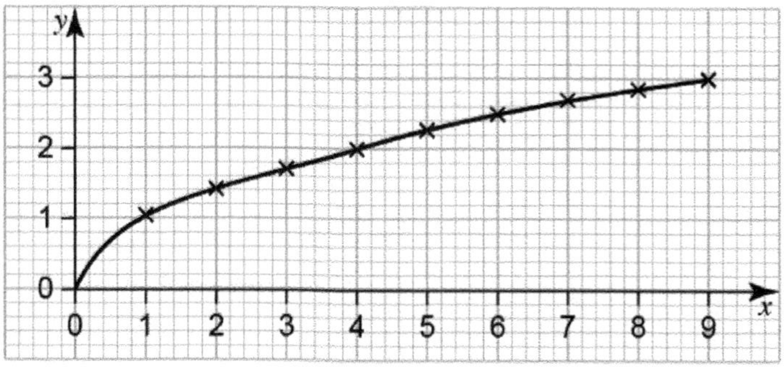

- Point out that if the graph is rotated 90° anticlockwise and reflected in the vertical axis it looks like a graph of $y = x^2$
- Now look at the graph of $y = \sqrt{x}$ at the start of section 16.6 and ask why it looks different.
- The scales are different so that more points can be shown. Square numbers have been used to plot points: (1, 1), (4, 2), (9, 3) and so on.
- Now look at the graph of $y = \frac{12}{\sqrt{x}}$ which is underneath the graph of $y = \sqrt{x}$ in section 16.6
- Make sure students can see how the values in the table have been calculated.
- The graph shows that if x is small, then $\frac{12}{\sqrt{x}}$ is large. For example, $\frac{12}{\sqrt{0.1}} = 37.9$
- Point out that $\frac{12}{0}$ has no value.
- Students can now do Exercise 16F.

Plenary

- Show this table of values:

x	0	1	2	3	4	5
$\sqrt{25 - x^2}$	5	4.90	4.58	4	3	0

- Write in the first two values on the bottom row and ask students to tell you the missing numbers.
- Ask students to use the table to draw a graph of $y = \sqrt{25 - x^2}$. They should use the same scale on both axes.
- When they have drawn the graph, tell them that it should be a quarter of a circle, centre the origin, radius 5.
- Ask students to draw this quarter of a circle with a pair of compasses and check the accuracy of their graph.

Homework and answers

Homework for this lesson is available in the homework folder. Consolidation and extension exercises are available in the further practice folder. Answers to the exercises can be found in the answers folder.

16.7 Exponential graphs

Syllabus references

Statement	Extended
E2.10	Draw and interpret graphs representing exponential growth and decay problems.

Collins references

- Student's Book pages 312–316

Learning aims

- To be able to draw and recognise the graphs of exponential growth and decay.

Key words

- exponential growth, exponential decay

Prior knowledge

Students will need to understand multipliers.

Students need to be familiar with drawing graphs.

Common mistakes and remediation

Students may expect their graphs to go through the origin.

Useful tips

Remind students that, when they are drawing these graphs, there will be no straight lines, just smooth curves.

Starter

- Tell the students that in 2017 there were estimated to be about 300 parakeets in a city. This number was expected to increase at the rate of 20% each year.
- Ask students how many parakeets they would expect to be in the city in 2018. (360)
- Ask students how they worked this out.
- Now ask for an estimate for the number of parakeets in the city in 2019.
- The calculated figure is 432. Discuss sensible rounding according to the context. In this case, it is appropriate to round the figure to the approximation 430.
- Ask for an estimated number for 2020. (520) Again, discuss rounding. If they are using rounded figures, should they use the rounded figure or the accurate one for the next calculation?
- Ask the students for an estimate for the year 2021.
- Refer back to previous work done on exponential growth. Remind students that they can calculate the number by multiplying 300 by 1.2^5 to give approximately 750.
- You may need to remind students who need more **support** about the multiplier 1.2.
  Now ask how many parakeets there will be by 2027. They should calculate 300×1.2^{10} which gives approximately 1860.

Main lesson activity

- Show students how to draw a graph from the information they have derived.
- Draw a table, as below, and fill in the numbers they have calculated.

2017	2018	2019	2020	2021	2022	2023	2024	2025	2026	2027
300	360	430	520		750					1860

- Put on the board a pair of axes, labelling the horizontal axis as Years, from 2015 to 2030, and the vertical axis as Population, from 300 to 2000.
- Plot the known points and draw a smooth curve through them.
- Explain that this shape shows *exponential growth*. Such a graph will always follow this type of curve.
- Ask students to estimate when the parakeet population will have tripled. (2023)
- Now talk through Example 6 in the Student's Book, which shows *exponential decay*. In this case, the value of one variable decreases as the other increases. The graph has a different form but is still a smooth curve.
- Ask the students who need more of a **challenge** to describe the differences between the graphs of exponential growth and decay – in terms of both the shape and the trends.
- The students can now do Exercise 16G.

Plenary

- Ask the students to discuss the different types of graph they have been working on, describing their shapes and their general equations.
- They should describe:
  - the graph of exponential growth as a steep positive curve, with the general equation $y = x^n$ where n is greater than 1
  - the graph of exponential decay, showing a steep negative curve, general equation $y = x^n$ where n is less than 1
  - a straight line, general equation $y = mx + c$
  - a quadratic graph showing a ∩- or ∪-shaped curve, general equation $y = ax^2 + bx + c$.
- They may also describe an exponential graph, showing a hyperbola, general equation $y = \frac{1}{x}$.
- As students mention each type of graph, ensure that they are aware of what the other letters stand for, other than just the x and y.

Homework and answers

Homework for this lesson is available in the homework folder. Consolidation and extension exercises are available in the further practice folder. Answers to the exercises can be found in the answers folder.

16.8 Estimating gradients

Syllabus references

Statement	Extended
E2.12	Estimate gradients of curves by drawing tangents.

Collins references

- Student's Book pages 316–319

Learning aims

- To estimate the gradient at a point of a curve.

Key words

- gradient, tangent

Prior knowledge

Students need to know how to find the gradient of a straight line and be able to draw quadratic, cubic, reciprocal and exponential graphs.

Common mistakes and remediation

Students often make arithmetic errors when minus numbers are involved. Plenty of worked examples and repeated practice should reduce the frequency of this happening.

Useful tips

Tell students to remember that if a tangent is horizontal, then its gradient is zero.

Starter

- Provide the students with a set of straight-line graphs, and ask them to find, by drawing, the value of their gradients.

Main lesson activity

- Ask students to draw the graph of $y = x^3 - x^2 - 4x + 4$ taking values of x from –3 to +3.
- Ask students to discuss, in pairs, the 'steepness' of this curve. They should say something like it starts off quite steep, then becomes less steep until it is flat and so on.
- Lead them to deduce that, unlike with straight lines, the gradient of a curve varies from point to point.
- Ask them to describe the gradient at each part of the curve, i.e. positive gradient, zero gradient, negative gradient, zero gradient, positive gradient.
- Explain that we can only find a gradient of a curve at a particular point, and that this itself is an estimate. To make this estimate we need to draw a *tangent* to the curve at that point. Explain that a tangent is a straight line that touches the curve at one point and as we know how to find the gradient of a straight line we can find the gradient of the tangent. Work through the text and Example 7 to illustrate this.
- Students can now do Exercise 16H.

Plenary

- Refer to question 6 of Exercise 16H.
- Ask students to look at their answer to part **c** and decide where else on the complete graph of $y = \frac{5}{x}$ would the gradient be the same. Tell them they will need to use their knowledge of the graphs of reciprocal functions.

  (From the symmetry of reciprocal graphs, students should deduce that the gradient will be the same at (–4, –1.25).)

Homework and answers

Homework for this lesson is available in the homework folder. Consolidation and extension exercises are available in the further practice folder. Answers to the exercises can be found in the answers folder.

17.1 Patterns in number sequences

Syllabus references

Statement	Core
C2.7/E2.7	Continue a given number sequence or pattern. Recognise patterns in sequences.

Collins references

- Student's Book pages 322–324

Learning aims

- Recognise how number sequences are built up.
- Find the next terms in a sequence.

Key words

- sequence, term, difference, consecutive

Prior knowledge

Students should know how to recognise number patterns.

Common mistakes and remediation

Students find sequences accessible and usually do well. Errors are mainly related to their method, so just emphasise the need to take care.

Useful tips

Students who need more **support** might find it helpful to write the sequences in tables, listing the term number with each term.

Starter

- Explain that a number sequence is an ordered set of numbers with a rule that allows them to find every number in the sequence. Each number is a *term* and it has its own place in the order.
- The first step in finding the general rule for a sequence is to find the link between the terms. This is the *term-to-term rule*. Often the easiest way to find this link is to look at the differences between consecutive terms.
- Write the numbers 2 and 4 on the board and ask students what the next two numbers in the pattern could be. The most likely answer is 6 and 8. Write this on the board. Ask students how the pattern is building up. (+ 2)
- Now write 2, 4, 8, 16, … on the board and ask students how this is building up. (× 2) Then write 2, 4, 7, 11, … on the board and ask how this is building up. (+ 2, + 3, + 4, …)
- Repeat with the starting numbers 1, 5, … and ask students for different series starting 1, 5, ….
- Ask students to describe how each series is building up.

Main lesson activity

- Ask students to suggest some sequences and to describe how they are building up.
- Ask first of all for series that increase by a fixed amount, such as + 3. Then ask for series that decrease by a fixed amount, such as – 4.
- Ask for series that increase by a different amount each time: + 1, + 2, + 3 or + 2, + 4.
- Then ask for sequences that involve multiplication by a fixed number each time, such as × 2. As these get large very quickly, keep the starting numbers simple or allow calculators.
- Now ask for sequences that involve division by a fixed number each time, such as ÷ 10. As these soon start producing answers that are decimals, often recurring, it is a good idea to keep to ÷ 2, ÷ 5 or ÷ 10.
- The majority of students can usually see how these patterns build up mentally.
- Use Example 1 to demonstrate using differences. This may confuse students who need more **support**, so take it slowly if necessary. However, the concept of differencing is important when dealing with quadratic sequences so students who need more of a **challenge** will need to understand it.
- Students can now do Exercise 17A.

Plenary

- Write some 'silly' sequences on the board and ask for the next term. For example:

  31, 28, 31, 30, 31, 30, ... 31 (days in the months)

  O, T, T, F, F, ... T (starting letters of the numbers, one, two, three, ... to ten)
- Ask students if they can find any 'silly' or unusual sequences for a future lesson.

Homework and answers

Homework for this lesson is available in the homework folder. Consolidation and extension exercises are available in the further practice folder. Answers to the exercises can be found in the answers folder.

17.2 The nth term of a sequence

Syllabus references

Statement	Core
C2.7/E2.7	Recognise patterns in sequences, including the term-to-term rule, and relationships between different sequences. Find and use the nth term of the following sequences: linear, simple quadratic, simple cubic.

Collins references

- Student's Book pages 324–329

Learning aims

- Recognise how number sequences are built up.
- Generate sequences, given the nth term.
- Find an algebraic rule for the nth term of a sequence.

Key words

- coefficient, square number, cube number, triangular number, linear sequence, nth term, quadratic sequence, cubic sequence

Prior knowledge

Students should know how to recognise number patterns and how to find the differences between consecutive terms.

Common mistakes and remediation

When finding the 10th term, students often just double the value of the 5th term. Emphasise the correct method.

Useful tips

Students who need more **support** might find it helpful to write the sequences in tables, listing the term number with each term.

Starter

- Give students a start number and an answer, e.g. 3 and 16. Ask them to give a rule of the form 'multiply by a number then add or subtract a number' to get from 3 to 16, e.g. × 3 + 7. Write this as a linear expression in n, e.g. $3n + 7$.
- Ask students for similar rules and ask them to write them as linear expressions. This should give rules such as $5n + 1$, $2n + 10$, $4n + 4$, $6n - 2$, $7n - 5$.

Main lesson activity

Finding the rule

- Discuss the rules written down for the sequences students have met so far. Establish that they contain a single term in the variable n and a constant term. They do not use squares or higher powers. Explain that these are linear expressions. Ask students to suggest similar linear expressions in n.

- When students suggest linear expressions, such as $3n + 2$, ask the class to substitute $n = 1, 2, 3, \ldots$ into this expression and write the results in a list: 5, 8, 11, 14, 17, ...
- Ask students if they can see any connections with previous work. They should see that the expression generates a series with a constant difference between pairs of adjacent terms. This constant difference is a *coefficient*, and this type of series is a linear sequence.
- In such an expression, n is a variable, and its value determines the value of the nth term in the sequence. If $n = 4$, this determines the value of the 4th term, and so on.
- Repeat with another linear expression. Work through Example 2 as a further example of a linear expression. Students can now do Exercise 17B.

Finding the nth term of a linear sequence

- Taking any of the sequences used at the beginning of the lesson, write the linear expression and the series on the board and ask students if they can see any connections. The difference and the coefficient of n is the most obvious.
- Students may also spot that the first term of the sequence is the difference plus or minus the constant term.
- Work through this section, which reinforces finding the nth term of a linear sequence and the work covered at the beginning of the lesson, and go through Example 3.

Quadratic and cubic sequence

Start by revisiting the triangular numbers from question 5 in Exercise 17A: 1, 3, 6, 10, ...
Draw a sequence of diagrams to show how these numbers are obtained. Now put two triangles of the same size together to make a rectangle.
So two triangles cover 1 × 2, 2 × 3, 3 × 4, 4 × 5 dots. Ask students how many dots two triangles with side length n dots will cover: $n(n + 1)$. So the nth triangle number is $n(n + 1)/2$.

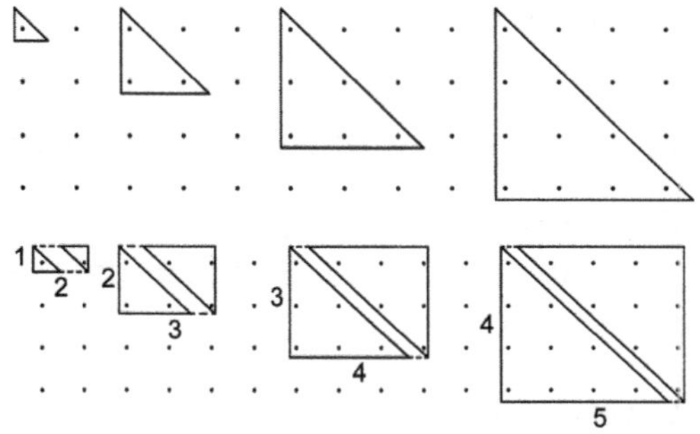

- Show students that not all series are linear by working through this section and Examples 4 and 5. Reinforce with further examples.
- Students can now do Exercise 17C.

Plenary

- Write the expression n^2 on the board. Ask students how this is different from other expressions. Explain that this is a non-linear expression as it has a power in it.
- Ask students what sequence is generated by substituting $n = 1, 2, 3, \ldots$ into this expression. It gives 1, 4, 9, 16, 25, ... which students should recognise as the square numbers.
- Write the sequence 2, 4, 8, 16 on the board. Ask students how this is building up. Ask if they can find an nth term. It will help to write it as 2, 2 × 2, 2 × 2 × 2, ...
- The nth term should be recognisable as 2^n.
- In a similar way, do this for the nth term of 10, 100, 1000, 10 000, ...

Homework and answers

Homework for this lesson is available in the homework folder. Consolidation and extension exercises are available in the further practice folder. Answers to the exercises can be found in the answers folder.

17.3 General rules from patterns

Syllabus references

Statement	Core
C2.7/E2.7	Recognise patterns in sequences, including the term-to-term rule, and relationships between different sequences. Find and use the nth term of the following sequences: linear, simple quadratic, simple cubic.

Collins references

- Student's Book pages 329–334

Learning aims

- Find the nth term from practical problems.

Key words

- rule, patterns

Prior knowledge

Students should be able to substitute into simple algebraic expressions and have some experience in continuing a sequence.

Common mistakes and remediation

Students often make the mistake of attempting to answer the question in their heads rather than writing down the sequence. Suggest that they sift through the words to get to the mathematics, write down the sequence and concentrate on the numbers rather than words such as 'fences' or 'pentagons'.

Useful tips

Always set up a table of results. This will help students to spot the pattern.

Starter

- Remind students how to find the next term and the nth term of a sequence. Provide two or three for students to practise, e.g. 7, 9, 11, 13, 15, ... 7, 17, 27, 37, ... 7, 107, 207, 307, ...
- Recap on the triangular numbers and how to find the nth term – this is repeated here from section 17.2 TG.
- Draw a sequence of diagrams to show how these numbers are obtained:

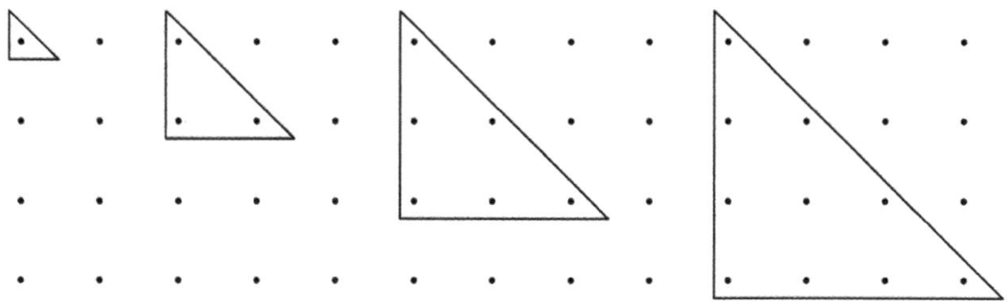

- Now put two triangles of the same size together to make a rectangle:

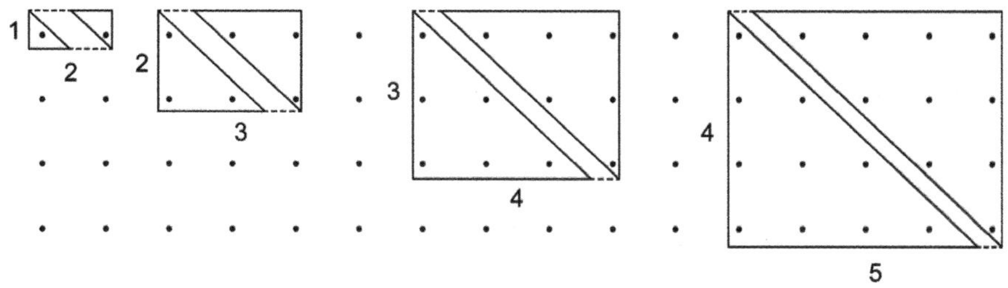

So two triangles cover 1 × 2, 2 × 3, 3 × 4, 4 × 5 dots.

- Ask students how many dots two triangles with side length n dots will cover: $n(n + 1)$
So the nth triangle number is $n(n + 1)/2$.

Main lesson activity

- Work through Example 6. The main point to emphasise is putting results in a table.
- Extend the table for a few columns (or rows). Test a generalisation to make sure the rule works.
- Initially, some students who need more **support** may need help with writing down a sequence.
- Now students can complete Exercise 17D. **Support** students by allowing them to work with a partner if you feel it is necessary. Each question in Exercise 17D requires setting up a table and spotting the nth term. Once students find the nth term, they should test it. Students who need more **support** may need help with question 9 on the Sierpinski triangle. They may see the patterns but they may not see the nth terms.

Plenary

- Draw the following patterns on the board

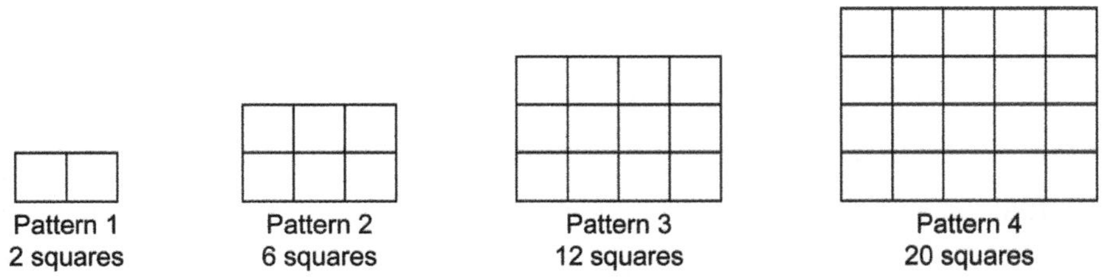

- Ask students how many squares there will be in pattern 5. (30)
- They may not spot this at first. You may need to prompt them.
- Ask how they can work it out. (The pattern goes up 4, 6, 8, 10, etc.)
- Ask if they could give a rule for the nth term.
- Look at the width (number of squares) of each block: 2, 3, 4, 5, … $n + 1$.
- Look at the height (number of squares) of each block: 1, 2, 3, 4, … n.
- The number of squares is: 1 × 2, 2 × 3, 3 × 4, 4 × 5, $n \times (n + 1)$.

Homework and answers

Homework for this lesson is available in the homework folder. Consolidation and extension exercises are available in the further practice folder. Answers to the exercises can be found in the answers folder.

17.4 Further sequences

Syllabus references

Statement	Extended
E2.7	Find the nth term of exponential sequences.

Collins references

- Student's Book pages 334–337

Learning aims

- To be able to find the general term for an exponential sequence.

Key words

- exponential sequence

Prior knowledge

Students need to be familiar with finding the nth term of linear and quadratic sequences.

Common mistakes and remediation

Students may mistake an exponential series for a quadratic series and vice versa.

Useful tips

Always look for the differences, to check for a pattern.

Starter

- Put on the board the sequence 3, 7, 11, 15.
- Ask the students to suggest the next few numbers. (19, 23, 27)
- Ask them how they found these numbers. Ask them for the general term. ($4n - 1$)
- Check that students remember this is a linear sequence.
- Say that we often use a letter to represent a sequence and subscripts to denote particular terms. So if we call this sequence T then the first term $T_1 = 3$ and $T_2 = 7$ and $T_7 = 27$ and so on.
- Write this on the board so that students can see how to do it correctly. Ask for some more terms such as T_5 and T_{10} (19 and 39)
- Say that we write the general term as $T_n = 4n - 1$
- Now write on the board the sequence 2, 5, 10, 17. Call it S. Ask students to tell you S_1 and S_3 (2 and 10)
- Then ask for S_5 and S_6 and S_6 (26, 37, 50)
- Ask students to give you the general term S_n ($S_n = n^2 + 1$)
- If necessary, as a hint, remind students what the square numbers are.
- Check that the students remember this is a quadratic sequence.

Main lesson activity

- Write on the board the sequence 2, 4, 8, 16, 32 .... And call it *T*
- Remind students that they have seen this pattern of numbers before. (Exponential growth)
- Ask students for the general term. ($Tt_n = 2^n$)
- Write this on the board, explaining that this is an *exponential series.*
- Now write on the board the sequence *E* which starts 10, 12, 14.4, 17.28, ... and tell students that this is an exponential series.
- Ask for the next term E_5. (20.736) Ask students to explain how they calculated it.
- Show that each term in the sequence is multiplied by 1.2 to get the next term.
- Ask students to suggest what the general term will be. ($E_n = 10 \times 1.2^{n-1}$) Explain why the power is $n - 1$.
- For extra **challenge**, explore how starting at different points within the same sequence will alter the power. For example, 12, 14.4, 17.28 could be expressed as $E_n = 10 \times 1.2^n$.
- Write on the board the sequence 2, 6, 12, 20, ... Say that this is a *quadratic sequence.* Call it Q. Ask students how they know this is the case. (The second differences are all the same.)
- Ask students to look at the numbers and try to find the general term. (Look at factors.)
- Show how the sequence can be written as (1 × 2), (2 × 3), (3 × 4), (4 × 5), ...
- Help students to recognise that the general term can be written as $Q_n = n(n + 1)$.
- Talk students through the sequence 3, 8, 15, 24, 35, ... shown below Example 7. Make sure that students understand how to use the square numbers to find the pattern. Check they understand how to use subscript notation.
- Students can now complete Exercise 17E.

Plenary

- Remind students that they have now seen a number of different types of sequence and have learnt how to find the general pattern by finding the nth term.
- Ask for a general term for a linear sequence. (e.g. $5n + 1$)
- Ask for a general term for a quadratic sequence. (e.g. $n^2 + 3$)
- Ask for a general term for an exponential series. (e.g. 3^n)
- Ask the students how they can tell which is which, by looking at the numbers. They should respond:
  - a linear sequence has constant first differences
  - a quadratic sequence has constant second differences
  - an exponential sequence has a constant ratio between successive terms.

Homework and answers

Homework for this lesson is available in the homework folder. Consolidation and extension exercises are available in the further practice folder. Answers to the exercises can be found in the answers folder.

18.1 Using indices

Syllabus references

Statement	Core
C1.7/E1.7	Understand and use indices (positive, zero and negative integers) and the rules of indices.

Collins references

- Student's Book pages 340–342

Learning aims

- Use powers (also known as indices).

Key words

- index, indices, power, power 1, power 0

Prior knowledge

Students should know the square numbers up to 15 × 15 with their corresponding square roots.

Common mistakes and remediation

The most common mistake is to ignore powers of 1, e.g. giving $2^5 \times 2$ as 2^5 instead of the correct 2^6. Tell students they can avoid this error if they write the power 1 into their copy of the question.

Useful tips

When explaining what powers are, avoid saying, e.g. x^n is x multiplied by itself n times.

Demonstrate that:

x^2 is x multiplied by itself once (not twice)

x^3 is x multiplied by itself twice

and so on. It is safer to think of x^n as being n 'lots' of x multiplied together.

It might be useful to display a list of square numbers and cube numbers, for students who need more **support**.

$1^2 = 1$, $2^2 = 4$, $3^2 = 9$, $4^2 = 16$, $5^2 = 25$, $10^2 = 100$

$1^3 = 1$, $2^3 = 8$, $3^3 = 27$, $4^3 = 64$, $5^3 = 125$, $10^3 = 1000$

Starter

- Show students the puzzle below and ask them to find the number in the middle. They must multiply the number in each square by the number in the next square, and write the answer in the circle between the two squares.
- They should then add the numbers in the circles and write the answer in the diamond in the middle. (35 035)
- Explain to students that powers are a shorter way of writing repeated multiplications.
- Now introduce cubes of numbers. Ask students the value of 10 × 10 × 10 (1000). Ask students how they can write it using powers (10^3).

- Ensure students know that this is read as '10 cubed' or 'the cube of 10' or '10 to the power 3'.
- Repeat for the other cubes that students are expected to know:
  $1^3 = 1$, $2^3 = 8$, $3^3 = 27$, $4^3 = 64$, $5^3 = 125$.

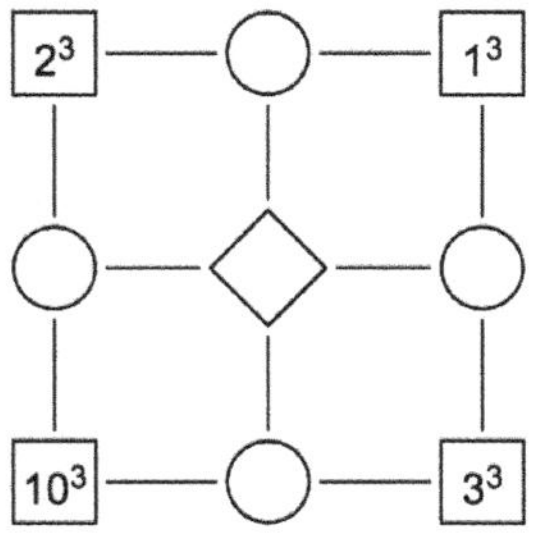

Main lesson activity

- Explain that the lesson is about working out the values of powers of numbers. Go around the class asking students to start from 1 and keep doubling the previous answer. (1, 2, 4, 8, 16, 32, ...)
- See how far students can get before making an error.
- Ask students to imagine a chess or draughts board with 64 small squares, then to imagine putting 1 cent on the first square, 2 cents on the second square, 4 cents on the next square and so on. Without discussing it, ask students to write down how much money they think will be put on the 64th square (over $90 thousand million million – this is not a misprint!). Students will most likely underestimate the answer. Use this example to explain how quickly powers grow.
- Now look at Example 1. Explain that a power defines how many times to multiply a number by itself, but the language needs care; 4^5 means $4 \times 4 \times 4 \times 4 \times 4$ and is expressed as: '4 to the power of 5'. However, it is not 4 multiplied by itself five times, which would be 4^6. When working with powers of 10, the power is the same as the number of zeros in the calculated answer, e.g. $10^3 = 1000$.

Working out powers on your calculator, Two special powers

- Show students how to use the x^2 button, the x^3 button and the power button on a calculator.
- Now ask students to input the number 7 into the calculator, press =, then press × 7 and repeatedly press the 'equals' button. This will display powers of 7: 7, 49, 343 and so on. For students who need more **support**, start with the number 2, to give the powers of 2. Link this to the work on doubling and the money on the chess board.
- Finally, check that students know that any number to the power 1 is itself and any number to the power zero is equal to 1.
- Now students can do Exercise 18A. Students who need more of a **challenge** could complete this as a class activity, answering the questions orally. Students who need more **support** will probably need to use pen and paper.

Plenary

- Discuss solutions to question 13 of Exercise 18A. Students will probably have used a calculator to find the answers, but explore with them how they could do this question using pencil and paper.

Homework and answers

Homework for this lesson is available in the homework folder. Consolidation and extension exercises are available in the further practice folder. Answers to the exercises can be found in the answers folder.

18.2 Negative indices

Syllabus references

Statement	Core
C1.7/E1.7	Understand and use indices (positive, zero and negative integers) and the rules of indices.
C2.4/E2.4	Use and interpret positive, negative and zero indices.

Collins references

- Student's Book pages 342–344

Learning aims

- Use negative indices.

Key words

- negative index, reciprocal

Prior knowledge

Students should understand the work on positive indices as covered in Lesson 18.1.

Common mistakes and remediation

Students often struggle to remember 'reciprocal' and so it is good practice to use the word as often as possible to establish familiarity. Some students will assume a negative power implies a negative answer, for example, they think 3^{-2} means $-3^2 = -9$, when in fact, it is $\frac{1}{3} \times \frac{1}{3} = \frac{1}{9}$ or 0.111 recurring.

Useful tips

Keep reminding students that if you move a power from top to bottom, or vice versa, then the sign changes. Emphasise that it does not mean that the answer is negative.

Starter

- Check students' understanding of the work covered in the previous lesson, by asking them to work out the following:

  a 3^5 **b** 8^2 **c** 4^4 **d** 0.5^3 **e** $(-2)^6$ **f** $(-3)^3$

  Tell them to confirm their answers on a calculator, using the power button.

- Now ask students to write the following as powers of 2 and powers of 4.

  a 4 **b** 16 **c** 64

  Again, they should confirm their answers on a calculator.

Main lesson activity

Negative powers (or negative indices)

- Ask students to use their calculators to answer:
  - $2^5, 2^4, 2^3, 2^2, 2^1, 2^0, 2^{-1}$ $(32, 16, 8, 4, 2, 1, \frac{1}{2})$
  - $3^5, 3^4, 3^3, 3^2, 3^1, 3^0, 3^{-1}$ $(243, 81, 27, 9, 3, 1, \frac{1}{3})$

  Discuss what is happening with each pattern.
- Reinforce the fact that power zero always gives an answer of 1 and that negative powers give the reciprocal of the equivalent positive power. Check that students understand the word 'reciprocal'. Work through Example 2.
- Now students can do Exercise 18B.

Plenary

- Review the work on positive, negative and zero indices by discussing the answers to questions 7, 8 and 10 of Exercise 18B.

Homework and answers

Homework for this lesson is available in the homework folder. Consolidation and extension exercises are available in the further practice folder. Answers to the exercises can be found in the answers folder.

18.3 Multiplying and dividing with indices

Syllabus references

Statement	Core
C1.7/E1.7	Understand and use indices (positive, zero and negative integers) and the rules of indices.
C2.4/E2.4	Use and interpret positive, negative and zero indices.

Collins references

- Student's Book pages 344–346

Learning aims

- Solve problems set in a real-life context.

Prior knowledge

Students need to understand the work covered in Lessons 18.1 and 18.2.

Common mistakes and remediation

Students often still apply the rules of indices when working with powers of different numbers or variables, e.g. they might say that $3^5 \times 4^2 = 12^7$. Highlighting this potential error and working through plenty of examples should help to address this.

Useful tips

When multiplying and dividing with indices, advise students to deal with numbers and indices separately, for example $12a^5 \div 4a^2 = (12 \div 4) \times (a^5 \div a^2) = 3a^3$.

Starter

- Ask students to copy the following table:

2 × 4 =	4 × 8 =	2 × 8 =	2 × 16 =

- Now ask students to fill in the answers and then re-write the calculations underneath using powers. They should produce the following table:

2 × 4 = 8	4 × 8 = 32	2 × 8 = 16	2 × 16 = 32
$2^1 \times 2^2 = 2^3$	$2^2 \times 2^3 = 2^5$	$2 \times 2^3 = 2^4$	$2 \times 2^4 = 2^5$

Main lesson activity

Rules for multiplying or dividing numbers in index form

- Refer to the example used in the starter and repeat for other base numbers, e.g. 4, 5 and 6 (note that 3 and 10 are used in the Student's Book). Ask students if they can see any connection.
- Now do the same with powers of 2 for division, e.g. $32 \div 8 = 4$ and $2^5 \div 2^3 = 2^2$.
- Again ask students to spot the rule or connection for this form.
- Write on the board:

  $2^2 \times 2^3 = 2^5$, $3^2 \times 3^3 = 3^5$, $4^2 \times 4^3 = 4^5$

  and ask students for the answer to $x^2 \times x^3$ (x^5).
- Repeat for a division, e.g. $4^5 \div 4^3 = 4^2$, $3^5 \div 3^3 = 3^2$, $2^5 \div 2^3 = 2^2$
- Then ask for the answer to $x^5 \div x^3$ (x^2).
- Write on the board: $2^3 \times 2^2 = (2 \times 2 \times 2) \times (2 \times 2) = 2^5$
- Go through the examples in the Student's Book so that students understand the two rules of indices, i.e. how to multiply them and how to divide them. Students will see that a quicker way of working is to add the indices. Similarly, when dividing, the rule is to subtract the indices. Students need to understand that these rules only work when the numbers they are multiplying or dividing are expressed as powers of the same algebraic unknown.
- For students who need more **support**, you may need to give more examples such as:

  $8^3 \times 8^4 = (8 \times 8 \times 8) \times (8 \times 8 \times 8 \times 8) = 8^7$

  $8^4 \div 8^3 = (8 \times 8 \times 8 \times 8) \div (8 \times 8 \times 8) = 8^1 = 8$

  For students who need more of a **challenge** include mixed multiplication and division, e.g. $2^7 \times 2^3 \div 2^4$
- Write $a^3 \times a^3 \times a^3 \times a^3 \times a^3$ on the board and ask students how they can simplify it (a^{15}).
  Now write $(a^3)^5$ and ask them to comment. (It is the same.)
- Establish the rule that if you raise a power to a power you can simply multiply the powers.
- Show students the examples in the Student's Book. They can then do Exercise 18C.

Plenary

- Ask students how to work out:

  $$\frac{3a^4b^6 \times 8ab^7 \times (2b)^2}{2a^2b^2}$$

- Remind them of BIDMAS/BODMAS (the order of operations), so that they work out $(2b)^2$ first.
  Having done that, encourage them to work out one term at a time, i.e. the number term, the 'a' term and then the 'b' term ($48a^3b^{13}$).

Homework and answers

Homework for this lesson is available in the homework folder. Consolidation and extension exercises are available in the further practice folder. Answers to the exercises can be found in the answers folder.

18.4 Fractional indices

Syllabus references

Statement	Extended
E1.7 **E2.4**	Understand and use indices (positive, zero, negative and fractional) and the rules of indices.

Collins references

- Student's Book pages 347–351

Learning aims

- Use fractional indices of the form $\frac{1}{n}$ and $\frac{a}{b}$.

Prior knowledge

Students need to understand the work covered in Lesson 18.3.

Advise students that when working with fractional indices they should deal with the denominator of the fraction first.

Common mistakes and remediation

Students often make mistakes when simplifying expressions involving fractional indices. Advise them to first simplify the expression using the rules of indices, and then write the simplified form without fractional indices.

Useful tips

Advise students that when working with fractional indices they should deal with the denominator of the fraction first.

Starter

- Revise the rules of indices:
  - When you multiply powers of the same number or variable, add the indices.
  - When you divide powers of the same number or variable, subtract the indices.
  - When you raise a power to a further power, multiply the powers.
- Work through a few examples orally.

Main lesson activity

Indices of the form $\frac{1}{n}$

- Ask students for the value of $\sqrt{5} \times \sqrt{5}$. (5)
- Repeat for other numbers, e.g. $\sqrt{6} \times \sqrt{6}$. (6)
- Then using algebra for $\sqrt{x} \times \sqrt{x}$. (x)
- Now write $x^{\frac{1}{2}} \times x^{\frac{1}{2}}$ and remind students of the rules for indices (x^1 or simply x).
- Write these expressions, aligning those in the second row under those in the top row.

  $\sqrt{x} \times \sqrt{x} = x \qquad x^{\frac{1}{2}} \times x^{\frac{1}{2}} = x$
- Ask students what this tells them $\left(x^{\frac{1}{2}} = \sqrt{x}\right)$.
- Repeat, using cube roots and the power $\frac{1}{3}$, in order to establish that $x^{\frac{1}{3}} = \sqrt[3]{x}$.
- Remind students that for negative indices they will need the reciprocal, e.g. $10000^{-\frac{1}{4}} = \dfrac{1}{10000^{\frac{1}{4}}} = \dfrac{1}{10}$
- Show students the examples in the Student's Book. Then let them do Exercise 18D. Students who need more **support** should concentrate on question 1.

Indices of the form $\frac{a}{b}$

- Work through the two examples in the Student's Book with positive powers, explaining the significance of each part of the power. Then go through the four steps in Example 4, looking at each part of the power in turn.

  Step 1: Rewrite the calculation as a fraction by dealing with the negative power.

  Step 2: Take the root of the base number given by the denominator of the fraction.

  Step 3: Raise the result to the power given by the numerator of the fraction.

  Step 4: Write the answer as a fraction.
- Remind students that showing their working is important.
- Point out that the steps need not be followed too rigidly but that they will make fewer mistakes if they deal with the denominator of the fraction before the numerator.
- Go through Example 5. Now students can do Exercise 18E.

Plenary

- Ask students to use powers of other numbers so that the value of every answer is 4.

  Students who need more **support** will need examples to help them start, e.g. 2^2, $8^{\frac{2}{3}}$, $\left(\frac{1}{2}\right)^{-2}$
- Compile a list of answers.
- Repeat for an answer of 9.

Homework and answers

Homework for this lesson is available in the homework folder. Consolidation and extension exercises are available in the further practice folder. Answers to the exercises can be found in the answers folder.

19.1 Direct proportion

Syllabus references

Statement	Extended
E2.8	Express direct proportion in algebraic terms and use this form of expression to find unknown quantities.

Collins references

- Student's Book pages 354–359

Learning aims

- Solve problems where two variables have a directly proportional relationship (direct variation).
- Work out the constant of proportionality.

Key words

- direct variation, direct proportion, constant of proportionality

Prior knowledge

Students should know squares, square roots, cubes and cube roots of integers and be able to substitute values into algebraic expressions and solve simple algebraic equations.

Common mistakes and remediation

Some students find the constant k and then stop, without completing the question. Encourage them to check that they have answered the question. Students also assume that all relationships are linear. They should highlight or underline words such as square, cube, square root and cube root, ensuring that they use them in their equations.

Useful tips

Direct variation means that as one variable increases, the connected variable increases. All graphs showing direct proportion must pass through the origin. Students who need more **support** will find this topic challenging and should concentrate on linear relationships first. Students who need more of a **challenge** may cover the whole lesson plan and work through both exercises.

Starter

- Give an oral test on squares, square roots, cubes and cube roots. For example, 3^2, 3^3, $\sqrt{16}$, $\sqrt[3]{8}$, etc. Students should also know the corresponding square roots and cube roots.
- Make a table showing squares and ask students to tell you the answers.

x	1	2	3	4	5	6	7	8	9	10	11	12	13	14	15
x^2															

- Repeat for cubes.

x	1	2	3	4	5	10
x^2						

Main lesson activity

- Display the table below and ask students for the connections between y and x. ($y = 2x$)

x	1	2	3	4	5
y	2	4	6	8	10

- Ask students for the value of y when $x = 150$. (300) Explain that as x increases, y increases, and as the graph of $y = 2x$ passes through the origin, they are directly proportional.
- Ask students to think of examples of two variables that are directly proportional (e.g. age of children and height). Now plot the points above on a graph to show students that it is a linear relationship.
- Show students how to obtain the equation if they cannot spot it. Write: y is directly proportional to (or varies directly as) x, so using symbols $y \propto x$. Point out that this means y is a constant multiple of x so $y = kx$, where k is called the constant of proportionality. Now explain that as students know some values for x and y they can find k. For example, using $x = 1$ and $y = 2$ gives $2 = k \times 1$, so $k = 2$ and $y = 2x$.
- Now repeat for other tables. For example:

x	5	7	9
y	25	35	45

$y = 5x$

- With students, look at Example 1, explaining that this is the formal method. Go through the four steps.
- Repeat with other examples using different letters for the variables for the students who need more **support**, e.g. C is directly proportional to d and when $C = 36$, $d = 9$, work out the value of C when d = 7 (28).
- Now students can complete Exercise 19A. Students who need more **support** may need to work with a partner to complete the exercise.

Direct proportions involving squares, cubes, square roots and cube roots

- If possible, demonstrate pouring a cup of water slowly onto a piece of paper. Ask students what happens to the surface area as more water is poured and the radius of the circle increases. (Surface area increases.)
- Explain that the relationship this time is again direct proportion, but the connection is between the area and the square of the radius. Write $A \propto r^2$ so $A = kr^2$. Ask students to tell you the value of k. (π)
- Now discuss Example 2 with students.
- Show students who need more **support** an example using cube roots, for example, y is directly proportional to (or varies directly with) the cube root of x. When $x = 1000$, $y = 5$. Work out the value of y when $x = 8$. (1)
- Now students can complete Exercise 19B. Students who need more **support** can work in pairs if necessary.

Plenary

- Prepare students for the next section on inverse variation by asking them to think of variables where, as one increases, the other decreases, e.g. number of people building a wall and time taken.

Homework and answers

Homework for this lesson is available in the homework folder. Consolidation and extension exercises are available in the further practice folder. Answers to the exercises can be found in the answers folder.

19.2 Inverse proportion

Syllabus references

Statement	Extended
E2.8	Express inverse proportion in algebraic terms and use this form of expression to find unknown quantities.

Collins references

- Student's Book pages 359–361

Learning aims

- Solve problems where two variables have an inversely proportional relationship (inverse variation).
- Work out the constant of proportionality.

Key words

- inverse variation, inverse proportion

Prior knowledge

As in the previous lesson, students will find that knowing squares and square roots, cubes and cube roots of integers speeds up their working of many of the questions. They should be able to substitute values into algebraic expressions and solve simple algebraic equations.

Common mistakes and remediation

Students ignore the 'inversely' and treat it as direct proportion or they incorrectly rearrange the equation, after substituting, to find the value of the required variable. Point out these errors as students make them and provide extra examples for practice.

As in the previous lesson, students often ignore words such as square, cube, square root and cube root. They should highlight or underline these words and the word 'inverse' or 'indirect'.

Useful tips

Inverse variation means that as one variable increases the connected variable decreases.

To avoid this seeming more difficult for students than direct variation, point out that the steps are the same, but this time they are working with reciprocals.

Starter

- Set the following problems:
  - If Lee (or pick a student) hired a taxi to take him ice-skating, it might cost him $30.
  - If Lee and a friend went, how much would it cost them each?
  - If Lee and two friends went, how much would it cost them each?
  - If Lee and three friends went, how much would it cost them each?
  - If Lee and four friends went, how much would it cost them each?
  - In other words, the more people who use the taxi, the less it costs each of them – this is inverse proportion. Put the information in a table.

People in taxi	1	2	3	4	5
Cost per person	$30	$15	$10	$7.50	$6

- Sketch the graph to show inverse proportion.
- Explain that this is a reciprocal graph.

Main lesson activity

- Explain that the only difference between direct and indirect variation is that indirect variation uses reciprocals.
- Make a table to illustrate this. Ask students to complete it for square and cube roots.
- Carefully work through each step in Example 3.
- Now students can do Exercise 19C. Students who need more **support** can work in pairs if necessary.

Direct variation		Indirect variation	
$y \propto x$	$y = kx$	$y \propto \frac{1}{x}$	$y = \frac{k}{x}$
$y \propto x^2$	$y = kx^2$	$y \propto \frac{1}{x^2}$	$y = \frac{k}{x^2}$
$y \propto x^3$	$y = kx^3$	$y \propto \frac{1}{x^3}$	$y = \frac{k}{x^3}$

Plenary

- Students may not understand why inverse proportion graphs all have similar shapes when the functions are different. Ask students to plot the graphs for each of the following tables on the same grids in order to demonstrate the differences. Use scales from 0 to 6 for both axes.

$y = \frac{6}{x}$

x	1	2	3	4	5	6
y	6	3	2	0.5	1.5	1

$y = \frac{6}{\sqrt{x}}$

x	1	2	3	4	5	6
y	6	4.2	3.5	3	2.7	2.4

$y = \frac{6}{x^2}$

x	1	2	3	4	5	6
y	6	1.5	0.7	0.4	0.2	0.2

Homework and answers

Homework for this lesson is available in the homework folder. Consolidation and extension exercises are available in the further practice folder. Answers to the exercises can be found in the answers folder.

20.1 Graphical inequalities

Syllabus references

Statement	Extended
E2.6	Represent and interpret linear inequalities in two variables graphically.

Collins references

- Student's Book pages 364–366

Learning aims

- Show a graphical inequality.

Key words

- boundary, region, included, origin, dashed line, solid line

Prior knowledge

Students will need to know how to draw straight-line graphs.

Common mistakes and remediation

Students often do not draw the correct line, or they are confused about whether the line should be dotted or solid. They also choose the incorrect side of the line to the one required. Testing a point to check should help to eliminate these mistakes. Also, students often plot points on the wrong axes. They may also shade the wrong side of the line or not define clearly which side is required. To help to prevent these mistakes, work through some examples and point out the danger areas.

Useful tips

Tell students: Whatever the inequality, draw as though the inequality is an '=' sign. Then, if the sign is > or <, draw a dotted line, and if the sign is $\leqslant$ or $\geqslant$, draw a solid line. Think carefully about whether you want the region above or below the line.

Starter

- Write $2x + 3y = 6$ on the board. Remind students that the x-axis is the line $y = 0$, so, covering up the y term and solving the remaining linear equation, $2x = 6$ results in the intersect on the x-axis (3, 0). Similarly, the y-axis is the line $x = 0$, leading to $y = (0, 2)$.
- Repeat with other similar lines if necessary. Note that there is no need for negative values, although there is no reason why it cannot be covered for students who need more of a **challenge**. Students who need more **support** will prefer positive values.

Main lesson activity

- Ask students to sketch the graph of $2x + 3y = 6$. Ask which side of the line would satisfy $2x + 3y \leqslant 6$.
- The natural instinct, as the sign is 'less than or equal to', is to go for the region under the line. This is correct. But ask students how they can be sure. The answer is to test a point. Suggest to students that there is an 'obvious' choice of a point to test.
- If they do not come up with the answer, tell them that the origin is the best point to test if it isn't on the line, so $2 \times 0 + 3 \times 0 \leqslant 6$. This means that the origin is on the side required.
- Make sure students realise that the answer to $2x + 3y \leqslant 6$ is a region and that it is infinitely big.
- Repeat with the inequality $x > 2$. Ask why this is different. Students should realise that there is only one variable and the inequality is a 'strict' inequality.
- Sketch the region and point out that it is still a region, and that the way we deal with the strict inequality is to draw the boundary line as a dashed line.
- Make sure students are aware of the steps involved in drawing a graphical inequality, as follows.
    - Step 1: Draw the boundary line (dashed or solid as appropriate).
    - Step 2: Test a point not on the line to establish which side of the line is required.
- A cause of confusion is which side to shade. There is no established convention so students should be careful to read instructions fully.
- Work through Example 1 to make sure students have understood the principles.
- Students can then do Exercise 20A. Tell them to note the hint provided.

Plenary

- Provide some simple inequalities such as $x \geqslant 2$, $y < 3$, $x > y$, $y \leqslant 2x + 1$ on the board, and ask volunteers to explain the steps they should take to show these graphically.

Homework and answers

Homework for this lesson is available in the homework folder. Consolidation and extension exercises are available in the further practice folder. Answers to the exercises can be found in the answers folder.

20.2 More than one inequality

Syllabus references

Statement	Extended
E2.6	List inequalities that define a given region.

Collins references

- Student's Book pages 367–371

Learning aims

- Show a region that satisfies more than one graphical inequality.

Key words

- required

Prior knowledge

Students will need to understand the work covered in Lesson 20.1.

Common mistakes and remediation

Students often do not draw the correct lines, or they are confused about whether the line should be dotted or solid. They also fail to identify the correct region as the one required. Testing a point to check should help to eliminate these mistakes.

Useful tips

Tell students that when they have to show a region that satisfies more than one inequality, it is clearer to shade the regions not required, leaving the required region blank.

Starter

- Return to the plenary from Lesson 20.1 where you were asked to explain how to show the following inequalities graphically:

  $x \geqslant 2$ $\quad y < 3$ $\quad x > y$ $\quad y \leqslant 2x + 1$
- Ask students to now show these inequalities on separate grids. Remind them that they need to think carefully about whether or not the lines should be solid or dashed.

Main lesson activity

- Introduce the idea that some regions may be bound by more than one line.
- Refer back to the starter activity and ask students to re-draw the graphs, showing the inequalities $x \geqslant 2$ and $y < 3$, but this time draw them on the same grid. Ask students to identify the region that satisfies both these inequalities. Explain that in such a situation, it is clearer to shade the regions that are **not** required, so that the required region is left blank. Remind students to test a point to check that they have identified the required region correctly.
- Now ask students to show the inequality $y \leqslant 2x + 1$ on the same grid, and identify the region that satisfies all three inequalities. Ask them to test a point to ensure they have identified the region correctly.
- Now work through Example 2 with the students to consolidate their understanding.
- Now students can complete Exercise 20B, noting the hint provided. Students who need more **support** will need guidance.

Plenary

- Show students this graph:

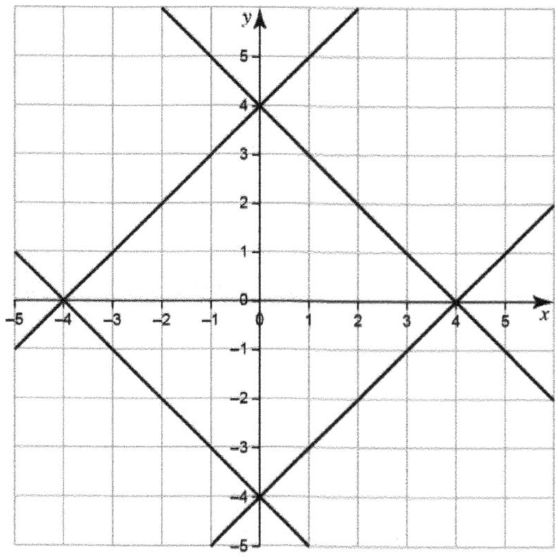

- Ask them for the equation of each of the four lines
- The answers are $y = x + 4$, $y = x - 4$, $x + y = 4$and $x + y = -4$
- You may get alternative answers such as $y = 4 - x$
- Now ask for inequalities that describe the points inside the square or on the perimeter
- The answers are $y \leqslant x + 4$, $y \geqslant x - 4$, $x + y \leqslant 4$and $x + y \geqslant -4$
- Students may suggest other correct answers such as $y \leqslant 4 - x$ instead of $x + y \leqslant 4$. Show that different possible versions are equivalent.

Homework and answers

Homework for this lesson is available in the homework folder. Consolidation and extension exercises are available in the further practice folder. Answers to the exercises can be found in the answers folder.

21.1 Function notation

Syllabus references

Statement	Extended
E2.13	Understand functions, domain and range and use function notation.

Collins references

- Student's Book pages 374–375

Learning aims

- Recognise and use function notation in either $f(x) = ...$ or $f: x \rightarrow ....$ form.

Key words

- function

Prior knowledge

Students should be familiar with evaluating functions in the form y = (an expression involving x) when drawing graphs or working with formulae.

Common mistakes and remediation

Students may revert to writing $y = ...$ rather than $f(x) = ...$ They should be encouraged to use the correct notation when writing this function.

Useful tips

Emphasise that writing f(2) = ... is a concise way of writing 'if $x = 2$ then $f(x) = ...$' and they should use the former.

The letter f is commonly used but any letter can be used for a function.

Starter

- Write 3 → 6 on the board without comment and then beneath it write 5 →. Ask students to suggest what the next number should be.
  The most likely replies are 8 (add 3) and 10 (multiply by 2). Do not ask for explanations at this stage but encourage different possible answers.
- After you have received a number of suggestions write the correct answer 5 → 12.
  Again, do not explain yet, but write underneath 8 →.
  Repeat the exercise and then write in 8 → 21.
- Ask students to give more possible pairs and record correct ones (the rule is subtract 1 and multiply by 3). Ask students to indicate if they understand the rule. When most of them do understand the rule, ask for explanations.
- There are two simple ways: –1 then × 3 or × 3 then –3. Try to elicit both from students and discuss why they are equivalent.
- Ask for an algebraic formulation if you are not given one. If we start with x then the answer is either $3(x - 1)$ or $3x - 1$.

Main lesson activity

- Draw a rectangle and label it f. Explain that this is a function machine. If you put a number in you will get a number out.

  It actually does the calculation in the starter.

- Say that we write this as $f(x) = 3(x - 1)$ and f is called a function. We could of course write $f(x) = 3x - 3$ as an equivalent form.

  Make sure students see the similarity with the starter activity.

  This makes it easy to show the start and finish numbers together: f(3) = 6, f(5) = 12, and so on.

  To practise, draw another box, label it g, and write $g(x) = \frac{x + 4}{2}$

- Ask for g(6), g(1), g(–2), g(4.5), etc. The answers are 5, 2.5, 1, 4.25.
- Students who need more **support** may need a further example to practise the new notation.
- When students are confident in using the notation they can do Exercise 21A.

Plenary

- Write on the board f(8) = 12 and ask students to write down three different possible expressions for $f(x)$.
- Each student should check the suggestions of their neighbour to make sure they are correct. Discuss any disputed answers.
- Students who need more of a **challenge** could be given restrictions such as only use subtraction; only use division.

  ($f(x) = 20 - x$ is one possible answer. $f(x) = \frac{96}{x}$ is another.)

Homework and answers

Homework for this lesson is available in the homework folder. Consolidation and extension exercises are available in the further practice folder. Answers to the exercises can be found in the answers folder.

21.2 Domain and range

Syllabus references

Statement	Extended
E2.13	Understand functions, domain and range and use function notation

Collins references

- Student's Book pages 375–377

Learning aims

- Students are able to identify the domain and range of a function.

Key words

- domain, range

Prior knowledge

Students should understand function notation.

Students should be familiar with the use of brackets { } to describe sets.

Common mistakes and remediation

Students often have difficulty remembering the difference between domain and range. Use the words frequently to reinforce meaning and tell students to think of a graph of the function.

Useful tips

Imagine a function pictorially like this:

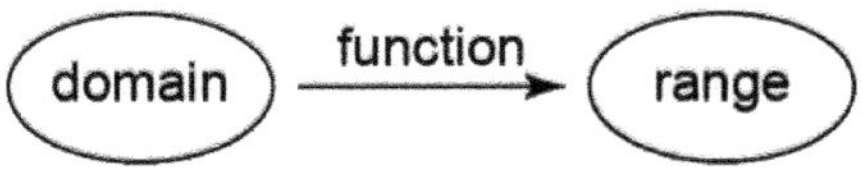

Starter

- Write down $\{x: 3 \leqslant \sqrt{x} \leqslant 6\}$ and ask students to tell you some numbers in this set. It can be any number between 9 and 36 inclusive. This is just to make sure students remember this set notation.
- Now write down $\mathrm{f}(x) = x^2$; $\mathrm{g}(x) = \sqrt{x}$; $\mathrm{h}(x) = \frac{1}{x}$
- Ask for the values of $\mathrm{f}(9)$ and $\mathrm{g}(9)$ and $\mathrm{h}(9)$
- The answers are 81, 3 and $\frac{1}{9}$
- Ask for the values of $\mathrm{f}(0)$ and $\mathrm{g}(0)$ and $\mathrm{h}(0)$
- The answers are $\mathrm{f}(0) = 0$ and $\mathrm{g}(0) = 0$ but $\mathrm{h}(0)$ does not exist. You cannot divide by 0.
- Now ask for the values of $\mathrm{f}(-4)$ and $\mathrm{g}(-4)$ and $\mathrm{h}(-4)$
- This time $\mathrm{f}(-4) = 16$ and $\mathrm{h}(-4) = -0.25$ but $\mathrm{g}(-4)$ does not exist. You cannot find the square root of a negative number.
- Ask for a sketch of the curve $y = x^2$ or look at it on graphic software. It only exists if x is 0 or positive. There is no line for negative values of x
- Finally ask for a sketch of $y = \frac{1}{x}$ or look at it on graphic software. There is a discontinuity where x = 0 because it is not defined for that value.

Main lesson activity

- Say you want to look in more detail at the function $\mathrm{f}(x) = \sqrt{x-2}$
- Ask students to give you some values of the function without using a calculator. For example, $\mathrm{f}(6) = \sqrt{6-2} = 2$ and $\mathrm{f}(102) = \sqrt{100} = 10$
- Establish that x can be 2 or greater than 2 but it cannot be less that 2.
- We say that the **domain** of the function is $\{x: x \geqslant 2\}$. The domain is the set of all the values that the variable x can take.
- What can we say about the values of $\mathrm{f}(x)$? It can be any non-negative number. We say that the **range** of the function is $\{y: y \geqslant 0\}$
- Tell students to look at the graph of $y = \sqrt{x-2}$in section 21.2. This shows the domain and range of the function on the x-axis and y-axis respectively.
- Use a graph plotter to show a graph of $y = \frac{2}{x+5}$ or sketch a copy (right).

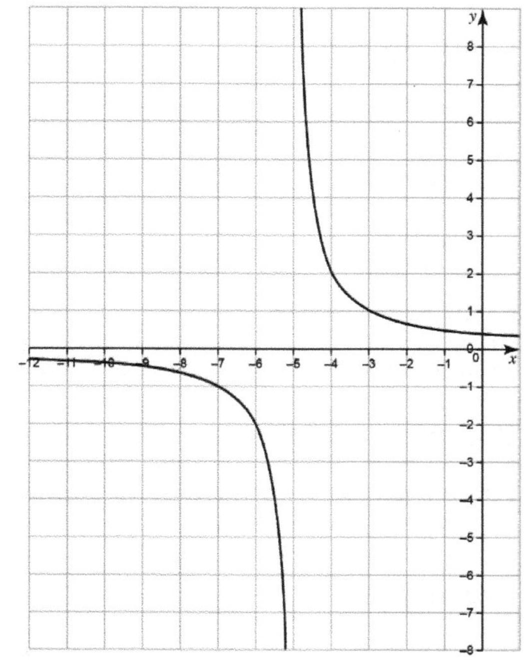

- If shows that x can be any value except −5 , and that y can be any value except 0.
- The domain of $\mathrm{f}(x) = \frac{2}{x+5}$ is $\{x: x \neq -5\}$and the range is $\{y: y \neq 0\}$
- Now look at Example 1 in section 21.2
- Here, the domain of the function $\mathrm{f}(x) = x(x-4)$ is given as $\{x: 0 \leqslant x \leqslant 6\}$
- The graph shows that the lowest value of y on the graph is −4 and the largest value is 12 and it can take any value in between those two.
- This means that the range of $\mathrm{f}(x)$is $\{y: -4 \leqslant y \leqslant 12\}$
- Students can now complete Exercise 21B.

Plenary

- Write down this function $\mathrm{f}(x) = x(20-x)$
- Say that the domain is the whole number line. Ask for the range of the function.
- It is $\{y: y \leqslant 100\}$. The best way to see that (10, 100) is a maximum point is to sketch a graph.

Homework and answers

Homework for this lesson is available in the homework folder. Consolidation and extension exercises are available in the further practice folder. Answers to the exercises can be found in the answers folder.

21.3 Inverse functions

Syllabus references

Statement	Extended
E2.13	Understand and find inverse functions $f^{-1}(x)$.

Collins references

- Student's Book pages 377–378

Learning aims

- Understand the concept of an inverse function and be able to find an expression for it.

Key words

- inverse

Prior knowledge

Students should be familiar with function notation.

Common mistakes and remediation

When reversing a sequence of operations it is easy to make algebraic mistakes. The best way to avoid this is to work systematically, reversing the operations one at a time and setting out the results clearly.

Useful tips

Writing $y = f(x)$ and manipulating the resulting formula may need some practice before students become confident with it.

Emphasise that f^{-1} does not mean $\frac{1}{f}$, as it does for indices. Make sure students do not confuse these two different uses of the same notation.

Starter

- Write down $f(x) = 2x + 5$. Then write down $f(a) = 13$ and ask for the value of a. (The answer is 4.) Ask students to explain how they know. (Because $2 \times 4 + 5 = 13$ is the most likely answer.)
- Repeat with $f(a) = 20$, then 30, then 0, then –15. The values of a are 7.5, 12.5, –2.5 and –10.
- Discuss how we find the value of a each time. In fact we are solving an equation, such as $2a + 5 = 13$ in the first case.
- Repeat the exercise with $g(x) = \frac{x-3}{4}$.
- Ask for a if $g(a) = 2$, then 5, then 1, then 4.5. The answers are 11, 23, 7 and 21.
- For Students who need more **support**, go through the solution of each corresponding equation in detail.
- Students who need more of a **challenge** could be asked to express their method algebraically – for example, the value of a is $4g(a) + 3$.

Main lesson activity

- Use $f(x) = 2x + 5$ as an example. This was used in the starter. Write down: $1 \to 7$, $4 \to 13$, $2 \to 9$
- Explain that the function that reverses the directions of the arrows is called the inverse of $f(x)$ and the notation for that is $f^{-1}(x)$. Explain that this does not mean $\frac{1}{f(x)}$ as it would for indices.

  We can write: $f^{-1}(7) = 1$
  $f^{-1}(13) = 4$
  $f^{-1}(9) = 2$
- Challenge the students to complete the expression, $f^{-1}(x) = \ldots\ldots\ldots\ldots$
- Using their experience of the starter they should be able to see that this is $\frac{x-5}{2}$.
- This can be checked by seeing that it gives the correct answer if x is 7, 13 or 9. Students who need more **support** may need to check other values too to become familiar with the idea.
- Repeat with $g(x) = \frac{x-3}{4}$ and lead the students to $g^{-1}(x) = 4x + 3$.

  Explain the following method that can be used to find the inverse of a function.
  Suppose $f(x) = 5(x - 7)$
  Write $y = 5(x - 7)$

  Rearrange to make x the subject:

  Divide by 5 $\frac{y}{5} = x - 7$

  Add 7 $\frac{y}{5} + 7 = x$

  Rewrite $x = \frac{y}{5} + 7$

  Now replace y by x on the right hand side to get $f^{-1}(x) = \frac{x}{5} + 7$.
- Demonstrate that this gives the right answers with particular values. For example, $f(10) = 15$ and $f^{-1}(15) = 10$.
- Students who need more **support** can find the idea of changing the letter confusing. It may be necessary to go through further examples to clarify this.
- When ready, students can start on Exercise 21C. In the exercise some functions are self-inverse, that is to say, $f(x) = f^{-1}(x)$. **Challenge** students by asking to find more self inverse functions.

Plenary

- Write down $f^{-1}(x) = 5x - 2$.
- Ask students to find $f(x)$.
  Students may try to do this by trial and improvement. In fact the inverse of $f^{-1}(x)$ is $f(x)$, and the same method of writing $y = 5x - 2$ and rearranging can be used to find $f(x) = \frac{x+2}{5}$.

Homework and answers

Homework for this lesson is available in the homework folder. Consolidation and extension exercises are available in the further practice folder. Answers to the exercises can be found in the answers folder.

21.4 Composite functions

Syllabus references

Statement	Extended
E2.13	Form composite functions as defined by gf(x) = g(f(x)).

Collins references

- Student's Book pages 378–381

Learning aims

- Understand how two functions can be combined to form a new composite function.

Key words

- composite

Prior knowledge

Students should be familiar with f(x) notation for functions.

Common mistakes and remediation

fg(x) means first g, then f, and students may make the mistake of assuming it is the other way round. Emphasise the point that it is the reverse order and if the order changes, so does the answer.

Starter

- This is an opportunity to revise the work done on inverses in the previous lesson.
- Write up the following and ask students to match each function with its inverse.

  a $(x) = 2x + 3$ **b** $(x) = \frac{x+2}{3}$ **c** $(x) = \frac{3-x}{2}$

  d $(x) = \frac{x-3}{2}$ **e** $(x) = 3x - 2$ **f** $(x) = 3(x + 2)$

  g $(x) = 3 - 2x$ **h** $(x) = 2 + 3x$ **i** $(x) = \frac{x}{3} - 2$

  j $(x) = \frac{x-2}{3}$

  Answers: a and d; b and e; c and g; f and i; h and j.

- Students could work in pairs on this exercise and be asked to justify their answers.
- Make the point that it does not matter which function you start with. For example, start with a and the inverse will be d; start with d and the inverse will be a.
- c and g are slightly harder, and with students who need more **support** the rearrangement of the corresponding formula may need more explanation.

Main lesson activity

- Write down $f(x) = 10 - x$ and $g(x) = 4x$.
- Starting with the numbers 1, 3, 5 and 0 as values for x, ask students to show values of $f(x)$. Put the numbers in a column like this:

	f	
1	→	9
4	→	6
5	→	5
0	→	10

- Now ask students to apply g to the answers to get this diagram:

	f		g	
1	→	9	→	36
4	→	6	→	24
5	→	5	→	20
0	→	10	→	40

- Show that the results could be written as g(f)(1) = 36, and so on.
- We can write this more simply as gf(1) = 36 and so on. Emphasise that this means first apply f to 1 and then apply g to the answer. The order is 'first f, then g'.
- Show that algebraically we could write $gf(x) = 4(10 - x)$. Demonstrate that this works with the numbers in the table.
- Ask the students to draw a similar table with the same starting numbers, but this time to first apply g, then f. The completed table should look like this:

	g		f	
1	→	4	→	6
4	→	16	→	–6
5	→	20	→	–10
0	→	0	→	10

  This shows that the order makes a difference; $gf(x)$ and $fg(x)$ are not the same.

  Show that algebraically $fg(x) = 10 - 4x$.
- Work through Example 2. Students can now do Exercise 21D.

Plenary

- Write $f(x) = 3(x)$. Ask students to find an expression for $ff(x)$.
- The answer is $ff(x) = 9x$. Some sample values should convince students.
- Write $g(x) = x + 2$. Ask students to find $gg(x)$. It should be clear that this is $x + 4$.
- Finally write $h(x) = 3x + 2$ and ask students to find an expression for $hh(x)$. Challenge them to make the expression as simple as possible.
- The answer is $9x + 8$. Students may deduce this from a table of values. It can also be done by simplifying $3(3x + 2) + 2$ and students who need more of a **challenge** could be asked to do this.

Homework and answers

Homework for this lesson is available in the homework folder. Consolidation and extension exercises are available in the further practice folder. Answers to the exercises can be found in the answers folder.

21.5 More about composite functions

Syllabus references

Statement	Extended
E2.13	Form composite functions as defined by $gf(x) = g(f(x))$.

Collins references

- Student's Book pages 381–383

Learning aims

- Students should be able to form and manipulate expressions for compound functions.

Prior knowledge

Students should have already been introduced to the notation for composite functions. They should also be familiar with the multiplication of two linear brackets.

Common mistakes and remediation

The most common error is in the algebraic manipulation of expressions. Expressions can be checked by substituting numerical values and seeing if the answers are correct.

Useful tips

The technique of substituting one expression into another is the important one for students to understand in this lesson.

Starter

- Display the following:

  $f(x) = x + 5$ $g(x) = 3x$ $h(x) = 2x - 3$
- Ask students to work in pairs to find expressions for the following:

  $ff(x)fg(x)$ $gf(x)$ $fh(x)$ $hf(x)$
- The answers are $x + 10$; $3x + 5$; $3(x + 5)$; $2x + 2$; $2x + 7$.
- Ask students to give their answers and ask them to justify their answers each time. If necessary clarify the algebra for the last two:

  $fh(x) = (2x - 3) + 5 = 2x - 3 + 5 = 2x + 2$

  $hf(x) = 2(x + 5) - 3 = 2x + 10 - 3 = 2x + 7$

Main lesson activity

- Write down $f(x) = x^2 - 1$ and $g(x) = 3x + 1$.
- Ask for an expression for $fg(x)$.
- Students need to understand why 'doing g' and then 'doing f' gives the result $fg(x) = (3x + 1)^2 - 1$. Explain that we are substituting $g(x)$ for x on the expression for $f(x)$ and make sure that students are clear what that means.
- Now simplify the expression by multiplying out the bracket:
  $fg(x) = (3x + 1)^2 - 1 = 9x^2 + 6x + 1 - 1 = 9x^2 + 6x$
- Students who need more **support** may need to be reminded how to multiply out the squared bracket.
- Now ask students for an expression for $gf(x)$ and simplify it.
- Ask a volunteer to go through a solution.
- The answer should show 'first f, then g' like this:
  $gf(x) = 3(x^2 - 1) + 1 = 3x^2 - 3 + 1 = 3x^2 - 2$
- If students are unconvinced, work through a numerical example.
- Students can now do Exercise 21E.

Plenary

- Write down $f(x) = 4(x - 5)$.
- Ask the students to find $f^{-1}(x)$.
- Confirm that the answer is $\frac{x}{4} + 5$.
- Now ask students to find and simplify an expression for $f^{-1}f(x)$.
- Ask a volunteer to show a solution.
- The answer is that $f^{-1}f(x) = x$.
- Algebraically: $f^{-1}f(x)\frac{4(x-5)}{5} = x - 5 + 5 = x$
- Ask students why this is the case. (Answer: Since f^{-1} is the inverse of f, we will always end up with the number we started with. Start with x, finish with x.)
- Ask students to find $ff^{-1}(x)$ in the same way.
- This time $ff^{-1}(x) = 4\left(\left(\frac{x}{4} + 5\right) - 5\right) = 4\frac{x}{4} = x$

Homework and answers

Homework for this lesson is available in the homework folder. Consolidation and extension exercises are available in the further practice folder. Answers to the exercises can be found in the answers folder.

22.1 The gradient of a curve

Syllabus references

Statement	Extended
E2.12	Use the derivatives of functions of the form ax^n. Estimate gradients of curves by drawing tangents.

Collins references

- Student's Book pages 386–388

Learning aims

- Calculate the gradient of a quadratic curve.

Key words

- gradient

Prior knowledge

Students need to be able to draw quadratic graphs and solve simple linear equations.

They should also be able to calculate the gradient of a straight line drawn between two points.

Common mistakes and remediation

Making an error in differentiating a quadratic equation is relatively rare. Some students forget that the constant term differentiates to 0.

Another common mistake is to write $\frac{dy}{dx}$ using partial derivatives rather than with normal 'd's.

Useful tips

Remind students that the gradient of the straight line $y = mx + c$ is m and that the gradient of a line $y = c$ is 0 as that line is parallel to the x-axis.

Starter

- Ensure students know how to calculate the gradient of a line between two points. Ask them to calculate the gradient between the following pairs of coordinates:

  (6, 2) and (8, 10) $\left(\frac{8}{2} = 4\right)$

  (5, 4) and (9, −10) $\left(-\frac{14}{4} = -3.5\right)$

Main lesson activity

- Let the students know that they are now studying a branch of calculus known as differentiation. This was developed in the seventeenth century, by Newton in England and independently by Leibniz in Germany, to help describe the motion of planets for astronomical purposes.
- If you have access to a graphing software or dynamic geometry software it is very useful to show a quadratic graph and how the gradient of a chord on the curve approaches the tangent as the moving point A gets closer to the fixed point P. The moving image is far more powerful that the written one!
- Discuss with students the following table of results obtained in the Student's Book:

x-coordinate	3	4	8
Gradient	3	5	−1

- Ask your students if they can find an equation linking the x-coordinate with the gradient. It looks as though the gradient is three times the x-coordinate minus 3, which is indeed the case.
- If time permits you might want to repeat this with $y = x^2 + 2x + 1$, in which case the gradient will be $2x + 2$.
- Now introduce the general result that if $y = ax^2 + bx + c$, $\frac{dy}{dx} = 2ax + b$.
- Students can now do Exercise 22A.

Plenary

- Ask students to produce a mini-poster, size A5, that explains what you should do to differentiate $y = ax^2 + bx + c$. Allow just five minutes.
- Students should then pass their mini-poster to a neighbour for checking and comments. The best can be displayed on the wall.

Homework and answers

Homework for this lesson is available in the homework folder. Consolidation and extension exercises are available in the further practice folder. Answers to the exercises can be found in the answers folder.

22.2 More complex curves

Syllabus references

Statement	Extended
E2.12	Use the derivatives of functions of the form ax^n where a is a rational constant and n is a positive integer or zero, and simple sums of not more than three of these. Apply differentiation to gradients.

Collins references

- Student's Book pages 388–390

Learning aims

- Differentiate simple polynomials with non-negative powers of x.
- Calculate the gradient at various points on polynomial curves.

Key words

- derivative, differentiation

Prior knowledge

Students need to have completed Section 22.1, and be able to substitute into and solve equations.

Common mistakes and remediation

Making an error in differentiating a simple polynomial equation is not common. Some students forget that the constant term differentiates to 0. Others may forget to multiply the power by the coefficient or possibly to reduce the power by 1. Ask students to check their answers regularly at the start of any exercise and seek help if they are consistently making mistakes.

Useful tips

Remind students that they need to take care when finding powers of negative numbers especially when using a calculator. Substituting −3 into x^2 means they should enter (−3) and square that: if they enter −3 then press the square button −9 will result.

Starter

- Ensure students know how to differentiate simple quadratics by asking them to differentiate the following:

  $y = x^2 + 4x + 6 \qquad \frac{dy}{dx} = 2x + 4$

  $y = 3x^2 - 8x + 5 \qquad \frac{dy}{dx} = 6x - 8$

  $y = 5x - 4 + 0.5x^2 \qquad \frac{dy}{dx} = x + 5$

- Now ask them to state the value of x when the gradient is 0.
  (−2, 4/3 and −5 respectively)

Main lesson activity

- Write the table of results for y and its derivatives from the Student's Book on the board and ask students if they can make a conjecture about the general result. Let them know that this is a conjecture and that if they continue with their study of mathematics they will be able to prove the result.
- Discuss the result that if $y = x^n$, then $\frac{dy}{dx} = nx^{n-1}$ and ask them to use this result to differentiate the following:

  $y = x^6$ $\quad\quad$ $\frac{dy}{dx} = 6x^5$

  $y = x^{101}$ $\quad\quad$ $\frac{dy}{dx} = 101x^{100}$
- Now ask students what they think would be the result if a is a constant and $y = ax^n$ $\left(\frac{dy}{dx} = anx^{n-1}\right)$.
- Ask them to use this result to differentiate the following:

  $y = 4x^3$ $\quad\quad$ $\frac{dy}{dx} = 12x^2$

  $y = -0.2x^{10}$ $\quad\quad$ $\frac{dy}{dx} = -2x^9$
- Students can now do Exercise 22B.

Plenary

- Ask for a volunteer to explain how to do question 8 or 9 from Exercise 22B. If nobody volunteers then you can sigh and do it for them.

Homework and answers

Homework for this lesson is available in the homework folder. Consolidation and extension exercises are available in the further practice folder. Answers to the exercises can be found in the answers folder.

22.3 Turning points

Syllabus references

Statement	Extended
E2.12	Apply differentiation to gradients and stationary points (turning points). Discriminate between maxima and minima by any method.

Collins references

- Student's Book pages 391–393

Learning aims

- Find the turning points on a polynomial curve.
- Identify the turning points as a maximum or a minimum.

Key words

- turning point, maximum, minimum

Prior knowledge

Students need to have worked through Sections 22.1 and 22.2 and be able to differentiate the more complex curves of Section 22.2.

They should also be able to solve simple quadratic equations by factorising and have met turning points in Section 16.3.

Common mistakes and remediation

Since differentiation is still fairly new to the students they might make some simple slips as described earlier: making an error in differentiating a simple polynomial equation is not common. Some students forget that the constant term differentiates to 0. Others may forget to multiply the power by the coefficient or possibly to reduce the power by 1. Ask students to check their answers regularly at the start of any exercise and seek help if they are consistently making mistakes.

Another mistake they might make is in solving a resulting quadratic equation: the questions as set will give whole number solutions or simple fractions.

Useful tips

Remind students that a quick sketch should help them determine whether the turning point is a maximum or minimum. None of the questions at this level will involve asymptotes.

Remind them about the work they did on turning points in Section 16.3 if they have forgotten this.

Starter

- To remind students about turning points, sketch some polynomial curves on the board with one, two, three or even four turning points and ask them to identify the type of turning point.

Main lesson activity

- Draw a smooth curve with two turning points on the board. Mark the turning points with A and B.
- Ask the students what they can say about the gradient of a tangent on the curve to the left of point A, at point A and to the right of it. (It will be either positive, 0, negative, or negative, 0, positive, depending on whether you have sketched a curve like that at the start of Section 22.3 in the Student's Book, or a 180° rotation of it.)
- Discuss with students the conditions for a maximum (the gradient should go positive, 0, negative as a point on the curve goes from the left to the turning point and then to the right of it) and a minimum (the gradient should go negative, 0, positive as a point on the curve goes from the left to the turning point and then to the right of it).
- Go through Example 3 on the board with your students.
- Students can now do Exercise 22C.

Plenary

- Ask students to produce a mini-poster, size A5, that explains how to determine whether a turning point is a maximum or minimum. Allow just five minutes.
- Students should then pass their mini-poster to a neighbour for checking and comments. The best can be displayed on the wall.

Homework and answers

Homework for this lesson is available in the homework folder. Consolidation and extension exercises are available in the further practice folder. Answers to the exercises can be found in the answers folder.

23.1 Angle facts

Syllabus references

Statement	Core
C4.6/E4.6	Calculate unknown angles and give simple explanations using the following geometrical properties: sum of angles at a point = 360°, sum of angles at a point on a straight line = 180°, vertically opposite angles are equal.

Collins references

- Student's Book pages 408–410

Learning aims

- Calculate angles on a straight line and angles around a point and use opposite angles.

Key words

- angles at a point, angles on straight lines, opposite angles, vertically opposite angles

Prior knowledge

Students need to be able to solve simple linear equations.

Common mistakes and remediation

Students sometimes think that the angle sum on a straight line is 200° or that there are 380° in a complete turn.

Useful tips

It is useful to have these key angle facts displayed in the classroom: angles on a line add up to 180°; angles around a point add up to 360°; opposite angles are equal.

Starter

- Ask students to do some mental arithmetic involving complements of 90, 180 and 360. For example, say a number, such as 35, and ask for the complement to 90 (55), 180 (145) and 360 (325).

Main lesson activity

- Draw the diagram opposite on the board. Explain to the students that $a + b = 180°$. Say that angles on a straight line add up to 180°.

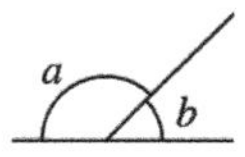

- Explain that this helps to find the size of an angle on a diagram that is not drawn accurately. For example, draw this diagram opposite and ask students to find the value of b.

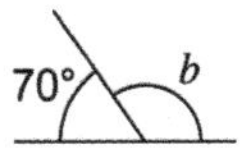

  Angles on a straight line add up to 180°.

  So $b + 70 = 180°$

  $b = 180° - 70°$

  $b = 110°$

- Encourage students always to write down an equation first. Students who need more **support** may find it easier just to work out 180° – 70° on a calculator.

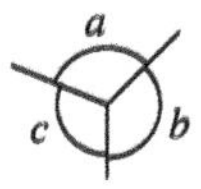

- Draw this diagram on the board. Explain to students that $a + b + c = 360°$.
  Say that angles around a point add up to 360°.

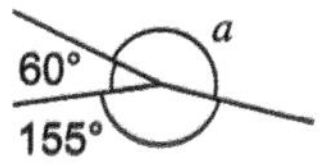

- Explain that this helps to find the size of an angle on a diagram that is not drawn accurately. For example, draw this diagram and ask students to find the value of a.

  Angles around a point add up to 360°.

  So $a + 60° + 155° = 360°$

  $a = 360° - 60° - 155°$

  $a = 145°$

- Students who need more **support** may find it easier just to work out 360° – 60° – 155° on a calculator.
- Draw this diagram on the board.
  Ask students what they can say about a and b.

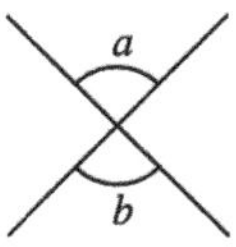

- Remind students that the angle sum on a straight line is 180°. Lead them to see that both a and b can combine with the same one of the unmarked angles to form angles on a straight line.
- Explain to students that $a = b$. Tell them that a and b are opposite angles and are equal. Tell students that we can also refer to them as vertically opposite angles.
- Draw this diagram and ask students to find the value of a.
  (Opposite angles are equal so $a = 124°$.)

- Students can now do Exercise 23A. Students who need more **support** may omit some of the more difficult questions.

Plenary

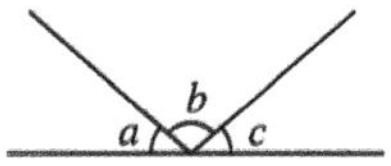

- Draw this diagram on the board and ask students:
- If $a = 45°$ and $b = 110°$, calculate the value of c.
- If $a = x$, $b = 2x$ and $c = 3x$, calculate the value of x.
- If $a = b = c$, find the value of a.

Homework and answers

Homework for this lesson is available in the homework folder. Consolidation and extension exercises are available in the further practice folder. Answers to the exercises can be found in the answers folder.

23.2 Parallel lines

Syllabus references

Statement	Core
C4.1/E4.1	Use and interpret the geometrical terms line, parallel, interior and exterior angles.
C4.6/E4.6	Calculate unknown angles and give geometric explanations for angles formed within parallel lines: corresponding angles are equal, alternate angles are equal, co-interior (supplementary) angles sum to 180°.

Collins references

- Student's Book pages 410–413

Learning aims

- Find angles in parallel lines.

Key words

- corresponding angles, alternate angles, allied angles, interior angles

Prior knowledge

Students should know how to calculate angles on a straight line, angles at a point, opposite angles and angles in a triangle. They should also know how to solve simple linear equations.

Common mistakes and remediation

Mistakes in this topic often result from students guessing the size of angles. This will indicate that they do not thoroughly understand the definitions of the various types of angle relationships.

Students often measure the angles on diagrams supplied in the questions. Remind them that diagrams are never drawn accurately in questions involving parallel lines and other topics involving measurement.

Useful tips

Always check that acute or reflex angles are correct on diagrams. It is useful to have these key angle facts displayed in the classroom: corresponding angles are equal; alternate angles are equal; allied angles (or interior angles) add up to 180°.

Starter

- Make sure that students have tracing paper and protractors.
- Students who need more **support** can use tracing paper to do this activity.
- Students who need more of a **challenge** should use a protractor to measure the angles carefully.
- Ask students to draw two parallel lines and a third line that crosses them both.
  They should then label the eight angles that are formed.

  Ask students to measure or trace one of the angles and then identify all the other angles that are equal to it.
- Students who need more of a **challenge** can continue with the activity, to establish the other equivalences for themselves. They should also be able to establish that allied angles add up to 180°.

Main lesson activity

- Explain to students that there are three types of angles in parallel lines. It is important that students become familiar with the mathematical terminology and use it in their working.
- Point out that corresponding and alternate angles are either both acute or both reflex and that for allied angles, one must be acute and one must be reflex unless they are both right angles.
- Students can now do Exercise 23B. Students who are having difficulty may be advised not to attempt all the questions.
- Students who need more **support** should be encouraged to use tracing paper to check that the size of corresponding and alternate angles are the same.

Plenary

- Draw this diagram on the board.

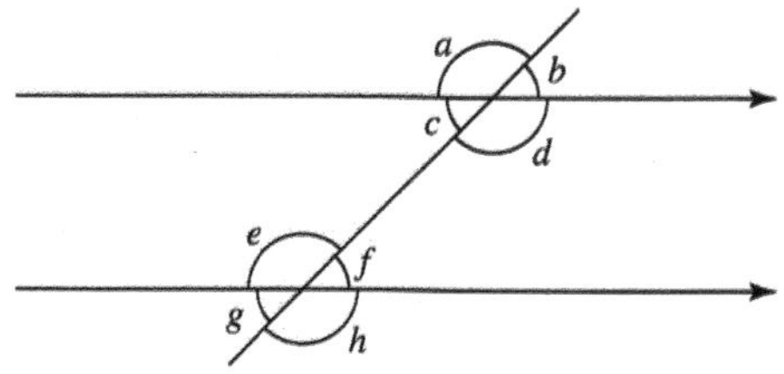

- Invite students to name a pair of corresponding angles, a pair of alternate angles and a pair of allied angles.

Homework and answers

Homework for this lesson is available in the homework folder. Consolidation and extension exercises are available in the further practice folder. Answers to the exercises can be found in the answers folder.

23.3 Angles in a triangle

Syllabus references

Statement	Core
C4.1/E4.1 C4.6/E4.6	Use and interpret the vocabulary of triangles. Calculate unknown angles and give simple explanations using the following geometrical property: angle sum of a triangle = 180°.

Collins references

- Student's Book pages 414–416

Learning aims

- Calculate the size of angles in a triangle.

Key words

- equilateral triangle, isosceles triangle, right-angled triangle

Prior knowledge

Students need to be able to solve simple linear equations.

Common mistakes and remediation

Students often think that the sum of the angles in a triangle is 200° or 360°. It is therefore useful to display facts about the different types of triangle.

Useful tips

It is useful to have these key triangle facts displayed in the classroom: an acute-angled triangle has three acute angles; an obtuse-angled triangle has one obtuse angle; a right-angled triangle has one right angle; a scalene triangle has sides all of different lengths; an isosceles triangle has two sides equal; the angles opposite the equal sides are equal; an equilateral triangle has three sides equal and three angles of 60°; the angles in a triangle add up to 180°.

Starter

- Ask students, 'What types of triangle are there? What can you say about the angles in a triangle?'
- Many should recollect most of the information they will require for this topic.

Main lesson activity

- Draw the triangle opposite on the board. In any triangle $a + b + c = 180°$.
  The sum of the angles in a triangle is 180°.

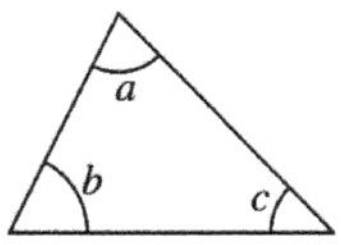

- Now do an example with an obtuse-angled triangle. Draw this triangle on the board. Ask students to calculate the value of a.

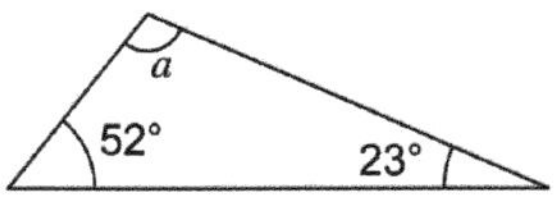

Angles in a triangle add up to 180°.

$$
\begin{aligned}
\text{So } a + 52° + 23° &= 180° \\
a &= 180° - 52° - 23° \\
a &= 105°
\end{aligned}
$$

- Remind students that an angle in a triangle can be obtuse and that a triangle with three different angles is called a scalene triangle.
- Point out to students that there are other different types of triangle:

An equilateral triangle
Three sides of equal length and three equal angles. Each angle is 60°.

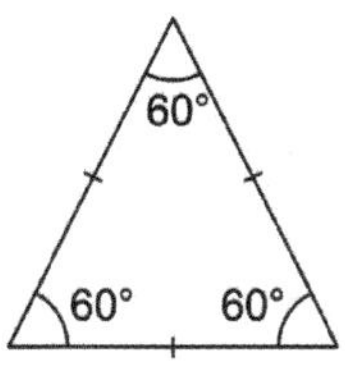

An isosceles triangle
Two sides of equal length and two equal angles - the equal angles are opposite the equal sides.

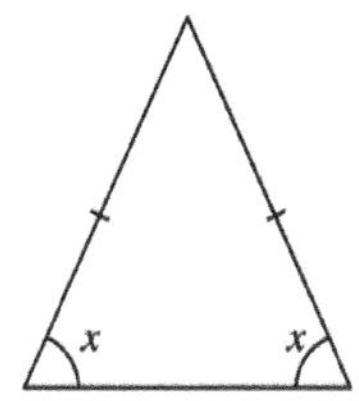

A right-angled triangle
one angle is 90° and therefore a + b = 90°.

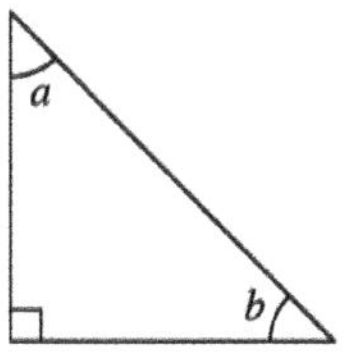

- Students who need more of a **challenge** can be shown a proof, using this diagram, that the exterior angle in a triangle is equal to the sum of the two opposite interior angles. a, b and c are the three *interior angles* of the triangle. x is an exterior angle of the triangle.

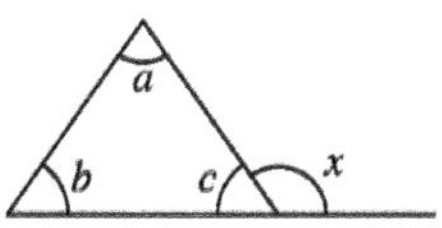

$a + b + c = 180°$ (angles in a triangle)

$x + c = 180°$ (angles on a line)

So $a + b = x$

- Students can now do Exercise 23C. In this exercise, question 6 is a problem-solving (PS) question.

Plenary

Ask students these questions.

- 'What do you understand a right-angled triangle to be?'
- 'What is an acute-angled triangle?' (One in which each angle is less than 90°.)
- 'What is an obtuse-angled triangle?' (One in which one angle is greater than 90°.)
- 'Can you draw an acute-angled isosceles triangle?'
- 'Can you draw an obtuse-angled isosceles triangle?'
- 'Can you draw a right-angled isosceles triangle?'

Homework and answers

Homework for this lesson is available in the homework folder. Consolidation and extension exercises are available in the further practice folder. Answers to the exercises can be found in the answers folder.

23.4 Angles in a quadrilateral

Syllabus references

Statement	Core
C4.6/E4.6	Calculate unknown angles and give simple explanations using the following geometrical property: angle sum of a quadrilateral = 360°. Calculate unknown angles and give geometric explanations for angles formed within parallel lines: corresponding angles are equal, alternate angles are equal, co-interior (supplementary) angles sum to 180°.

Collins references

- Student's Book pages 416–419

Learning aims

- Work out the sizes of angles in quadrilaterals.

Key words

- quadrilateral, parallelogram, rhombus, kite, trapezium

Prior knowledge

Students should know that the sum of the angles in any triangle is 180°, and that the sum of the angles in a quadrilateral is 360°. Students should also know that when a transversal cuts parallel lines, the alternate angles are equal, corresponding angles are equal, and allied (or interior) angles are supplementary (have a sum of 180°).

Common mistakes and remediation

Some students may be confused between alternate and corresponding angles. Constant reference to them during the lesson would be beneficial.

Note: Remind students that they should not refer to alternate angles as Z angles and corresponding angles as F angles.

Useful tips

Remind students that they should not assume diagrams are drawn accurately unless it is clearly stated that they are. Therefore, students should never try to measure angles in diagrams.

Starter

- Draw on the board a pair of parallel lines cut by a transversal.

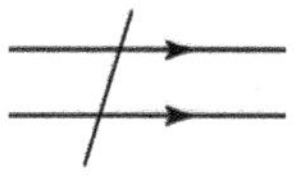

- Ask a student to mark a pair of alternate angles on the diagram.
  Repeat for a pair of corresponding angles and a pair of allied angles.
- Encourage students who need more **support** to copy the following into their notes.

Angles in parallel lines

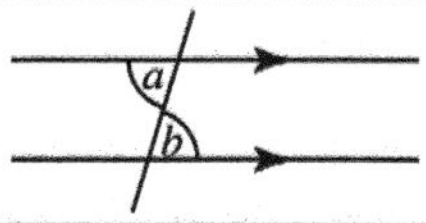	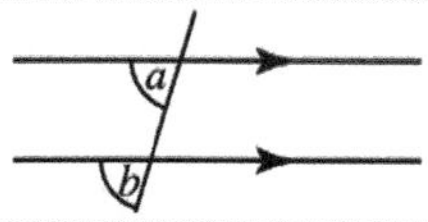	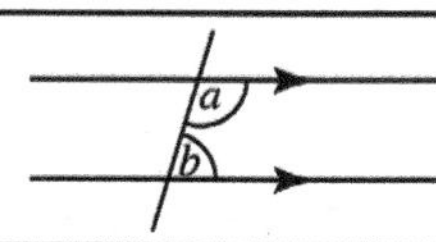
a and b are equal a and b are alternate angles	a and b are equal a and b are corresponding angles	$a + b = 180°$ a and b are allied angles

Main lesson activity

- Remind students that the sum of the angles in a quadrilateral is 360°. Go through Example 1.
- Draw a parallelogram on the board. Go through the geometrical properties of the parallelogram and encourage students to record all the facts for future reference. Go through Example 2.
- Remind students who need more **support** about the mathematical conventions for parallel lines, equal sides and naming angles. Remind students who need more of a **challenge** that the diagonals of the parallelogram create alternate angles and vertically opposite angles.
- Draw a rhombus on the board. Go through its geometrical properties and encourage students to record all the facts for future reference. Rhombuses are often called diamonds.
- Draw a kite on the board. Go through its geometrical properties and encourage students to record all the facts for future reference.
- Draw a trapezium on the board. Go through its geometrical properties and encourage students to record all the facts for future reference.
- Tell students who need more of a **challenge** that if the sloping sides are equal, the shape is an *isosceles trapezium*.
- Briefly revise the symmetry of the four quadrilaterals, since this is often a problematic area for students who need more **support**.

Shape	Number of lines of symmetry	Order of rotational symmetry
Parallelogram	0	2
Rhombus	2	2
Kite	1	0
Trapezium	0	0

- Now students can complete Exercise 23D. If necessary, let students who need more **support** work on the questions in pairs.

Plenary

- Draw on the board an isosceles triangle, a trapezium, a parallelogram, a kite and a rhombus (without marking identical angles and sides).
- Ask students (in turn) to mark pairs of identical angles or sides on the diagrams until all possibilities have been covered.

Homework and answers

Homework for this lesson is available in the homework folder. Consolidation and extension exercises are available in the further practice folder. Answers to the exercises can be found in the answers folder.

23.5 Regular polygons

Syllabus references

Statement	Core
C4.1/E4.1	Use and interpret the vocabulary of polygons.
C4.6/E4.6	Calculate unknown angles and give simple explanations using the following geometrical properties: sum of angles at a point = 360°, sum of angles at a point on a straight line = 180°, vertically opposite angles are equal, angle sum of a triangle = 180° and angle sum of a quadrilateral = 360°. Know and use angle properties of regular polygons.

Collins references

- Student's Book pages 419–421

Learning aims

- Calculate the exterior angles and the interior angles of a regular polygon.

Key words

- polygon, regular polygon, external angles, pentagon, hexagon, octagon, square

Prior knowledge

Students should know how to calculate angles on a straight line and angles at a point.

Common mistakes and remediation

A common error is to calculate the exterior angle as 360° – (interior angle). Ensure that students learn the fact that an exterior angle = 180° – its interior angle.

Useful tips

It is helpful to have available a collection of regular polygons or templates. Students who need more **support** may find it helpful to extend the side of a polygon and identify the interior angle (inside the polygon) and the **ex**terior angle (found by **ex**tending the side).

Starter

Ask: 'What is a three-sided shape called? a four-sided shape? a five-sided shape? a 10-sided shape?'

Students who need more of a **challenge** might ask about polygons with more than 10 sides (e.g. 11 sides: *hendecagon*, 12 sides: *dodecagon*, 20 sides: *icosagon*). They could do their own research into this.

Main lesson activity

- Define a regular polygon as one in which all the sides are equal and all the angles are equal.
- Draw a regular pentagon and show students how to divide it into five isosceles triangles by drawing lines from the centre to each vertex. From this, deduce that each interior angle is 108°.
- Ask students to do the same for an equilateral triangle, a square, a regular hexagon and a regular octagon, and record their results in a table.
- Ask students to work out the size of the interior angles of a regular nonagon and a regular decagon without drawing the shapes.

- Define an exterior angle of a polygon. Make sure students know that the sum of an interior angle and its exterior angle is 180°. Ask them to find the size of each exterior angle of the regular polygons and record their results in their tables. Tell them to go on to work out the sum of the exterior angles in each of the regular pentagons. A completed table is given below.

SHAPE	**NUMBER OF SIDES**	**INTERIOR ANGLE**	**EXTERIOR ANGLE**	**SUM OF EXTERIOR ANGLES**
Equilateral triangle	3	60°	120°	3 × 120° = 360°
Square	4	90°	90°	4 × 90° = 360°
Regular pentagon	5	108°	72°	5 × 72° = 360°
Regular hexagon	6	120°	60°	6 × 60° = 360°
Regular octagon	8	135°	45°	8 × 45° = 360°
Regular nonagon	9	140°	40°	9 × 40° = 360°
Regular decagon	10	144°	36°	10 × 36° = 360°

- Students who need more **support** should notice that the formula in words for finding each exterior angle and interior angle in a regular polygon is:

$$\text{exterior angle} = \frac{360°}{\text{number of sides}} \qquad \text{interior angle} = 180° - \text{exterior angle}$$

- Students who need more of a **challenge** should write down the algebraic formulae for any regular polygon:

$$E = \frac{360°}{n} \qquad I = 180° - E = 180° - \frac{360°}{n}$$

where E is the exterior angle, I is the interior angle and n is the number of sides.

- Conclude that, when finding the size of an interior angle of a regular polygon, it is easier to find the size of the exterior angle first. Go through Example 3.
- Students can now do Exercise 23E.

Plenary

- Discuss the formula for finding the exterior angle of a regular pentagon through the following approach.
- Draw a regular pentagon and identify the exterior angles. Now starting at *A*, for example, and walking round the pentagon clockwise, you would finally turn through one complete revolution i.e. 360° before returning to *A*. Therefore, since there are five identical exterior angles, each of these must be 360° ÷ 5 = 72°.

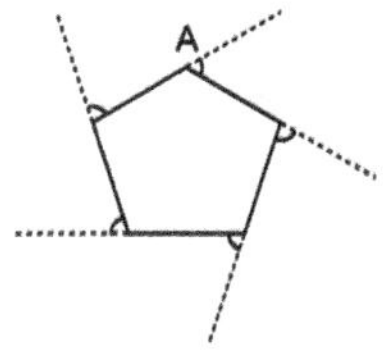

Homework and answers

Homework for this lesson is available in the homework folder. Consolidation and extension exercises are available in the further practice folder. Answers to the exercises can be found in the answers folder.

23.6 Irregular polygons

Syllabus references

Statement	Extended
E4.6	Know and use angle properties of irregular polygons.

Collins references

- Student's Book pages 422–424

Learning aims

- Work out the size of interior angles and exterior angles in an irregular polygon.

Key words

- irregular polygon, heptagon, nonagon, decagon

Prior knowledge

Students should know how to calculate angles on a straight line, and that the sum of the interior angles of a triangle is 180°.

Common mistakes and remediation

Students often mistakenly think that the sum of an exterior angle and an interior angle is 360°. Throughout the lesson, reinforce that it is 180°.

Useful tips

Students who need more **support** may find it helpful to extend the side of a polygon and identify the interior angle (inside the polygon) and the **ex**terior angle (found by **ex**tending the side).

Technically the nine-sided shape is known as an enneagon, rather than a nonagon. This is because the prefixes come from the Greek words, rather than from the Latin.

Starter

- Review the rules for finding the interior and exterior angles of a regular polygon.
- Display a table for regular polygons with headings 'Number of sides', 'Exterior angle' and 'Interior angle'. Ask individuals to complete rows of the table for shapes with three to 10 sides, and with n sides.
- Ask: 'What happens to the interior angle of a regular polygon as the number of sides increases? Why?'

Main lesson activity

- Explain to students that they are now going to look at irregular polygons in which the sides are not all the same length and the angles are not all the same size.
- Draw a quadrilateral and a pentagon on the board. Show students how to split them into triangles to find the sum of the interior angles. Ask students to do the same for a hexagon and a heptagon. They should use convex polygons, but not regular ones. Ensure that students draw all the diagonals from one vertex of the polygon.
- To help students who need more **support**, provide them with a worksheet of ready-drawn polygons.
- Ask students to work out the sum of the interior angles for octagons, nonagons and decagons without drawing the shapes. Ask them to record their results in a table.
- Students who need more **support** should notice the formula in words:

  Sum of interior angles in degrees = 180 × (number of sides – 2)

- Students who more of a **challenge** should write down the algebraic formula:

  $S = 180(n - 2)$, where S is the sum of the interior angles in degrees and n is the number of sides.

- Remind students that the sum of the **exterior** angles of **any** polygon is 360° but that their sizes may not all be the same, so to find a specific exterior angle, we use the fact that an exterior angle = 180° – the adjacent interior angle.
- Go through Example 4.
- Now students can do Exercise 23F. Advise students who need more **support** to do the earlier questions in pairs until they are confident enough to complete the exercise on their own.

Plenary

- Consider whether the results still hold for concave polygons. These are polygons in which at least one internal angle is greater than 180°.

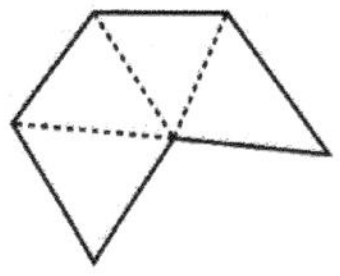

For example, a concave hexagon is shown here.

- It can still be split into four triangles so the sum of the interior angles is 4 × 180° = 720°, as for a regular hexagon.

Homework and answers

Homework for this lesson is available in the homework folder. Consolidation and extension exercises are available in the further practice folder. Answers to the exercises can be found in the answers folder.

23.7 Tangents and diameters

Syllabus references

Statement	Core
C4.1/E4.1 C4.7/E4.7	Use and interpret the vocabulary of a circle. Calculate unknown angles and give explanations using the following geometrical properties of circles: angle in a semicircle = 90°, angle between tangent and radius = 90°.

Collins references

- Student's Book pages 424–426

Learning aims

- Use tangents and diameters to find the sizes of angles in circles.

Key words

- radius, tangent, point of contact, diameter, semi-circle

Prior knowledge

Students need to know that angles in a triangle add up to 180° and that angles in a quadrilateral add up to 360°.

Common mistakes and remediation

Students often find it difficult to solve problems on angles in circles if they have to find more than one angle on a diagram. Advise them to mark on their diagram all the angles they know before proceeding with the question.

Useful tips

As some of the work involves drawing circles, make sure all students have the necessary equipment: a sharp pencil, compasses and a protractor.

Starter

- Ask students, 'What can you say about the angles in a triangle? What about an isosceles triangle? What can you say about the angles in a quadrilateral?'
- Draw some examples of triangles and quadrilaterals on the board where students have to find the missing angles.

Main lesson activity

- Explain to students that they will be meeting circle theorems that involve tangents and diameters. Explain the definition of a tangent and a diameter.
- A *tangent* is a line that touches a circle at a point, which is called *the point of contact*.
- A *diameter* is a line that joins any two points on the circumference of the circle and that passes through the centre of the circle. It divides a circle into two semi-circles.

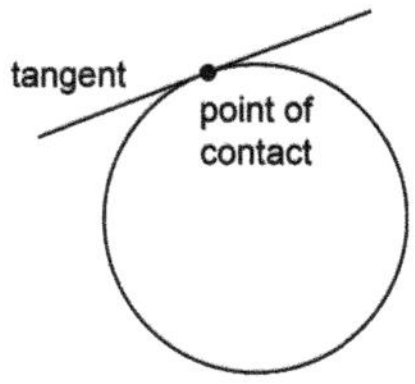

- Ask students to draw this diagram, with O as the centre of the circle. OA and OB are radii of the circle and XZ and YZ are tangents that touch the circle at A and B.

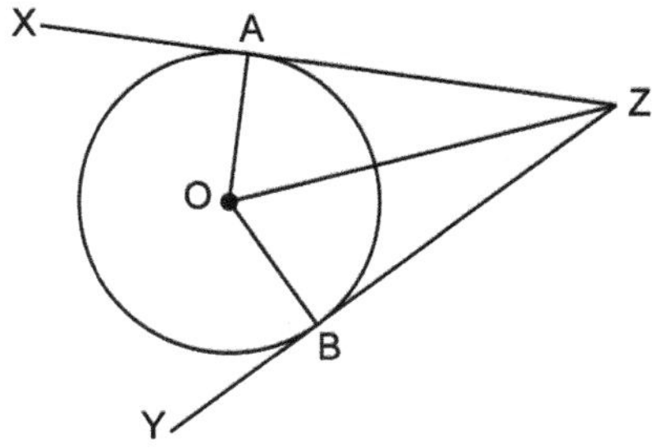

- Ask them to measure angle OAZ and angle OBZ and write down anything they notice. They should find that both angles are 90°.
- Explain that this fact is always true and leads to a circle theorem:

  'A tangent to a circle is perpendicular to the radius drawn to the point of contact.'

  This is usually abbreviated to: 'the tangent is perpendicular to the radius'.

- Ask students to draw this diagram. AB is a diameter of the circle.
- Ask students to measure angle ACB and write down anything they notice. They should find that the angle is 90°. Explain that this is always true and leads to another circle theorem:

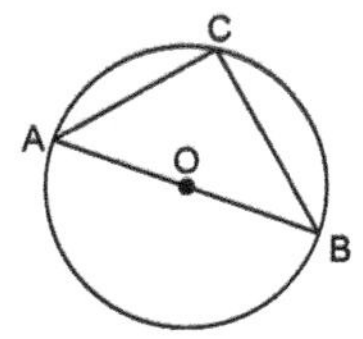

  'Every angle at the circumference of a semi-circle that is subtended by the diameter is a right angle (90°).'

  This is usually abbreviated to: 'angle in a semi-circle is 90°'.

- Go through Example 5.
- Now students can complete Exercise 23G. Tell them to note the hint with question 3.

Plenary

- Draw this diagram on the board. O is the centre of the circle.

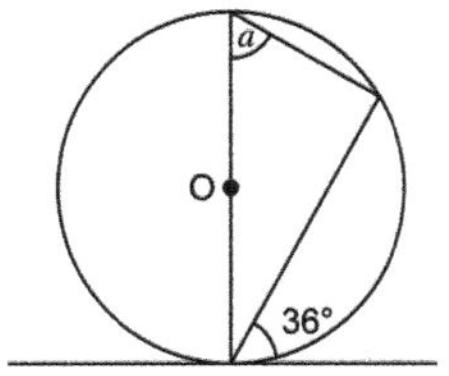

- Ask students to work out the value of a, giving reasons for their answers.

Homework and answers

Homework for this lesson is available in the homework folder. Consolidation and extension exercises are available in the further practice folder. Answers to the exercises can be found in the answers folder.

23.8 Angles in a circle

Syllabus references

Statement	Extended
E4.7	Calculate unknown angles and give explanations using the following geometrical properties of circles: angle at the centre of a circle is twice the angle at the circumference, angles in the same segment are equal.

Collins references

- Student's Book pages 426–430

Learning aims

- Work out the size of angles in circles.

Key words

- arc, circumference, segment, subtended

Prior knowledge

Students should know that the sum of the interior angles of a triangle is 180°. They should be familiar with circle terms such as chord, segment, circumference, diameter, radius and arc. Clarify the word 'subtends' with an example showing two points subtending angles at the centre and the circumference.

They should also be familiar with the angle properties of isosceles triangles.

Common mistakes and remediation

Students may find it difficult to solve problems about angles in circles if they have to find more than one angle on a diagram. Advise them to mark on their diagram all the angles they know before proceeding with the question.

Useful tips

As some of the work involves drawing circles, make sure all students have the necessary equipment: a sharp pencil, compasses and a protractor.

Starter

- Draw a suitably labelled circle and remind students of the main circle terms. For example, draw a tangent line to the circle and ask students to define precisely what a tangent line is.

Main lesson activity

- Ask students to draw this diagram, with O as the centre of the circle.

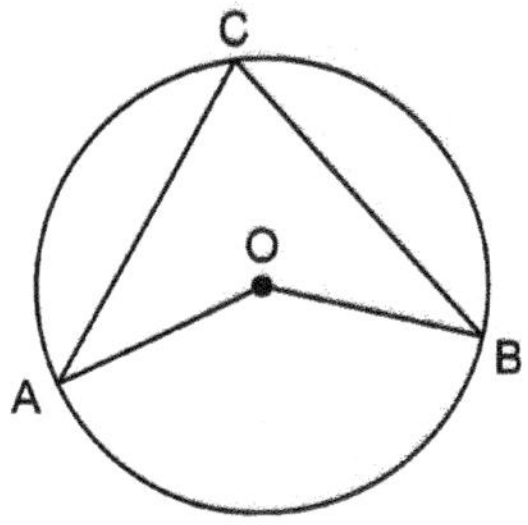

- Ask them to measure angle ACB and angle AOB and write down anything they notice about the two angles. They should find that the angle made at the centre is twice the angle made at the circumference.
- Explain that this is always true. The mathematical fact is another circle theorem: 'The angle at the centre of a circle is twice the angle at the circumference, subtended by the same arc.' This is usually abbreviated to: 'angle at the centre is twice the angle at the circumference'.
- Show students these two examples and ask them to write down the values of a and b.

1

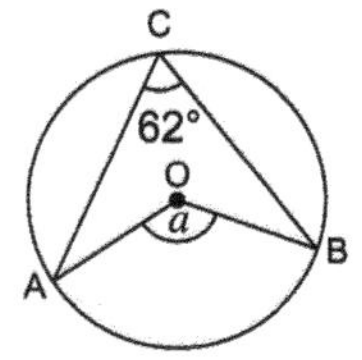

2

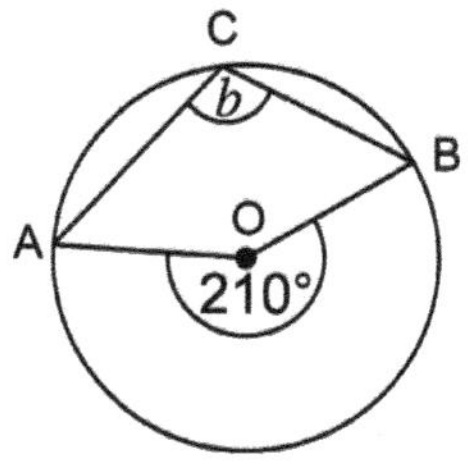

- Now ask students to draw this diagram.

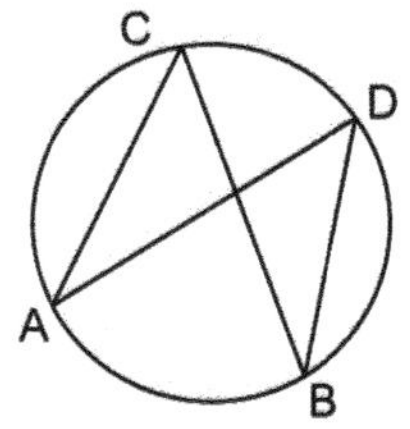

- Ask them to measure angle ACB and angle ADB and write down anything they notice. They should find that the angles are both the same. Explain that this is always true and leads to another circle theorem: 'Angles subtended at the circumference in the same segment of a circle are equal.'
- Explain that the segment is the part of the circle above the chord AB. This is usually abbreviated to: 'angles in the same segment are equal'.
- Go through Examples 6 and 7.
- Now students can do Exercise 23H. Go through the earlier questions with students who need more **support**, before they tackle questions 5–9.

Plenary

- Use the properties of an isosceles triangle to prove that the angles subtended at the circumference in the same segment of a circle are equal.

Homework and answers

Homework for this lesson is available in the homework folder. Consolidation and extension exercises are available in the further practice folder. Answers to the exercises can be found in the answers folder.

23.9 Cyclic quadrilaterals

Syllabus references

Statement	Extended
E4.7	Calculate unknown angles and give explanations using the following geometrical properties of circles: angle at the centre is twice the angle at the circumference, angles in the same segment are equal, opposite angles of a cyclic quadrilateral sum to 180° (supplementary).

Collins references

- Student's Book pages 430–432

Learning aims

- Find the size of angles in cyclic quadrilaterals.

Key words

- cyclic quadrilateral, opposite segment, supplementary

Prior knowledge

Students need to have covered and understood Lessons 23.7 and 23.8.

Common mistakes and remediation

Students find it difficult to solve problems about angles in circles if they have to find more than one angle on a diagram. Advise them to mark on their diagram all the angles they know before proceeding with the question.

Useful tips

As some of the work involves drawing circles, make sure all students have the necessary equipment: a sharp pencil, compasses and a protractor.

Starter

- Ask students to summarise the four circle theorems they have covered so far.

Main lesson activity

- Explain that students will now learn another circle theorem.
- Ask them to draw this diagram, with O as the centre of the circle.

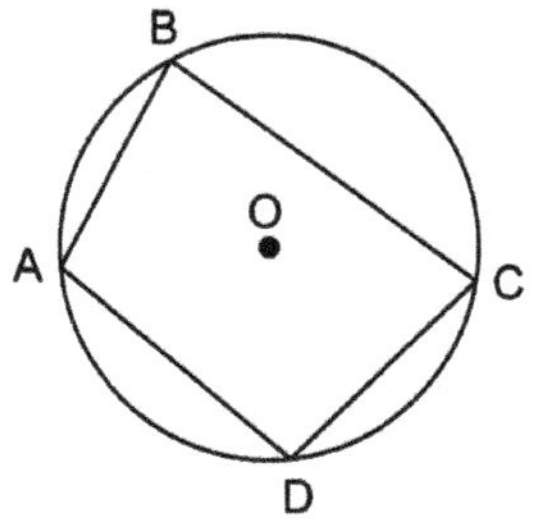

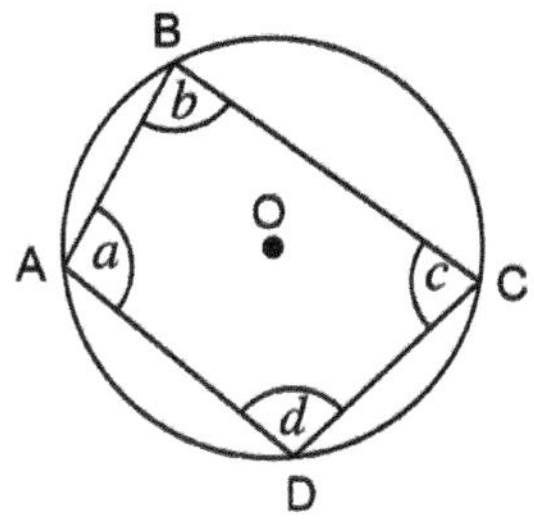

- Explain that the quadrilateral ABCD, with all its vertices on the circumference of a circle, is a *cyclic quadrilateral.*
- Ask students to measure the four interior angles of the quadrilateral and write down anything they notice. They should find that opposite angles add up to 180°.
- Explain that this is always true for cyclic quadrilaterals and leads to another circle theorem, which students can now copy into their notes: 'The sum of the opposite angles of a cyclic quadrilateral is 180°: $a + c = 180°$ and $b + d = 180°$.' This is usually abbreviated to: 'opposite angles in a cyclic quadrilateral add up to 180°'.
- Show students this example and ask them to work out the values of a and b in the diagram.

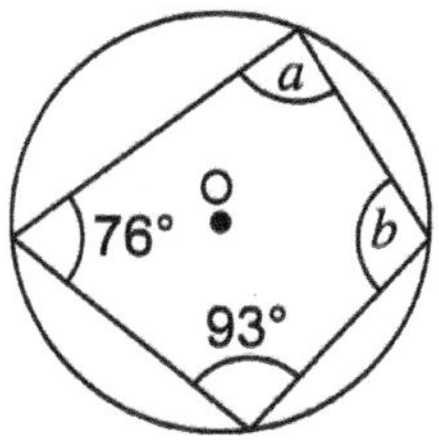

  $a + 93° = 180°$ so $a = 87°$ and

  $b + 76° = 180°$ so $b = 104°$

- Direct students who need more of a **challenge** to the proof of the circle theorem above that precedes Exercise 23I. Tell them that it involves applying a circle theorem that they have previously learned.
- Students can now do Exercise 23I. Go through the earlier questions with students who need more **support** before they tackle questions 7–9. Let them work in pairs if necessary.

Plenary

- Ask students if all quadrilaterals are cyclic. After discussion, explain that a quadrilateral is cyclic only if a pair of opposite angles adds up to 180°.

Homework and answers

Homework for this lesson is available in the homework folder. Consolidation and extension exercises are available in the further practice folder. Answers to the exercises can be found in the answers folder.

23.10 Alternate segment theorem

Syllabus references

Statement	Extended
E4.7	Calculate unknown angles and give explanations using alternate segment theorem.

Collins references

- Student's Book pages 432–435

Learning aims

- Find the size of angles in the alternate segment.

Key words

- tangent, chord, major segment, minor segment, major arc, minor arc, alternate segment

Prior knowledge

Students need to have covered and understood Lessons 23.7, 23.8 and 23.9.

Common mistakes and remediation

Students find it difficult to solve problems about angles in circles if they have to find more than one angle on a diagram. Advise them to mark on their diagram all the angles they know before proceeding with the question.

They sometimes find it hard to identify the alternate segment when the angle between the chord and tangent is obtuse, so it is worth drawing a variety of diagrams to cover this.

Useful tips

As some of the work involves drawing circles, make sure all students have the necessary equipment: a sharp pencil, compasses and a protractor.

Starter

- Ask students to summarise all the four circle theorems they have covered so far.

Main lesson activity

- Explain that students will now learn another circle theorem.
- Draw a circle on the board and add a tangent a chord from the point of contact. Ask students to name the parts of the circle. (Circumference, tangent, chord, major segment, minor segment, major arc, minor arc.)
- Now ask them to draw the following diagrams (you might want to have copies ready for them to use to save time).

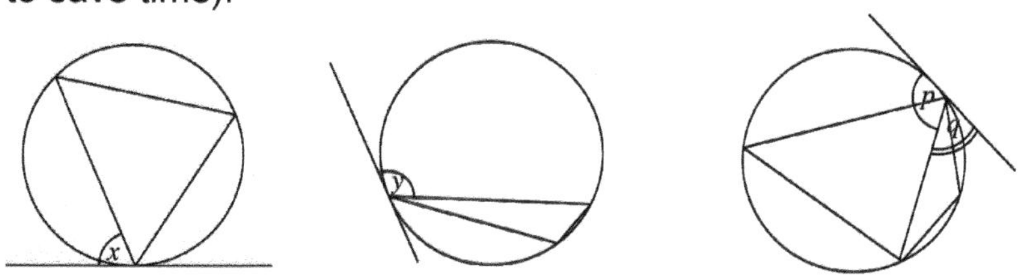

- Ask students to first identify the alternate segment (the one on the other side of the chord from the given angle), then to find the appropriate angle in that segment that corresponds to the lettered angle. They should write the letter in that angle.
- Check that in the third diagram the opposite angles are identified as p and q. Ask the students how this proves that the opposite angles in a cyclic quadrilateral add up to 180°. (p and q are shown as the angles on a straight line.)
- Work through Example 8 with your students.
- Students can now do Exercise 23J

Plenary

- Go through question 5 or 7 of Exercise 23J
- Recap on all the circle theorems met in this chapter.

Homework and answers

Homework for this lesson is available in the homework folder. Consolidation and extension exercises are available in the further practice folder. Answers to the exercises can be found in the answers folder.

24.1 Measuring and drawing angles

Syllabus references

Statement	Core
C4.1/E4.1 C4.2/E4.2	Use and interpret geometrical terms and vocabulary. Measure and draw lines and angles.

Collins references

- Student's Book pages 438–441

Learning aims

- Measure and draw an angle of any size.

Key words

- right angles, acute angles, obtuse angles, reflex angles, perpendicular, protractor

Prior knowledge

Students need to be able to use a semi-circular protractor to measure acute angles and obtuse angles.

Common mistakes and remediation

A common error is to read the angle on the wrong scale on a semi-circular protractor. Students who need more **support** are often unsure where to put the centre of the protractor. They may find it useful to turn the page so that one of the lines forming the angle is horizontal. Students may use the percentage (%) symbol instead of the degrees (°) symbol. Ensure students are aware of the difference.

Useful tips

Remind students to place the protractor so that the centre is over the vertex and the base line is over one of the 'arms' of the angle, and then to locate the zero on that arm and follow the scale round, to the other arm. Students who need more **support** may find it easier to use a circular protractor.

Starter

- Draw some angles on the board. Include angles up to 360°. Mark each angle with a circular arc. Ask students to say whether each angle is acute, obtuse or reflex and to estimate its size. Check by measuring.

Main lesson activity

- Draw an acute angle on the board. Show students how to use a semi-circular protractor to measure the size of the angle. Make sure they understand and that they use the degree (°) symbol correctly.

- Students who need more **support** could be shown how to use a circular protractor to measure the angle.
- Repeat for an obtuse angle. It may be useful to draw an obtuse angle in a different orientation, as on the right:

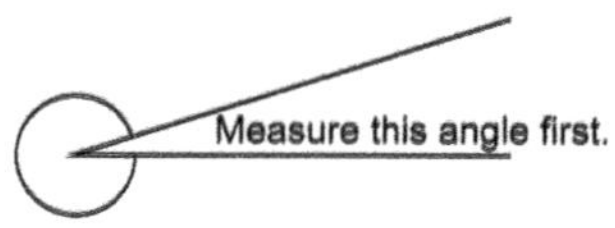

- Remind students that if they are unsure where to place the protractor when measuring angles, they can turn the page so that one of the lines appears horizontal.
- Repeat for a reflex angle. For students who need more **support,** using a semi-circular protractor, explain that it is easier to measure the acute or obtuse angle first and then subtract this from 360°.
- Show students the convention for naming angles. The acute angle at B, formed by the lines AB and BC, is written as angle ABC. This could also be written as angle CBA. Reflex angles are written in the same way. This is the reflex angle XYZ.

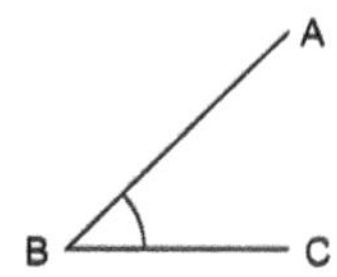

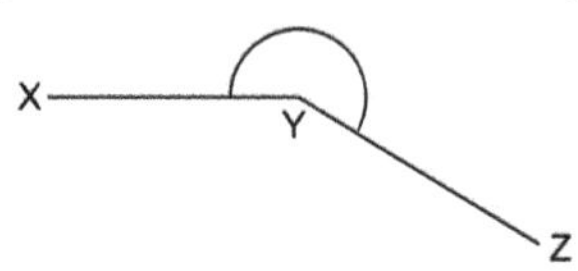

- Reiterate this work by going through Example 1.
- Students can now do Exercise 24A.

Plenary

- Ask each student to draw one acute, one obtuse and one reflex angle.
- Ask them to measure their angles, swap their diagrams with a neighbour, measure their neighbour's angles, and check each others' answers.

Homework and answers

Homework for this lesson is available in the homework folder. Consolidation and extension exercises are available in the further practice folder. Answers to the exercises can be found in the answers folder.

24.2 Bearings

Syllabus references

Statement	Core
C4.1/E4.1	Use and interpret geometrical terms and vocabulary.
C4.3/E4.3	Use and interpret three-figure bearings.

Collins references

- Student's Book pages 441–444

Learning aims

- Use a bearing to specify a direction.

Key words

- bearing, three-figure bearing

Prior knowledge

Students should know how to use a protractor to measure and draw angles.

Common mistakes and remediation

Students may not be clear which of the two points is the base one, from which the bearing of the other is taken. Remind students that bearings are always measured from north, so the first thing to do is add the north line through the point from which the bearing is to be taken.

Students who need more **support** could be encouraged to circle or underline the name of the point that the bearing is to be taken from.

Useful tips

Once students have established which is the base point from which to measure the bearing, and drawn in the north line through this point, they place the protractor with its centre over the point and its base line along the line pointing north. They identify the zero on this line, and read the angle clockwise.

Starter

- Draw a compass rose on the board. Starting from north as 0° and working clockwise, establish that the angles at E, S and W are 90°, 180° and 270°. Mark these on the diagram.
- Angles for NE, SE, SW and NW could also be given (45°, 135°, 225°, 315°).
- Now draw some lines from the centre of the rose and ask students who need more of a **challenge** to estimate the angles measured clockwise from north.

Main lesson activity

- Draw diagram **i** on the board. Inform the class that A and B represent two towns on a map and you need to know the direction of B from A. The best way is to give the direction as an angle. This angle is known as a bearing. Bearings are always measured in a clockwise direction from the north line, which is normally abbreviated to N.

i

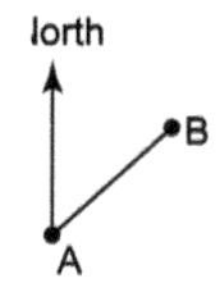

- Now redraw as diagram **ii**. Explain that the bearing of B from A is 70°. Students must be careful to get this the right way round and not confuse it with the bearing of A from B.

ii

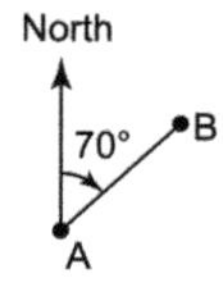

- Explain that a bearing is usually given with three digits and is referred to as a three-figure bearing. A two-digit bearing has a zero put in front of it, so due east is 090°. Now complete diagram **iii**.

iii

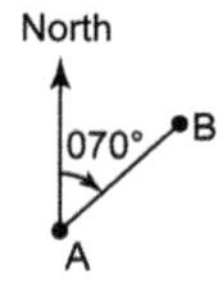

- Bearings may be obtuse or reflex angles, in which case students need to take especial care. Students who need more **support** could use a circular protractor. Emphasise this by going through the following examples. Diagram **iv**: the bearing of D from C is 125°. Diagram **v**: the bearing of E *from* F is 305°.

iv

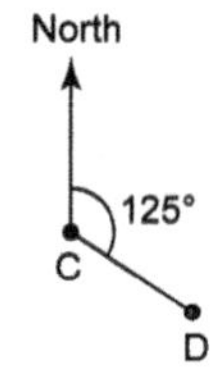

v

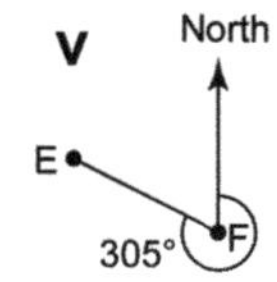

- Ask the class who might use bearings, e.g. orienteers, hikers, explorers, yachtsmen, airports, mapmakers.
- Explain that bearings are often used with a scale, e.g. an aircraft flies from X to Y on a bearing of 150° for a distance of 250 km. Draw diagram **vi** to show this journey. Scale: 1 cm represents 50 km (XY should be 5 cm long).

vi

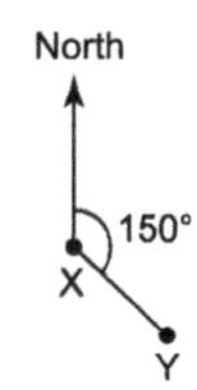

- Work through Example 2 with the students, referring them to the text above it.
- Students can now do Exercise 24B. Students who are having difficulty may be advised not to attempt all the questions.

Plenary

- Ask students, 'If B is on a bearing of 060° from A, what is the bearing of A from B?' (240°) Draw a diagram to demonstrate this.
- Repeat for several more bearings such as 045° (225°), 090° (270°), 160° (340°), 290° (110°).
- Tabulate the results so that students can see them clearly. Students who need more of a **challenge** may notice that there is a connection between the two bearings in each pair. There is always a difference of 180° between them.

Homework and answers

Homework for this lesson is available in the homework folder. Consolidation and extension exercises are available in the further practice folder. Answers to the exercises can be found in the answers folder.

24.3 Nets

Syllabus references

Statement	Core
C4.2/E4.2	Draw, use and interpret nets.

Collins references

- Student's Book pages 444–448

Learning aims

- Draw and recognise solids from their nets.

Key words

- net, cube, cuboid, prism, pyramid, vertex

Prior knowledge

Almost all cardboard packaging is made from a net, that is, made from one piece of cardboard. The students should know the names of solids (cube, cuboid, pyramid and prism) as well as shapes such as square, rectangle and triangle (equilateral, isosceles and right-angled).

Common mistakes and remediation

A common mistake is to draw the two sides that should join in the solid as different lengths. Students' drawings are often outside the maximum 2 mm error allowed.

Students who need more **support** often confuse 'accurate drawing' with 'sketch'. Check they understand the difference as they work through the exercise.

Useful tips

Students who need more **support** may have difficulty visualising a net of a solid. Having models available which they can hold will often help.

Starter

- Show the class a collection of cardboard boxes, for example cereal packets, triangular prism chocolate boxes or prepared ones. Ask the class how they would make the shapes from card.
- Most students should be familiar with the word 'net'. Remind the class that a net is a flat shape that can be folded into a solid.
- Open up one of the cuboid boxes to show the net and explain that it is made from six rectangles with three pairs the same size. Point out that when a net is drawn, it is not necessary to put on the tabs. The tabs are only drawn if the shape is to be made.
- Open up a different shape, such as the triangular prism, to show the class its net.

Main lesson activity

- It is useful to have a supply of centimetre-squared paper for this lesson. For students who need more **support** a collection of solids, including cubes, cuboids, square-based and pentagon-based pyramids, and triangular and hexagonal prisms, is also useful.

- Draw a 2 × 4 3 cuboid on the board. Show students how to draw an accurate net for the cuboid. Make sure they understand that there are three pairs of rectangles that make up the shape. Students who need more **support** could make sketches of them first.

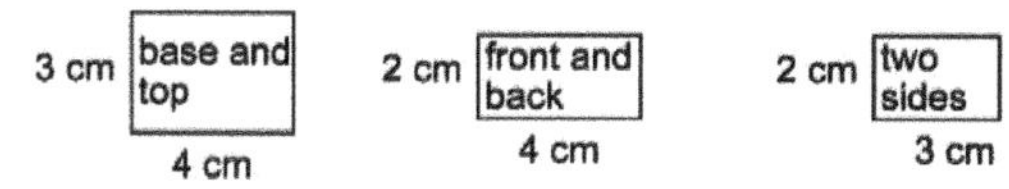

- Ask students to draw the net on centimetre-squared paper, using a pencil and ruler.
- Then students have drawn the net explain that the surface area of the net is the sum of the areas of each face. Ask them to find the surface area of the cuboid.

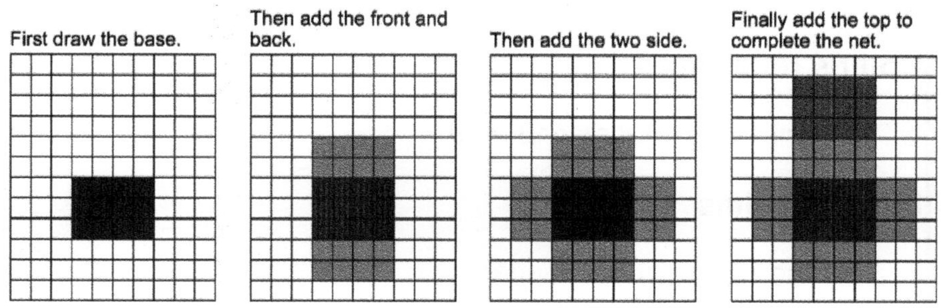

- They should get the answer of 52 cm^2 and they can find in several ways.
- The six faces come in pairs so the area is $2(3 \times 4) + 2(2 \times 4) + 2(2 \times 3) = 24 + 16 + 12 = 52$ cm^2
- Or they could say $2(12 + 8 + 6)$
- Or looking at the net there are three rectangles: $(2 \times 4) + (10 \times 4) + (2 \times 4) = 52$
- Ask students who need more of a **challenge** to describe different way to find the surface area.
- Students van also find the volume of the cuboid. It is the product of the three dimensions.
- In this case the volume is $3 \times 4 \times 2 = 24$ cm^3 and of course the numbers can be multiplied in any order. Make sure students use the correct units for area and volume.

- Students who need more of a **challenge** could be asked if there are other nets for the same cuboid.
- Explain to the students that sometimes only a sketch of a net is required.
- They will still need to use a pencil and ruler, but the measurements need not be accurate.
- Draw a sketch of a net for this square-based pyramid. The net is made from a square and four identical isosceles triangles. The net can be a square with a triangle on each edge.

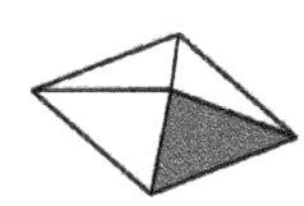

- Examples 3 and 4 show the ideas of this unit. They can be used to reinforce what you have done.
- Students can now do Exercise 24C. If necessary, remind students what an open box is when they do question 2.
- Students who need more **support** may find it helpful to work with a partner, and use squared paper, for question 7.

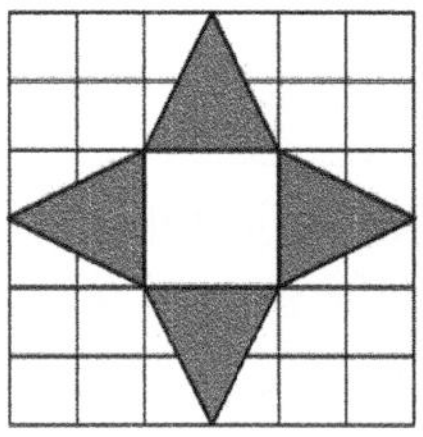

Plenary

- Show students a more complex solid, such as a hexagonal prism.
- Allow students to work in pairs or small groups and ask them to draw a sketch of the shape.
- You can also ask them to sketch a net of the solid.
- Ask the groups to compare their drawings. A display could be made of the students' work.
- Ask students who need more **challenge** why it is difficult to draw nets for cylinders and cones.

Homework and answers

Homework for this lesson is available in the homework folder. Consolidation and extension exercises are available in the further practice folder. Answers to the exercises can be found in the answers folder.

24.4 Congruent shapes

Syllabus references

Statement	Core
C4.1/E4.1	Use and interpret the geometrical term congruent.

Collins references

- Student's Book pages 448–449

Learning aims

- Recognise congruent shapes.

Key words

- congruent

Prior knowledge

No prior knowledge is needed.

Common mistakes and remediation

Students often do not recognise that identical shapes can be rotated and still be congruent. By using tracing paper as the method for finding congruent shapes, students should avoid this error.

Useful tips

Have a good supply of tracing paper available for this lesson.

Starter

- Ask students to draw, using squared paper, a shape with four squares that must touch, edge to edge. There are five different shapes possible.

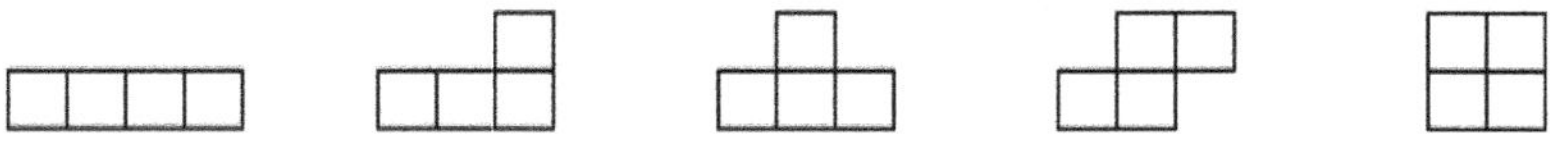

- Go round the class and collect examples. It is likely that students will have the same shape but in different orientations. Show examples of these.
- Make sure that students have an intuitive idea of congruency.

Main lesson activity

- Use the students' collected shapes to talk about those that are actually the same but are simply turned around or back to front. Introduce the word *congruent*.
- Ask the students to look at some of the building materials that may be present in the classroom – tiles, bricks or wooden blocks on the floor; discuss the fact that these are congruent and why these shapes have been used – so that they can all fit together properly.
- Draw two circles of different sizes on the board and ask if these two shapes are congruent. Some of the students who need more **support** may think that they are congruent since all circles are the same shape.
- Reinforce the idea that they are not congruent because they are not both exactly the same size. Congruent shapes are exactly the same size and shape.
- Introduce the word *similar* to students who need more of a **challenge**, explaining that shapes that are the same but different sizes, such as large and small circles, are called *similar* shapes.
- Encourage the students to use tracing paper. Emphasise that the best way to check whether two shapes are congruent is to trace one of them and see if the tracing fits exactly over the other one, putting it on top either way round. Students can work through Example 5 using tracing paper.
- Students can now do Exercise 24D.

Plenary

- Revisit the starter, this time using five squares. They can start by using their four-square shapes and deciding where they can add a square to each, remembering that there might be more than one possibility. After a time, collect the different shapes they have drawn and see which are congruent.
- There are just 12 different possible shapes to be made with five squares.

Homework and answers

Homework for this lesson is available in the homework folder. Consolidation and extension exercises are available in the further practice folder. Answers to the exercises can be found in the answers folder.

24.5 Similar shapes

Syllabus references

Statement	Core
C4.1/E4.1	Use and interpret the geometrical terms similar, scale factor.
C4.4/E4.4	Calculate lengths of similar shapes.

Collins references

- Student Book pages 449–452

Learning aims

- Show that two shapes are similar.
- Work out the scale factor between similar shapes.

Key words

- similar, enlargement, linear scale factor, corresponding angles, corresponding sides

Prior knowledge

Students should have covered work on scale factors and enlargements, simplifying ratios and solving equations.

Common mistakes and remediation

Students are often unsure of which side they need to find. Unless a missing length is already labelled with an algebraic unknown on the diagram, encourage students to identify the 'missing' length and label it, using a letter such as x, and mark all corresponding sides.

Useful tips

Calculating scale ratios involves solving equations of the form $\frac{x}{a} = \frac{b}{c}$

It would be helpful to show students that the solution is $x = \frac{b \times a}{c}$

Encourage students to describe similar shapes by naming the vertices in the corresponding order, in both.

Make sure that students who need more **support** redraw shapes so that they are in the same orientation, then draw arrows from each side in one shape to the corresponding side in the other.

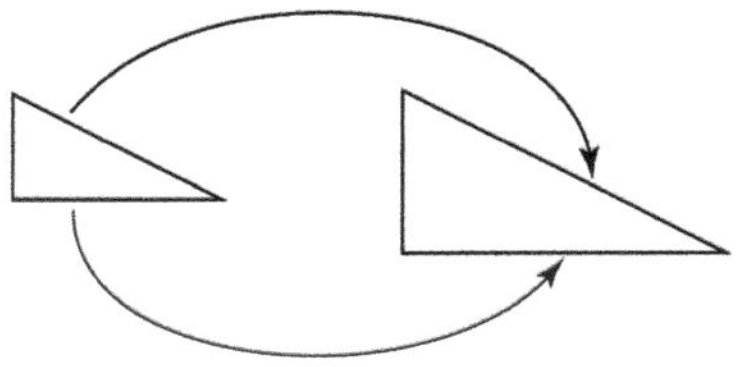

Starter

- Draw two triangles on the board: one that is 4 × 3 × 6 cm, and one that is 8 × 6 × 12 cm. Discuss the similarities and differences between the two triangles. Similarities: the angles are the same; corresponding angles are the same, for example, the smallest angle is opposite the smallest side. Differences: they are not congruent; the lengths of the sides in the second triangle are double those in the first; one is an enlargement of the other.

- Explain that the sides of the second triangle are an enlargement of the first by a scale factor of 2 and hence all the corresponding sides are in the same ratio of 1 : 2 (3 : 6 = 4 : 8 = 6 : 12). The two triangles are mathematically similar.
- Two shapes are similar if all the corresponding sides are in the same ratio, or one is an enlargement of the other. Stress that all corresponding angles in similar shapes are equal.
- Introduce the term linear scale factor to describe the ratio between the lengths of corresponding sides.

Main lesson activity

- Show the class how to find the missing length of a side in a pair of similar triangles. Draw two similar triangles on the board. Triangle ABC is similar to triangle XYZ. Find the length of the side marked x.

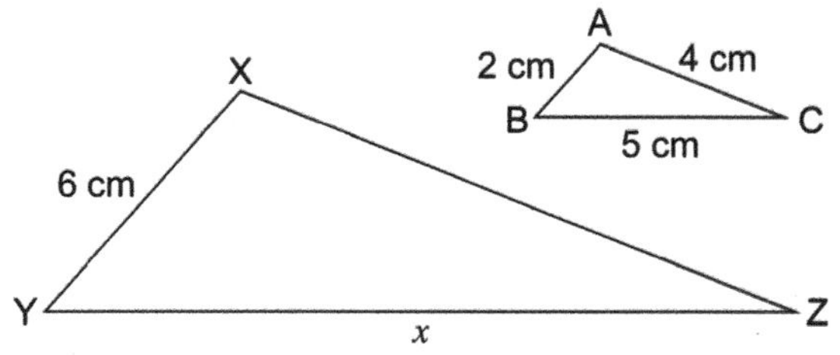

- Since the triangles are similar, triangle XYZ must be an enlargement of triangle ABC. Ask students to give the scale factor of the enlargement and say how they worked it out.
- Students who need more **support** can use this method to find x: the scale factor is 3, so $x = 5 \times 3 =$ 15 cm.
- Show students who need more of a **challenge** how to find x by using ratios or equations. The corresponding sides are in the same ratio, so $6 : 2 = x : 5$.
- Show students how this can be set up as an equation with x as a numerator:

  $\frac{x}{5} = \frac{6}{2}$ so $x = \frac{6 \times 5}{2} = 15$ cm
- Now extend this to other similar shapes and demonstrate that the same principle applies to these, i.e. corresponding angles are the same and all corresponding sides are in the same ratio.
- Work through Example 6 with students.
- Now students can continue with Exercise 24E.

Plenary

- Draw pairs of similar triangles on the board, but in different orientations. Ask students to provide the information to show similarity.
- Draw two rectangles on the board (3 × 5 cm and 6 × 10 cm). Ask students why they are similar. Ask for the linear scale factor of the enlargement. Then ask students to compare the areas and discuss what they notice.

Homework and answers

Homework for this lesson is available in the homework folder. Consolidation and extension exercises are available in the further practice folder. Answers to the exercises can be found in the answers folder.

24.6 Areas of similar shapes

Syllabus references

Statement	Extended
E4.4	Use the relationships between lengths and areas of similar shapes. Solve problems and give simple explanations involving similarity.

Collins references

- Student's Book pages 452–455

Learning aims

- Show that two shapes are similar.
- Work out the scale factor between similar shapes.

Key words

- area scale factor

Prior knowledge

Students should have covered work on scale factors, linear scale factors and enlargements, simplifying ratios and solving equations.

Common mistakes and remediation

The most common error students make is not using corresponding sides when forming the ratio.
We are using the ratio of corresponding sides in different shapes, not two sides in the same shape.

Useful tips

Students need to understand that a ratio of $a:b$ is equivalent to a multiplier of $\frac{b}{a}$

It would also be helpful to remind students how to calculate the area of a triangle:

$$\text{area} = \frac{1}{2} \times \text{base} \times \text{height}$$

$$A = \frac{1}{2}bh$$

Starter

- Draw two similar right-angled triangles on the board: one with a base of 5 cm and height of 3 cm, the other with a base of 10 cm and height of 6 cm.
- Establish that these are similar triangles and that the second triangle is an enlargement of the first by a linear scale factor of 2 because corresponding sides are in the ratio 1 : 2.
- Students calculate the area of each triangle and find that the smaller has an area of 7.5 cm^2 and the larger one of 30 cm^2.
- Ask what this tells them about the scale factor of their areas. (It is 4) The area of the larger shape is 4 times the area of the smaller shape.
- Introduce the term 'area scale factor'.

Main lesson activity

- Draw a pair of similar rectangles on the board. Make the first one 2 cm by 3 cm and the second one 6 cm by 9 cm.
- Ask for the ratio of corresponding sides (1 : 3) and the linear scale factor (3)
- Ask students that the areas of the rectangles (6 cm^2 and 54 cm^2)
- Ask for the area scale factor is. It is 9 because $54 \div 6 = 9$
- Use this example and the starter example to show that the area scale factor is the linear scale factor squared. Say that it is always true that if the sides of a shape are multiplied by k then the area is multiplied by k^2
- Give students further examples:
  - If the ratio of the sides of two similar shapes is 2 : 3 what is the linear scale factor? It is $1\frac{1}{2}$
  - What is the area scale factor? It is $\left(1\frac{1}{2}\right)^2 = 2\frac{1}{4}$
  - If the scale factor of corresponding sides is 2 : 5 what is the linear scale factor? It is $2\frac{1}{2}$ or 2.5
  - What is the area scale factor? It is $\left(2\frac{1}{2}\right)^2 = 6\frac{1}{4}$ or 6.25
  -
- From this lead students towards understanding that for two similar shapes, if the ratio of corresponding sides is $a : b$, then the linear scale factor is $\frac{b}{a}$ and the area scale factor is $\left(\frac{b}{a}\right)^2$
- Students can now do Exercise 24F.

Plenary

- Draw a simple shape. Say that the area is 20 cm^2.
- Ask for the area of a similar shape which has a linear scale factor of 2 or 3 or … (80 cm^2 and …)
- What is the area if the linear scale factor is 0.5? The answer is 5 cm^2
- Now say that you want to draw a similar shape with an area of 40 cm^2. What is the linear scale factor?
- Students may try to do this by finding a number whose square is 2. So, for example, 1.4 is close because $1.4^2 = 1.96$
- The exact answer is $\sqrt{2} = 1.41$ to 3 d.p.

Homework and answers

Homework for this lesson is available in the homework folder. Consolidation and extension exercises are available in the further practice folder. Answers to the exercises can be found in the answers folder.

24.7 Areas and volumes of similar solids

Syllabus references

Statement	Extended
E4.4	Use the relationships between surface areas and volumes of similar solids. Solve problems and give simple explanations involving similarity.

Collins references

- Student's Book pages 455–459

Learning aims

- Solve problems involving the surface areas and volumes of similar solids.

Key words

- volume scale factor

Prior knowledge

Students should be familiar with finding squares, cubes, square roots and cube roots of numbers on their calculators. They should also be confident in solving linear equations.

Common mistakes and remediation

Remind students that a ratio between corresponding sides similar shapes of $a:b$ is equivalent to a linear scale factor of $b \div a$ or $\frac{b}{a}$ and an area scale factor of $\left(\frac{b}{a}\right)^2$

Useful tips

Encourage students to label all diagrams carefully, showing which are the corresponding sides. It might be helpful to display the following information: for two similar solids, if the length ratio is a : b, then the area ratio is a^2 : b^2 and the volume ratio is a^3 : b^3.

Starter

- Ask students to draw a line 3 cm long. Now ask them to draw another line twice as long. Ask: What is the length of this line?
- Ask them to draw a square 3 cm by 3 cm and tell you what its area is, then draw another square with sides twice as long. Ask what the area of this square is. Why is the area not twice as big? Elicit that the lengths are doubled in both directions.
- Ask students to consider what they would expect the volume to be of cubes of sides 3 cm and 6 cm.

Main lesson activity

- Say that you are going to look at volumes of similar 3D solids. Draw two similar cuboids on the board.
- Stress that the cuboids are similar because all corresponding sides are in the same ratio. Here the length ratio is 1 : 2 and the linear scale factor is 2.

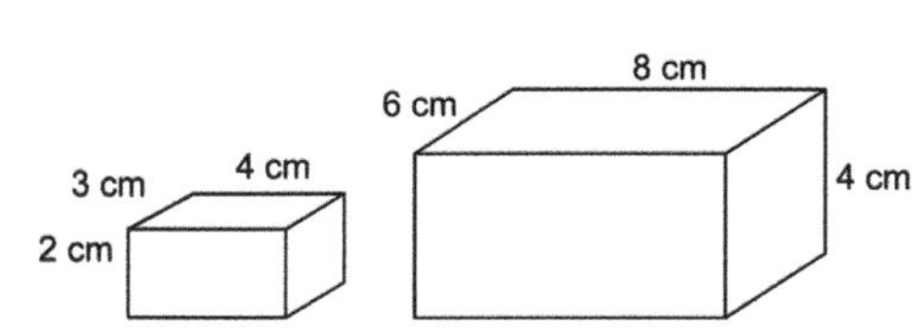

- The volumes of the two cuboids are 24 cm^3 and 192 cm^3. So the *volume ratio* is 24 : 192 = 1 : 8 and the volume scale factor is 8.
- Explain that for any similar solid, if the length ratio is $a : b$ then the volume ratio is $a^3 : b^3$
- If the length scale factor is k then the volume scale factor is k^3. In the example above the linear scale factor is 2 and the volume scale factor is $2^3 = 8$
- Use the example of two similar cans. The label on one can is 5 cm high and on the other can it is 8 cm high.
- The area of the label on the smaller can is 12 cm^2. Work out the area of the label on the second can. A sketch will make this clearer.
- The length ratio is 5 : 8, so the linear scale factor is $8 \div 5 = 1.6$
- Ask for the area scale factor. It is $1.6^2 = 2.56$
- The area of the label on the larger can is 8 × 2.56 = 20.5 cm^2 to 1 decimal place.
- Now say that the volume of the smaller can is 60 cm^3 What is the volume of the larger can?
- The length scale factor is 1.6 and so the volume scale factor is $1.6^3 = 4.096$
- The volume of the larger tin is $60 \times 1.6^3 = 246$ cm^3 o the nearest whole number.
- Point out that with a calculator you do not need to actually find the value of 1.6^3 ; the calculation can be done in a single sequence of steps.
- Examples 7 and 8 give more examples if you need them.
- Students can do Exercise 24G.
- When students have made progress on Exercise 24G look at Examples 9 and 10 which give more complicated examples.
- The important points here are the square root of the area factor is the linear scale factor, and the cube root of the volume scale factor is the linear scale factor.
- When students are confident about moving between a linear, area and volume scale factor they can start on Exercise 24H.

Plenary

- Ask students to summarise what they have learned in the lesson by completing this table (completed version shown next to it).

Length scale factor	Area scale factor	Volume scale factor
2		
	2.25	
	0.64	
		64

Length scale factor	Area scale factor	Volume scale factor
2	4	8
1.5	2.25	3.375
0.8	0.64	0.512
5	25	125

Homework and answers

Homework for this lesson is available in the homework folder. Consolidation and extension exercises are available in the further practice folder. Answers to the exercises can be found in the answers folder.

25.1 Constructing shapes

Syllabus references

Statement	Core
C4.2/E4.2	Measure and draw lines and angles. Construct a triangle given the three sides using ruler and pair of compasses only.

Collins references

- Student's Book pages 462–465

Learning aims

- Construct triangles using compasses, a protractor and a straight edge.

Key words

- construct, ruler, protractor, compasses, set square

Prior knowledge

Students should be able to draw and measure straight lines, and use a protractor to draw and measure angles accurately. They should also be able to open a pair of compasses and set them to a given radius and draw arcs from a fixed point.

Common mistakes and remediation

Most errors in this type of construction are due to poor accuracy in drawing triangles.
Remind students of this.

Useful tips

Tell the class that using poor equipment and not taking care with their drawings can make them inaccurate. Make clear to students that a ruler and compasses produces the most accurate drawing and they should learn this technique, rather than relying on using a protractor.

Students need to make sure their compasses are in good condition, with the joint tight but smooth. They should keep their pencils sharp, and rulers and protractors clean; not scratched or worn. Tell students that it is better to choose a ruler with a small border before the zero, since trying to align the point of the compasses to the very edge of a slippery piece of plastic can cause problems.

Starter

- Students will almost certainly have done this activity before, but it is a useful exercise in using compasses.
- Ask students to draw a circle of any radius, to fit the page.
- Now, keeping the compasses set to the same radius, students place the point anywhere on the circumference and draw an arc to cut the circle in two places. They repeat, but now using as centres the points where the arcs cut the circle. They do this until they have drawn six arcs. If they have drawn the arcs accurately, they will produce a flower pattern with sharp pointed petals.

Main lesson activity

- Have plenty of plain paper available, as students will use a lot in these lessons. Make sure they all have a pair of compasses fitted with a sharp pencil, a ruler, a protractor and another sharp pencil for drawing.
- Go through Example 1, the construction of a triangle with three sides known. Then ask students to construct a few more triangles, giving them the three sides each time. Emphasise to the students who need more **support** that they should always start by drawing the longest side as the base. Ask students to measure the largest angle (opposite the longest side), for them to compare their accuracy and to practise using a protractor.
- Now students can complete Exercise 25A. Students who need more **support** may work in pairs to complete these questions.

Plenary

- Ask students to draw a line 5 cm long. To do this, they should draw a straight line, then open their compasses so that the point is on zero on a ruler, and the pencil point is on 5 cm. They should mark a point near the end of the line they have drawn, and mark the 5 cm-length with an arc.
- Now ask students to draw arcs above the line, centred on each end of the 5 cm line they have marked. Then they should join each of the points on the line to the point where the arcs cross. Ask what type of triangle they have drawn. Let them measure all the angles to check (an equilateral triangle, 60°).

Homework and answers

Homework for this lesson is available in the homework folder. Consolidation and extension exercises are available in the further practice folder. Answers to the exercises can be found in the answers folder.

25.2 Scale drawings

Syllabus references

Statement	Core
C4.3/E4.3	Draw and interpret scale drawings.

Collins references

- Student's Book pages 466–469

Learning aims

- Read scales and draw scale drawings.

Key words

- scale drawing

Prior knowledge

Students should be able to measure and draw lines.

They should know that 100 cm = 1 m and 1000 m = 1 km.

They should be familiar with multiplying and dividing by powers of 10.

Common mistakes and remediation

Students may not be able to work out how many kilometres there are in, for example, 3 000 000 cm or how many centimetres there are in, for example, 0.5 km.

For students who need more **support**, initially only use questions with given units, for example,1 cm : 20 km. Remember that some students who need more **support** will require a great deal of practice with the 1 : 20 000 type question to be able to change it into 200 m or 0.2 km. You could use the 'drip feed' method and give one question (for example, change 1 000 000 cm to metres and kilometres) each lesson as an extra starter.

Useful tips

A selection of maps could be available to show students the different ways scales are shown.

Starter

- Check that students can convert between units readily and easily.
- Ask the class how they multiply whole numbers by 10, 100 and 1000. (Place 1 zero, 2 zeros and 3 zeros at the end.)
- Write on the board: 100 cm = 1 m and 1000 m = 1 km.
- Ask individual students how they would convert various units, for example, change 5 metres into centimetres, change 7 kilometres into metres, change 4 kilometres into centimetres (500, 7000, 400 000).
- Students who need more **support** may need some more practice at doing this.
- Students who need more of a **challenge** could be given examples that involve decimals, for example, change 3.8 metres into centimetres; change 6.5 kilometres into metres.

- If time permits, do a few examples involving division, for example, change 900 centimetres into metres; change 12 000 metres into kilometres.

Main lesson activity

- Draw two lines on the board. Inform students that the second line is three times longer than the first. Ask students to give different lengths for each line. Give an example, such as 5 cm and 15 cm, and remind students that the ratio of the lengths can be written as 1 : 3. Explain that this idea can be used with scale drawings.
- Ask students for examples of where scale drawings are used (architects, designers, DT, maps, sports). Explain that a scale drawing is a larger or smaller version of an original shape or object.
- Draw this diagram on the board. It represents the dimensions of the floor of a room. Explain that, to make a scale drawing of the room, first you need to choose a scale. A suitable scale would be one centimetre represents one metre. Lead students to make this scale drawing in their books. Point out how the scale is written by the side of the scale drawing.

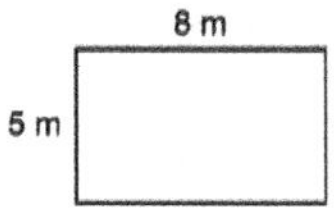

- Draw another rectangle on the board to represent a sports pitch. Ask: 'What scale could be used?' Possible answers include: 1 cm represents 10 m, 1 cm represents 20 m or 1 cm represents 40 m. Ask which one is most suitable. For ease of drawing in their books, students should suggest using 1 cm represents 20 m.

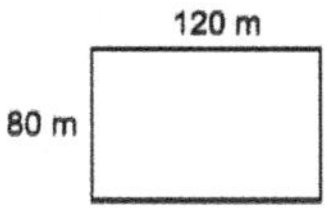

- Work through Example 2 with students, to make sure that they are confident with this topic.
- Students can now complete Exercise 25B.

Plenary

Write the following figures on the board and ask:

- 'How many centimetres are there in: 1 km, 2 km, 10 km, 0.5 km, 1.5 km?'
- 'How many metres are there in: 2000 cm, 10 000 cm, 15 000 cm, 4 000 000 cm?'

Ask:

- 'If a map has a scale of 1 cm to 100 km, how long is a road which is drawn as 4 cm long on the map?' (400 km)
- 'If a map has a scale of 1 cm to 100 m, how long is a fence which is 2.5 cm long on the map?' (250 m)
- 'The boundary of a state is 2430 km. How long will this be shown on a map with a scale of 2 cm to 500 km?' (9.72 cm)

Homework and answers

Homework for this lesson is available in the homework folder. Consolidation and extension exercises are available in the further practice folder. Answers to the exercises can be found in the answers folder.

26.1 Pythagoras' theorem

Syllabus references

Statement	Core
C6.1/E6.1 C6.2/E6.2	Know and use Pythagoras' theorem. Solve problems in two dimensions using Pythagoras' theorem and trigonometry.

Collins references

- Student's Book pages 472–475

Learning aims

- Calculate the length of the hypotenuse in a right-angled triangle.

Key words

- hypotenuse, Pythagoras' theorem

Prior knowledge

Students should know how to find the square and square root of a number, and it will be helpful if they are familiar with square numbers. Students should also know how to round numbers correctly, and they should have an appreciation of what is a suitable level of accuracy given the context of the problem involved.

Common mistakes and remediation

Perhaps the most common mistake students make is to forget to take the square root to find the final answer. They can generally avoid doing this if they pay attention to the layout of solutions. Students also try to apply the theorem in unsuitable situations, such as when the triangle in the question is not known to be a right-angled triangle. Emphasise that Pythagoras' theorem only applies to right-angled triangles.

Useful tips

Pythagoras' theorem may be familiar to students. If this is the case, this lesson will be revision.

Starter

- Display a target board of whole numbers and ask students to find the square numbers.
- Ask for any known square numbers that are missing from the board. Try to elicit the first 12 square numbers, with some other commonly known numbers such as 400, 625, 1 000 000.
- Ask for a definition of square numbers. Ask for squares of some simple decimal numbers such as 0.5 and 0.1.

Main lesson activity

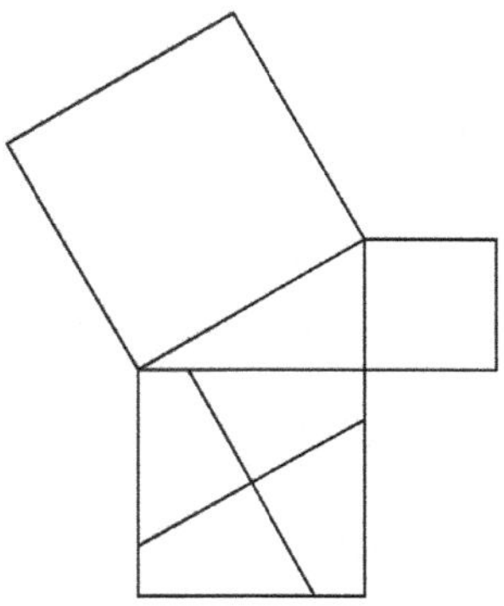

- Ask students if they have heard of Pythagoras' theorem, which concerns a triangle. They might remember it in the form $a^2 + b^2 = c^2$ or they might remember it as a fact about squares.
- Try to encourage a clear and precise description.
- Now ask how students know it is true. They are unlikely to recall a proof.
- Give students a copy of the dissection on the right. Make sure they understand that it shows squares on the sides of a right-angled triangle.
- To benefit students who need more **support**, the class can work in pairs to cut out the five smaller pieces and fit them into the largest square. This demonstrates Pythagoras' theorem. By labelling the sides of a right-angled triangle as a, b and c, deduce that $a^2 + b^2 = c^2$.
- Draw the 5.4 cm × 7.3 cm triangle on the board.

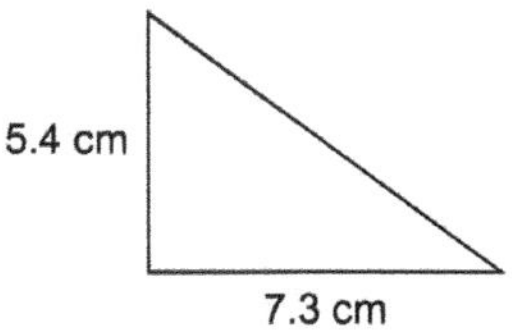

- Discuss how to find the missing side. Ask students to do the required calculation.
- Discuss the appropriate way to round the calculator answer. In this case, 1 decimal place is appropriate, because the lengths are given to that degree of accuracy. It gives 9.1 cm.
- If necessary, work through Example 1 with students.
- Now students can complete Exercise 26A. It may be useful to have a plenary during the lesson to look at question 10, as it is important for all students to think about it.
- Now show students how to find the length of one of the shorter sides.
- Draw a right-angled triangle with hypotenuse 13.6 cm and shortest side 8.4 cm and ask how the third side can be found.
- Make sure students understand that this can be done by subtraction. To 1 decimal place, the answer is 10.7 cm.
- Look at Example 2 with students.
- Now students can complete Exercise 26B. Students who need more **support** can work on the questions with a partner, possibly a student who needs more of a **challenge**. Provide support as needed.

Plenary

- Draw a right-angled triangle with a semi-circle on each side. Ask if Pythagoras' theorem is true for semi-circles as well as squares (yes). If students want to check a numerical example, give the sides as 6 cm, 8 cm and 10 cm.
- Encourage discussion, asking students to justify their answers. **Challenge** students with the question: 'What about other shapes?' (In fact, it is true for any similar shapes.)

Homework and answers

Homework for this lesson is available in the homework folder. Consolidation and extension exercises are available in the further practice folder. Answers to the exercises can be found in the answers folder.

26.2 Trigonometric ratios

Syllabus references

Statement	Core
C6.2/E6.2	Know and use the sine and cosine ratios for acute angles in calculations involving sides and angles of a right-angled triangle.

Collins references

- Student's Book pages 476–477

Learning aims

- Use the three trigonometric ratios.

Key words

- ratio, sine, cosine, tangent, opposite side, adjacent side

Prior knowledge

Students should be able to round off numbers to a given number of significant figures.

Common mistakes and remediation

Students need to use calculators efficiently and correctly. A common source of errors is if students do not have their calculators set to degrees. Ensure that each student is able to check and correct this on his or her calculator.

Useful tips

This lesson and the following lessons introduce the three trigonometric ratios and how to use them to make simple calculations. Students should be able to complete the exercises fairly quickly.

Tell students that 'soh-cah-toa' should help them to remember the three ratios: sine, cosine and tangent. (sine opposite/hypotenuse-cosine adjacent/hypotenuse-tangent opposite/adjacent).

You could suggest that students remember 'soh-cah-toa' as a mnemonic.

Starter

- Show students the diagram below on the right.
- Ask them about the angles at P. Are they all the same size? What can they say about their sizes? Ensure that you make the following points:
  - They are not all the same size.
  - The first one is 45°.
  - They get smaller as you go around the spiral.
  - If the spiral were to continue the angles would become smaller and smaller.

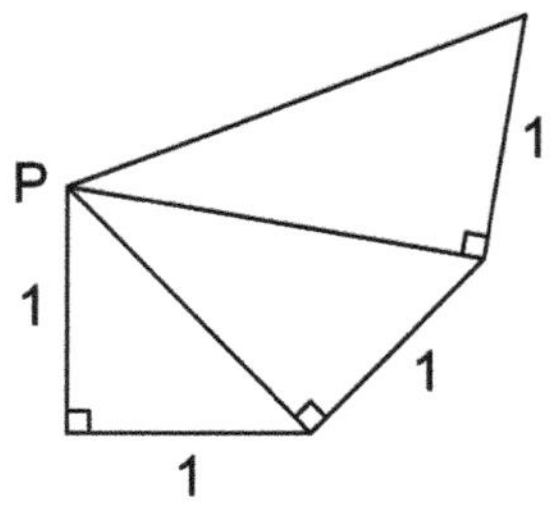

Main lesson activity

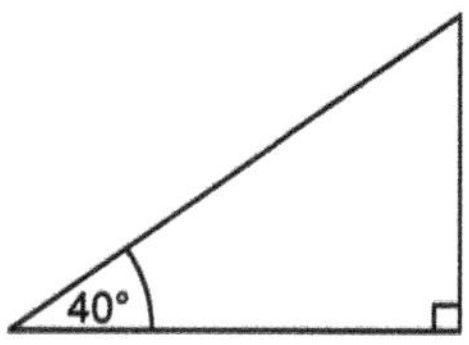

- Ask students to draw a triangle like the one on the right. It should be reasonably large but you should have a range of sizes for what follows.
- Ask them to label the 'Hypotenuse'. Point out that the other two sides are either opposite the 40° angle or adjacent to it. Students should label these 'Opposite' and 'Adjacent'.
- Then ask students to measure the three sides and write the lengths on their triangles.
- Collect the measurements from each student and add them to a table, ideally in a spreadsheet. Use these headings.

Opposite	Adjacent	Hypotenuse	O/H	A/H	O/A

- Use the spreadsheet to work out the ratios in the last three columns. Round them to 2 dp in the display. The sides should vary but the ratios will be similar. If any ratios are very different, check the accuracy of the measurements. Small diagrams can also lead to bigger variation.
- Use the spreadsheet to average the numbers in each column. These should be 0.64, 0.77 and 0.84. Discuss why these might be better values to use as the 'correct' value for these ratios.
- Now explain that these ratios are called the sine, the cosine and the tangent of 40°, that their abbreviations are sin, cos and tan, and are written sin 40°, and so on.
- Make sure students can use a calculator correctly and check against the values in the table. Ask them to find the ratios for other angles.
- Finally, explain that they will need to be able to work out calculations like 7.2 sin 35° or $\frac{7.2}{\sin 35°}$.
  Check to see that students who need more **support** can also do this efficiently.
- Work through Example 3 in the Student's Book.
- Now students can complete Exercise 26C, which will give them practice in these skills.

Plenary

- Return to the 40° triangle students drew earlier. Ask how this could be used to find the sine, cosine and tangent of 50° (third angle in triangle). Then ask the students to do the relevant calculations and check their accuracy by using the calculator.

Homework and answers

Homework for this lesson is available in the homework folder. Consolidation and extension exercises are available in the further practice folder. Answers to the exercises can be found in the answers folder.

26.3 Calculating angles

Syllabus references

Statement	Core
C6.2/E6.2	Know and use the sine, cosine and tangent ratios for acute angles in calculations involving sides and angles of a right-angled triangle.

Collins references

- Student's Book page 478

Learning aims

- Use the trigonometric ratios to calculate an angle.

Key words

- inverse

Prior knowledge

Students should be able to calculate values of trigonometric ratios to a given degree of accuracy.

Common mistakes and remediation

The notation $\sin^{-1} 0.6$ will be unfamiliar to students. It must not be written the wrong way around as $0.6 \sin^{-1}$. If necessary, give guidance on how students should use their calculators correctly.

Useful tips

Make sure students know that inverse means opposite in effect, or the reverse of, or vice versa.

Starter

- Draw a 3, 4, 5 triangle on the board. Mark the smallest angle x.
- Ask students to write down the sine, cosine and tangent of x.
- Mark the other angle y and repeat the exercise.
- Reinforce the three ratios, which students need to remember.

Main lesson activity

- Use the results from the starter and point out to students that they know that sin x = 0.6, so x is the angle whose sine is 0.6.
- Introduce the notation $x = \sin^{-1} 0.6$ and then show them how to find this using a calculator. Make sure they can use their particular model (37° or 36.9°).
- Ask students to try and get the same answer using $\cos^{-1}$ and $\tan^{-1}$.
- They should be able to work with angle y in the same way.
- Work through Example 4 to help students who need more **support**.
- Now students can complete Exercise 26D.

Plenary

- Ask students what they would expect to find if they calculated $\sin \sin^{-1} 0.6$. (0.6)
- What about $\cos^{-1} \cos 0.6$? (0.6)
- Can they suggest other similar results?
- Why does this happen?
- Now ask students to find $\sin \cos^{-1} 0.6$ (0.8) and try to explain the result.

Homework and answers

Homework for this lesson is available in the homework folder. Consolidation and extension exercises are available in the further practice folder. Answers to the exercises can be found in the answers folder.

26.4 Using sine, cosine and tangent functions

Syllabus references

Statement	Core
C6.2/E6.2	Know and use the sine, cosine and tangent ratios for acute angles in calculations involving sides and angles of a right-angled triangle.

Collins references

- Student's Book pages 479–483

Learning aims

- Find lengths of sides and angles in right-angled triangles using the sine and cosine functions.

Prior knowledge

Students should be familiar with the sine, cosine and tangent functions and be able to evaluate simple expressions involving them using their calculators.

Common mistakes and remediation

Students have the greatest difficulty when the unknown side is in the denominator. Encourage students to check whether their answer is reasonable in the context of the question. For example, if the hypotenuse is shorter than one of the other sides then an error has been made.

Useful tips

Write the following on the board, which may help students who need more **support**.

- *Opposite* is opposite to, or on the other side of, the angle θ.
- *Adjacent* is adjacent, or next to, the angle θ.
- *Hypotenuse* is the long side.

Starter

- Display these equations and ask students, working in pairs, to solve them.

  $\frac{x}{4} = 12$ $\quad \frac{x}{12} = 4$ $\quad \frac{12}{x} = 4$ $\quad \frac{4}{x} = 12$ $\quad (48, 48, 3, \frac{1}{3})$

- When they have finished, ask for answers, with justifications. If different answers are given, encourage students to come to an agreement.
- The last equation is the hardest. Ask for methods to use to solve it.
- One way is to rewrite as $4 = 12x$, so $x = \frac{4}{12} = \frac{1}{3}$. However, students may come up with other ideas.
- Check that students understand clearly by asking them to write another similar set of four questions using different numbers and to solve them. Ask one or two pairs with good examples to share them with the class.
- Show a triangle with angles 45°, 45° and 90°. Ask students to explain why the tangent of 45° must be 1. What can they say about the tangent of an angle bigger that 45°? (It will be larger than 1.)
- What is the largest value a tangent can have? (There is no limit.)

Main lesson activity

- Display the three diagrams on the right side by side.
- Label 'opposite' and 'hypotenuse' in each one.
- Ask students to write the sine ratio for each one.
- Under the diagrams, write:

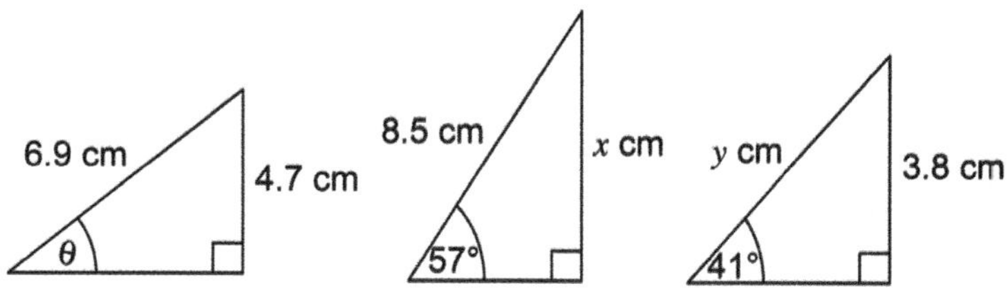

$$\sin \theta = \frac{4.7}{6.9};\ \sin 57° = \frac{x}{8.5};\ \sin 41° = \frac{3.8}{y}$$

- Ask students to solve each equation. Check that they all agree on the results.
- Refer back to the starter if students are unsure about the last equation.
- When students are clear about the three types of problem, they can complete Exercise 26E. Examples 5–7 give more guidance if students who need more **support** need it.
- Now show the triangle on the right.
- Ask students what ratio could be used to find x, which is the side adjacent to 62°, as the hypotenuse is known (cosine).
- Ask them to complete the question using cosine and a similar method to that used earlier in the lesson with sines (x = 39 cm).
- If this is clear to students, they can complete Exercise 26F. Examples 8 and 9 give further guidance for students who need more **support**.

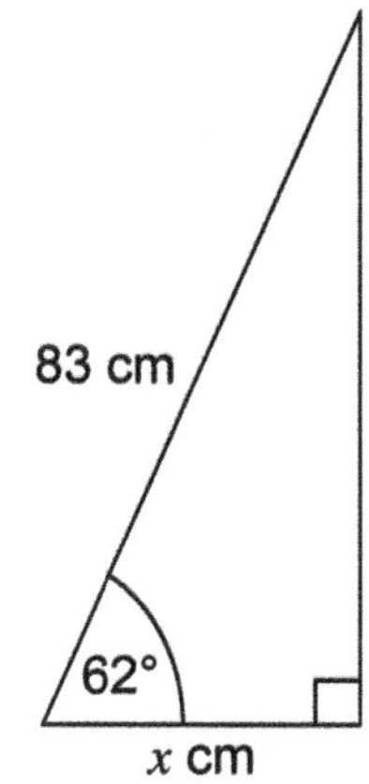

- Draw the two triangles on the right on the board.
- Tell students that tangents can be used to find the lettered angle and side.
- Ask for suggestions on the method and work through students' solutions.
- Work through Example 10 with students who need more **support**. Students who need more of a **challenge** should be sufficiently familiar with the method to move straight to the exercise.
- Now students can complete Exercise 26G.

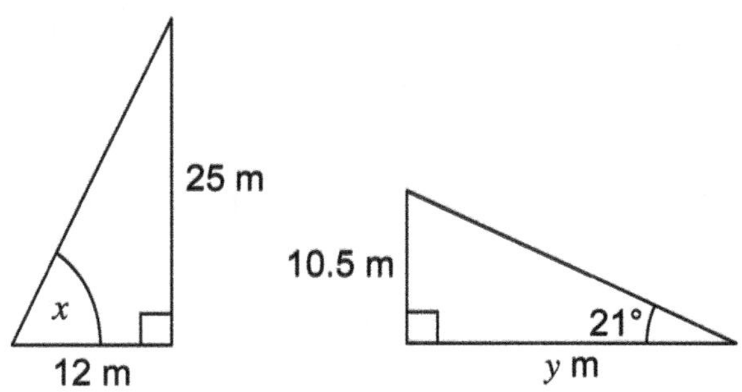

Plenary

- Look at the triangle with the angle of 62° again. Could students have used sine instead of cosine to find x? (Yes, by using the angle of 28° at the top of the triangle.)
- Ask them to check that this provides the same answer.
- Can you always use a sine instead of cosine? (Yes, in the questions they have seen so far.) Tell students that they should use whichever is more convenient.

Homework and answers

Homework for this lesson is available in the homework folder. Consolidation and extension exercises are available in the further practice folder. Answers to the exercises can be found in the answers folder.

26.5 Which ratio to use

Syllabus references

Statement	Core
C6.2/E6.2	Know and use the sine, cosine and tangent ratios for acute angles in calculations involving sides and angles of a right-angled triangle.

Collins references

- Student's Book pages 483–486

Learning aims

Decide which trigonometric ratio to use in a right-angled triangle.

Prior knowledge

Students should be familiar with calculating missing sides and angles of right-angled triangles using the trigonometric ratios sine, cosine and tangent.

Common mistakes and remediation

Selecting the correct ratio is the main stumbling block for students. A neat and clearly labelled diagram is the best way to make the correct choice.

Useful tips

Tell students that they should make sure they understand Examples 11–13 clearly. If students who need more **support** find they do not, then they should ask for your help. It might help if these students work with a student who needs more of a **challenge** to complete Exercise 26H.

Starter

- Draw the diagram below on the board.

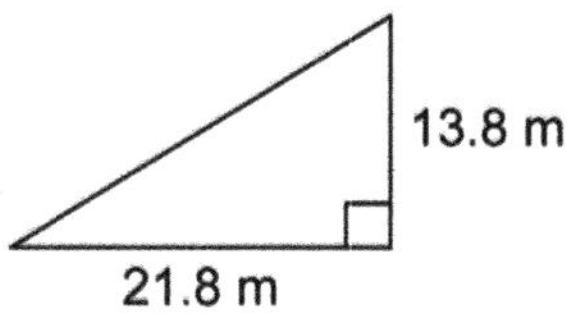

- Ask students, in pairs, to discuss how they can find the missing side and angles. Are there different ways to do it? Emphasise that they are looking for methods, not answers.
- After a few minutes, gather suggestions. The following are ideas that could be drawn from the discussion.
  - Use tangent to find one of the angles.
  - Angle sum of a triangle will give the third.
  - Use Pythagoras' theorem to find the third side.
  - Find the third side first then use sine or cosine to find an angle.
  - Find the hypotenuse first then use sine or cosine to find an angle.
- Make sure you mention Pythagoras' theorem to remind students of previous work.

Main lesson activity

- Draw the triangles from Examples 11–13 on the board. Explain that they must decide which ratio to use in each case.
- Model how to do this by first labelling the sides H, O, A in that order; looking to see which two sides are involved; writing down the appropriate equation. From then on the problem is similar to those in the preceding lessons.
- Point out that Example 11 lists the steps in detail.
- Now students can complete Exercise 26H, leaving question 4, which is a problem-solving question, for the plenary. The first question is about finding sides; the second about finding angles; the third is a mixture of these.

Plenary

- Look at question 4 of Exercise 26H.
- Ask students to give their explanations of parts **a** and **b**.
- Encourage clear reasoning and precise use of language. Students who need more **support** may need extra help in understanding the explanations.
- Use students' numerical examples from part **c** to reinforce what they have found.

Homework and answers

Homework for this lesson is available in the homework folder. Consolidation and extension exercises are available in the further practice folder. Answers to the exercises can be found in the answers folder.

26.6 Exact values of trigonometric functions

Syllabus references

Statement	Extended
E6.3	Know the exact values of: (a) sin x and cos x for x = 0°, 30°, 45°, 60° and 90°. (b) tan x for x = 0°, 30°, 45°, 60°.

Collins references

- Student's Book pages 486–489

Learning aims

- Recall and use the exact values of sine, cosine and tangent of 0°, 30°, 45°, 60° and 90°.

Key words

- no new keywords

Prior knowledge

Students should know how to find the sine, cosine and tangent of an angle in a right-angled triangle.

Common mistakes and remediation

Students can muddle the values of the trigonometric functions. Encourage them to draw the two triangles to show the ratios.

Useful tips

This topic is linked to the topic of surds. The ways that expressions involving surds can be manipulated can be revised here.

Starter

- Tell students not to use a calculator in this lesson.
- Draw this triangle and ask students to calculate to value of x.
- They should do this calculation: $x^2 = 5^2 - 4^2 = 25 - 16 = 9$, so $x = \sqrt{9} = 3$
- Do the same with this triangle and ask for the value of y.
- This time $y^2 = 4^2 - 3^2 = 16 - 9 = 7$ and so $y = \sqrt{7}$. This is the exact value of the length. Remind students that a calculator will give 2.64575… which is an approximation to the exact value.

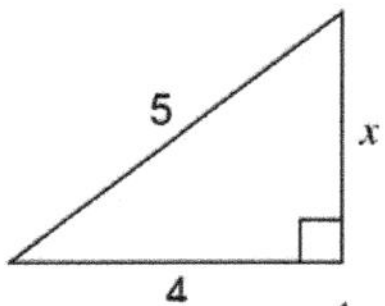

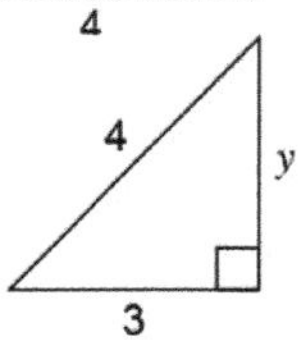

Main lesson activity

- Build up this diagram in stages.
- Draw an equilateral triangle ABC. Draw the perpendicular CD from C to AB and label the angles. Say that each side of the triangle is 2 units so AD = DB = 1. Label the sides as shown.
- Finally use Pythagoras' theorem in triangle ACD to find the length of CD.
- $CD^2 = AC^2 - AD^2 = 2^2 - 1^2 = 3$ and so $CD = \sqrt{3}$

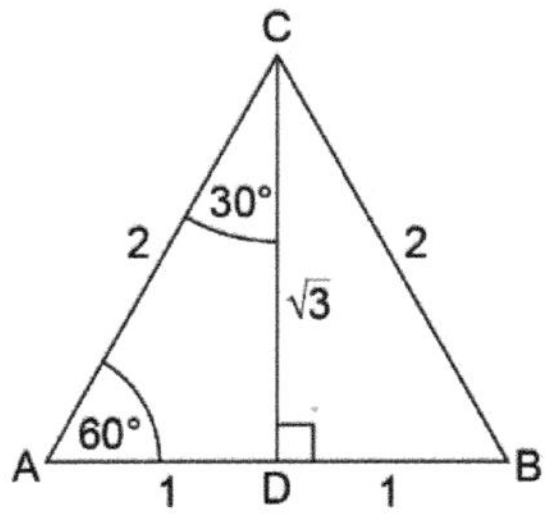

- Demonstrate how to use triangle ACD to find exact trigonometric ratios.

$$\sin 60^\circ = \frac{CD}{AC} = \frac{\sqrt{3}}{2} = \cos 30^\circ \qquad \sin 30^\circ = \frac{AD}{AC} = \frac{1}{2} = \cos 60^\circ$$

$$\tan 60^\circ = \frac{CD}{AD} = \frac{\sqrt{3}}{1} = \sqrt{3} \qquad \tan 30^\circ = \frac{AD}{CD} = \frac{1}{\sqrt{3}}$$

- Tell students they need to remember these exact values but they can use the diagram to help them. They should be able to sketch that if they need it.
- Now draw this right-angled equilateral triangle in stages.
- Say that EG and FG are 1 unit long. As students to explain why $EF=\sqrt{2}$
- This is Pythagoras again: $EF = \sqrt{1^2 + 1^2} = \sqrt{2}$
- Use the triangle to find more trigonometric ratios.
- $\sin 45^\circ = \cos 45^\circ = \frac{1}{\sqrt{2}}$ and $\tan 45^\circ = 1$

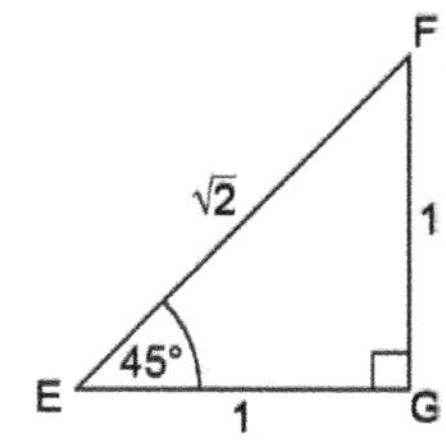

- Students need to remember these. Once again, the diagram will help.
- Some final values to remember: $\sin 0^\circ = \cos 90^\circ = \tan 0^\circ = 0$ and $\sin 90^\circ = \cos 0^\circ = 1$
- You can justify these values by imagining a right-angled triangle where one of the angle is very close to 0° and the other is very close to 90° and looking at the sides for each particular ratio.
- Finally, say that tan 90° has no value.
- Students can now do Exercise 26I.

Plenary

- Draw this diagram or ask the students to draw it. Start with triangle ABC, then draw CD, then AD, and finally DE and AE. If the students are drawing it, then the unit length can be any convenient length, such as 5 cm.

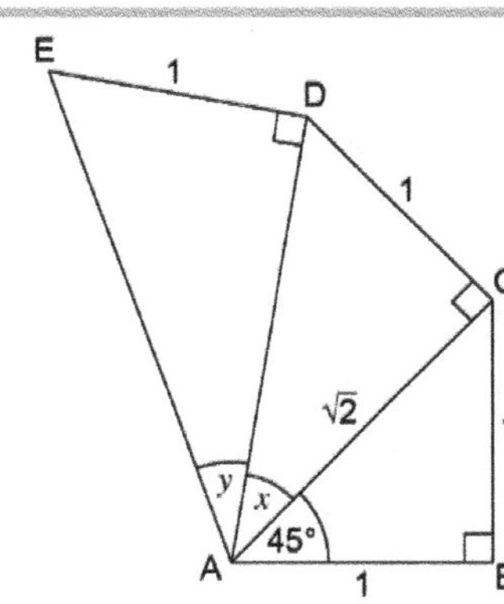

- Ask students to find the exact length of AD.
- They should see that $AD=\sqrt{AC^2 + CD^2} = \sqrt{2+1} = \sqrt{3}$
- Now ask for the sin, cos and tan of x.
- $\sin x = \frac{1}{\sqrt{3}}$ and $\cos x = \frac{\sqrt{2}}{\sqrt{3}}$ and $\tan x = \frac{1}{\sqrt{2}}$
- Finally ask students to find AE and the sin, cos and tan of y°.
- AE = 2, $\sin y^\circ = \frac{1}{2}$, $\cos y = \frac{\sqrt{3}}{2}$, $\tan y = \frac{1}{\sqrt{3}}$
- This shows that $y = 30^\circ$

Homework and answers

Homework for this lesson is available in the homework folder. Consolidation and extension exercises are available in the further practice folder. Answers to the exercises can be found in the answers folder.

26.7 Applications of trigonometric ratios

Syllabus references

Statement	Extended
E6.2	Know that the perpendicular distance from a point to a line is the shortest distance to the line. Carry out calculations involving angles of elevation and depression.

Collins references

- Student's Book pages 489–492

Learning aims

- Calculate the shortest distance of a point from a line.
- Solve problems using an angle of elevation or an angle of depression.

Key words

- angle of elevation, angle of depression

Prior knowledge

Students should be familiar with calculating missing sides and angles of right-angled triangles using the trigonometric ratios sine, cosine and tangent. They should be able to recognise which trigonometric ratio is required in each situation.

Common mistakes and remediation

Students often fail to find a right-angled triangle and do not use the correct ratio. Encourage students always to draw the relevant triangle and label the three sides.

Useful tips

Tell students that trigonometry questions are sometimes not straightforward triangles but are part of solving a practical problem in a real-life situation. Also refer to 'Common mistakes and remediation'..

Starter

- Lean a window pole, ladder or broom against a classroom wall. This will form a right-angled triangle: three sides and three angles. Explain that you want to find all the sides and angles, making as few measurements as possible.
- Ask: 'What is the least number needed? Which measurements will it be easiest to make?'
- Discuss these points with the class. They will probably agree that it will be easiest to measure two lengths. Make the two measurements they suggest and ask them to do the necessary calculations. Afterwards, discuss how accurate their answers will be. Real objects have thickness, which limits the accuracy of results. Students should think about this when answering questions and round off calculator answers sensibly.
- Now ask your students what the shortest distance from the right-angle to the pole is: check with a tape measure and show that it is in fact the perpendicular distance from the point to the line.

Main lesson activity

- Tell students that sometimes people need to know the shortest distance from a point to a line as this can save money. For example, a bridge engineer might need to put a strengthening rod between a point on a girder and another girder. Finding the shortest distance will save on the cost of the material used.
- Go over Example 14. Now students can complete Exercise 26J, questions 1 to 4. Students who need more **support** may need help with identifying the shortest distance to calculate. Check some answers before moving on.
- Explain that safety guidelines recommend the '1 in 4 rule' for ladders and sketch the drawing on the right on the board to illustrate it.

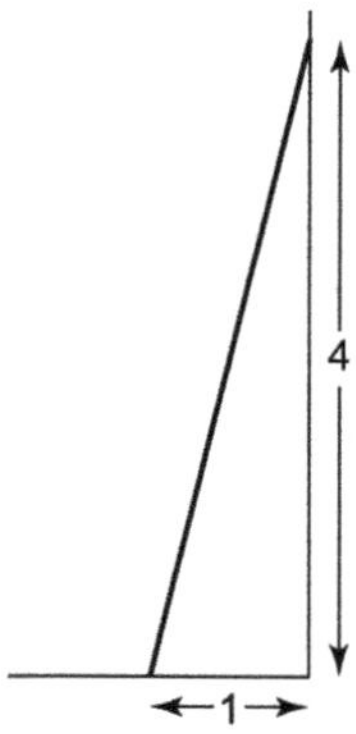

- Ask: 'What is the recommended safe angle for ladders?'
- Model the following procedure.
  - Identify the right-angled triangle.
  - Identify the angle you want.
  - Label the sides you know (O and A in this case).
  - Decide on the ratio you want (tangent in this case).
  - Write down the appropriate equation and solve it.
- Confirm that the angle with the ground is 76° and it is appropriate to give this to the nearest degree.

Angles of elevation and depression

- Ask: 'When you look up at an aircraft in the sky, can you tell how far away it is? Can you measure the distance? What can you measure?'
- Explain what is meant by an angle of elevation or an angle of depression.
- Model how to solve the problem in Example 15, paying particular attention to how a right-angled triangle can be found from the initial picture of the situation.
- Students can now start on Exercise 26J, questions 5 onwards. Students who need more **support** may need help with identifying the correct triangle and ratio to use.

Plenary

- Show the picture on the right of a person between two tall buildings.

- Ask: 'If the angle of elevation of the top of each building is the same, what can you say about the heights of the buildings above the eye-level of the person?' (If lines are added, it should be clear that the triangles are similar. One height above eye-level is double the other.)
- Then tell students that the angle of elevation is 60° and ask them to find the heights if the person's eyes are 1.6 m above ground level. (18.9 m and 36.2 m)

Homework and answers

Homework for this lesson is available in the homework folder. Consolidation and extension exercises are available in the further practice folder. Answers to the exercises can be found in the answers folder.

26.8 Problems in three dimensions

Syllabus references

Statement	Extended
E6.6	Carry out calculations and solve problems in three dimensions using Pythagoras' theorem and trigonometry, including calculating the angle between a line and a plane.

Collins references

- Student's Book pages 493–494

Learning aims

- Use trigonometric ratios and Pythagoras' theorem to solve more complex three-dimensional problems.

Prior knowledge

Students should be confident in applying trigonometric ratios and Pythagoras' theorem to solve problems in two dimensions.

Common mistakes and remediation

Remind students that if they round off their answers in any intermediate working, then the final answer may well be incorrect.

Useful tips

Make models of 3D shapes available to students, so that they can visualise lines and planes. Models made from geo-strips or similar are especially useful for students who need more **support**. It is much easier to explain the length of diagonals in cuboids and an angle between a line and a plane if there is a 3D model to demonstrate this.

Again, display the important information on a chart.

Pythagoras' theorem $c^2 = a^2 + b^2$

$\sin\theta = \frac{\text{O}}{\text{H}}$ $\cos\theta = \frac{\text{A}}{\text{H}}$ $\tan\theta = \frac{\text{O}}{\text{A}}$

Suggest that students always identify what information is required by highlighting it in the question. Encourage students to draw separate diagrams to help them work out the different stages in a problem.

Starter

- Demonstrate the idea of the angle between a line and a plane to the class. Ask students to try to define the angle between a line and a plane.
- It is crucial that they include in their definition a line that is perpendicular to the plane. Only students who need more of a **challenge** are likely to formalise their ideas in this way.

Main lesson activity

- Explain that the lesson is about using trigonometry in more complex 3D problems.
- Carefully explain all the stages in the working of an example, as follows. It is useful for them to have a model of a cuboid available, to refer to.
- ABCDEFGH on the right is a cuboid.

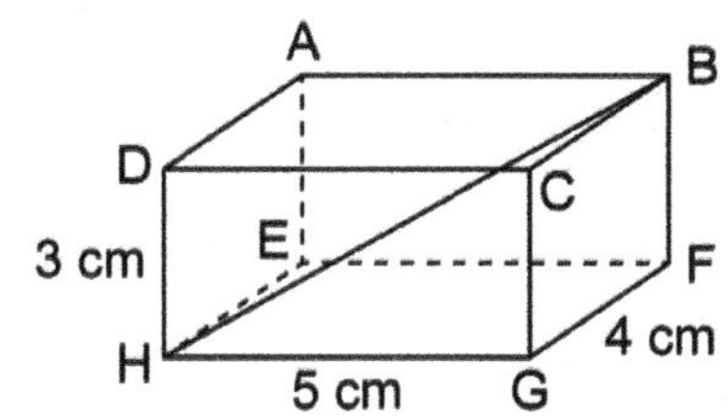

  1 Calculate the length of the diagonal BH.

  2 Calculate the angle between BH and the base EFGH.

  3 First look for a right-angled triangle with BH as one of its sides.

- The diagram on the right is triangle BFH.

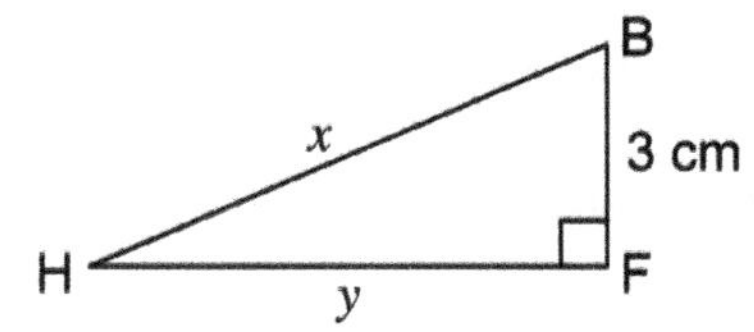

  1 Label the known side and the two unknown sides as x and y.

  2 Now to find y look for a right-angled triangle with HF as one of its sides.

- The diagram on the right is triangle FGH.

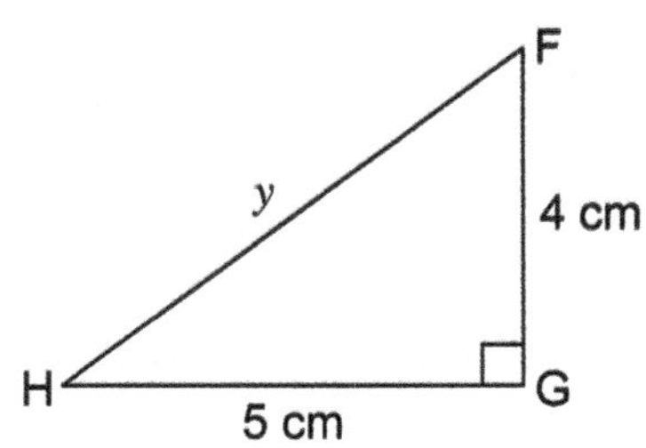

  1 Use Pythagoras' theorem to find y.

  $y^2 = 5^2 + 4^2$, so $y = \sqrt{41}$

  (There is no need to work this out.)

  2 Now find x. By Pythagoras' theorem:

  $x^2 = y^2 + 3^2 = 41 + 9 = 50$, so $x = \sqrt{50}$

  3 BH = 7.07 cm (3 sf)

- The angle between BH and the base EFGH is marked as θ on triangle BFH.

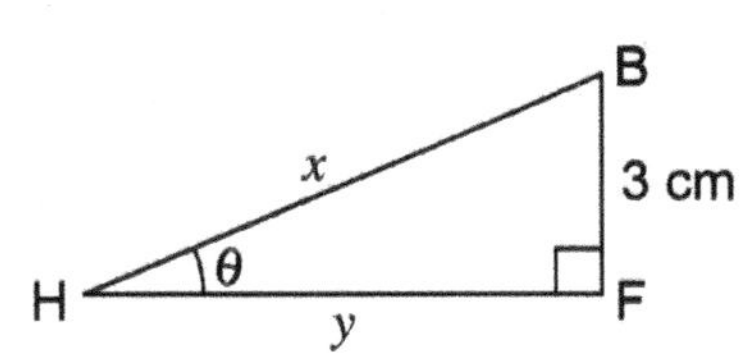

  1 Use any of the trigonometric ratios to find θ as all three sides are known.

  2 Use $\sin\theta = \frac{3}{x} = \frac{3}{\sqrt{50}} = 0.4243...$

  3 So $\theta = 25.1°$ (3 sf)

- Point out to students the importance of drawing clearly labelled diagrams.
- Use Example 16 to consolidate this work.
- Students can now do Exercise 26K. Students who need more of a **challenge** should work steadily through it.

Plenary

- Go through the exercise in class. Question 4, in particular, focuses directly on the angle between a line and a plane. Many students will find it difficult to pick out the required angle.

Homework and answers

Homework for this lesson is available in the homework folder. Consolidation and extension exercises are available in the further practice folder. Answers to the exercises can be found in the answers folder.

26.9 Sine and cosine of obtuse angles

Syllabus references

Statement	Extended
E6.5	Use the sine and cosine rules in calculations involving lengths and angles for any triangle.

Collins references

- Student's Book pages 495–496

Learning aims

- Find lengths of sides and angles in right-angled triangles using the sine and cosine functions.

Key words

- obtuse angle

Prior knowledge

Students should be familiar with the sine and cosine functions for acute angles and be able to evaluate simple expressions involving them using their calculators.

Common mistakes and remediation

Ensure that students remember the differences between acute, obtuse and reflex angles.

Useful tips

Again, display the important information on a chart.

Pythagoras' theorem $c^2 = a^2 + b^2$

$\sin \theta = \frac{\text{O}}{\text{H}}$ $\cos \theta = \frac{\text{A}}{\text{H}}$

Suggest that students always identify what information is required by highlighting it in the question. Encourage students to draw separate diagrams to help them work out the different stages in a problem.

Starter

- Sketch a right-angled triangle on the board and ask students to identify sin and cos of the two acute angles.
- Ask the class what a triangle with an obtuse angle might look like. Establish that a triangle with an obtuse angle cannot be right angled.
- Tell the class to choose any obtuse angle and work out its sine and cosine using their calculators. They will find that the sine is positive but the cosine is negative. Tell them that they will discover why this is during the lesson.

Main lesson activity

- Sketch the drawing from Student's Book page 470 on the board:

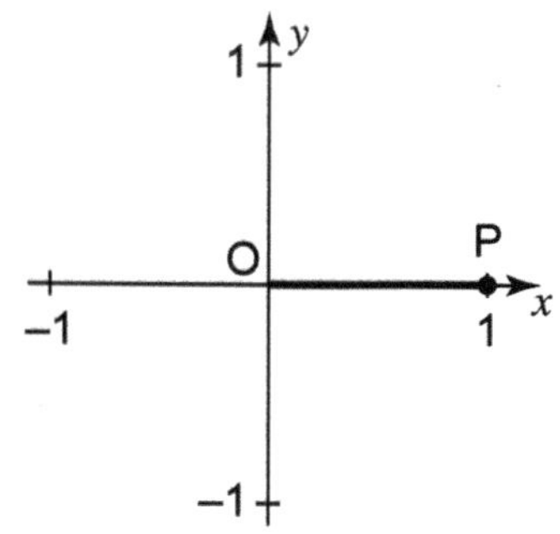

- Show how OP can rotate upwards to form a right-angled triangle and establish that the x-coordinate of P is cos θ, and the y-coordinate is sin θ.

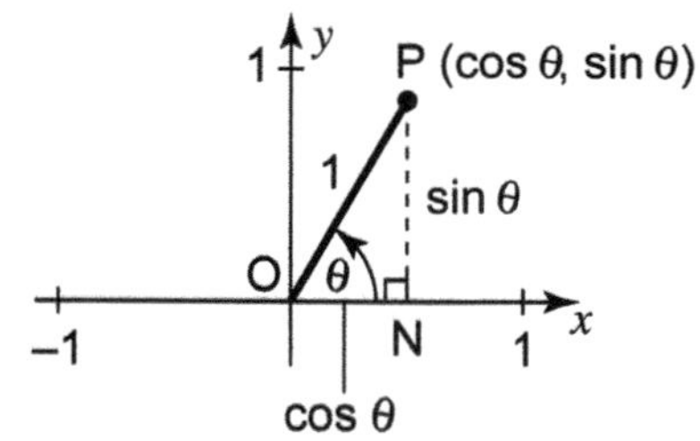

- Now show OP rotating into the next quarter of the graph.

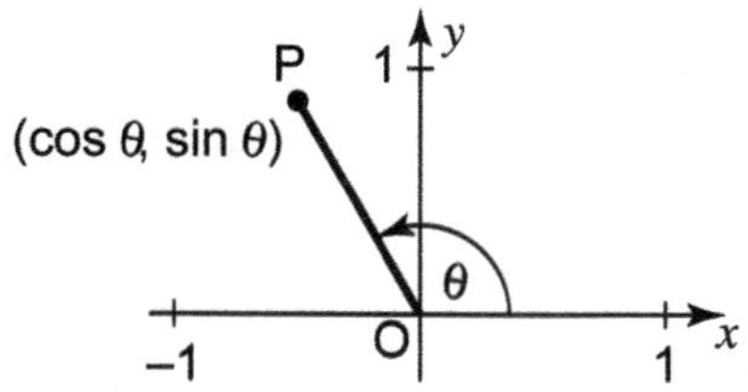

- The y-coordinate (sin θ) is still positive but the x-coordinate (cos θ) is now negative.
- Add in values to the diagram to illustrate this for students.

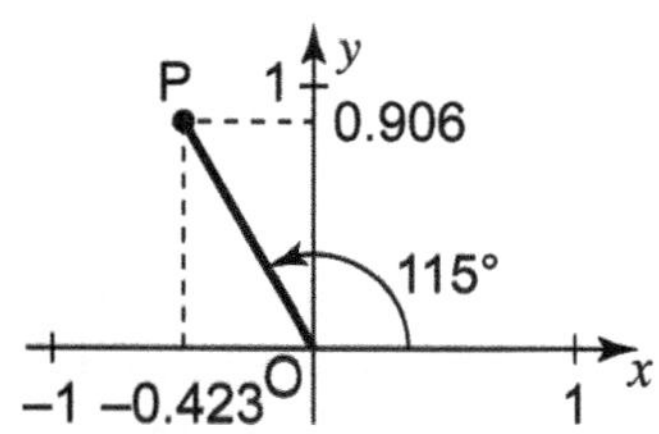

- Now point out that the angle adjoining θ on the x-axis must have the same sine as θ. Extend this to make the point that when the sum of two angles is 180°, their sines and cosines are the same value. The cosine is negative however.
- Students can now do Exercise 26L, referring back to this explanation in their books if necessary.

Plenary

- Go through the answers to Exercise 26L to ensure that all students have understood the concepts correctly.
- **Challenge** students by asking what they think the sines and cosines of a reflex angle might be – positive or negative? (Both will be negative.)

Homework and answers

Homework for this lesson is available in the homework folder. Consolidation and extension exercises are available in the further practice folder. Answers to the exercises can be found in the answers folder.

26.10a The sine rule

Syllabus references

Statement	Extended
E6.5	Use the sine and cosine rules in calculations involving lengths and angles for any triangle. Use the formula area of a triangle $= \frac{1}{2}ab \sin C$.

Collins references

- Student's Book pages 497–500

Learning aims

- Use the sine rule to find sides and angles in any triangle.

Key words

- sine rule

Prior knowledge

Students should know how to use sine, cosine, tangent and Pythagoras' theorem to calculate the angles and sides of right-angled triangles.

Common mistakes and remediation

Remind students that if they round off their answers in any intermediate working then the final answer could well be incorrect. Students who need more **support** may need to show intermediate working (to 4 sf). Show students who need more of a **challenge** how to enter a calculation directly into their calculators.

Useful tips

It would be useful to provide students with printed sheets of the various forms of the sine rules. Students who need more **support** may need to refer to them constantly. Encourage students to draw separate diagrams to help them work out the different stages in a problem.

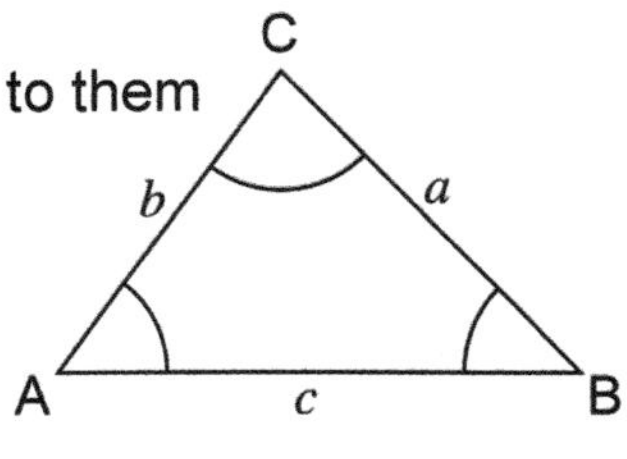

The sine rule: $\frac{a}{\sin A} = \frac{b}{\sin B}$ and $\frac{\sin A}{a} = \frac{\sin B}{b}$.

Starter

- To illustrate the sine rule, draw the triangle on the right on the board.
- Ask: 'How might you find the length of side b? Why can't we use sin, cos or tan here?' (only used in right-angled triangles).
  'What happens if we 'make' a right-angled triangle by dropping a perpendicular from vertex C? Can you see how we might find the length of line AC?'
- This should lead to the fact that $b \sin 50° = 5 \sin 65°$ and, therefore, that
  $\frac{b}{\sin 65°} = \frac{5}{\sin 65°}$

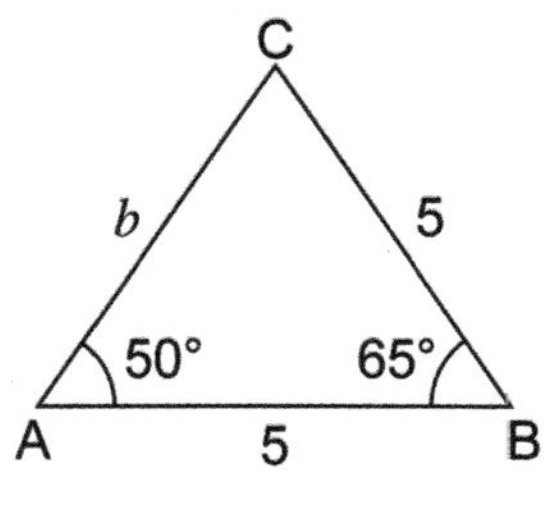

- Point out that the equal ratios are the lengths of the sides divided by the sines of the opposite angles. Is this always true? Investigate by forming a general triangle and thus deriving the sine rule.

Main lesson activity

- Explain how to use the sine rule. Say that, in navigation and engineering, problems may need to be solved which might involve triangles that do not contain right angles. Draw 'the general triangle' on the board. Explain how the sides and angles are labelled:
  - capital or upper-case letters for vertices or angles
  - lower-case letters for sides.

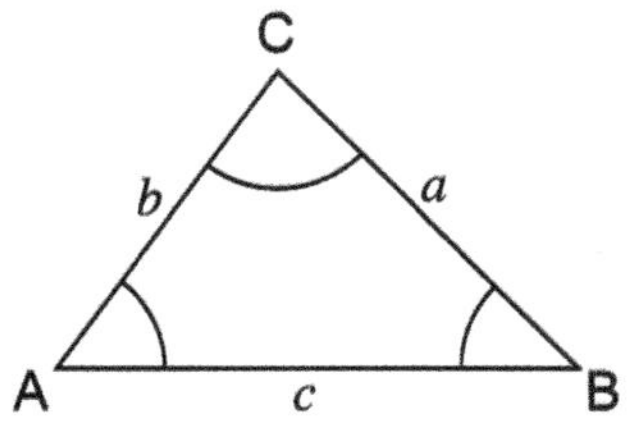

- Triangles can be solved like equations, provided that three of the values in the triangle are known.
- Introduce the sine rule:

$$\frac{a}{\sin A} = \frac{b}{\sin B} = \frac{c}{\sin C} \text{ or } \frac{\sin A}{a} = \frac{\sin B}{b} = \frac{\sin C}{c}$$

- Explain that the first form is used to find a missing length and the inverted form is used to find a missing angle. When solving triangles, students only need to use part of the formula.
- Go through an example to show how to apply the sine rule.
  Calculate the size of angle A.

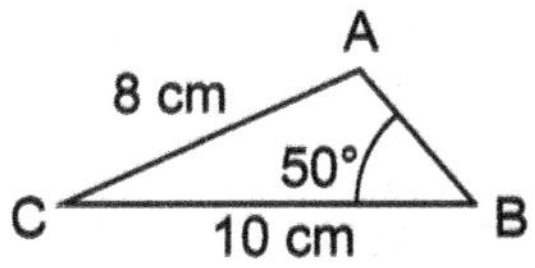

- To find an angle, use the inverted form of the sine rule: $\frac{a}{\sin A} = \frac{b}{\sin B}$
- Substitute the known values into the formula: $\frac{\sin A}{10} = \frac{\sin 50°}{8}$
- Rearranging gives $\sin A = \frac{10 \sin 50°}{8}$

  (Demonstrate how to evaluate this on a calculator.)
- So $A = 73.2°$ (3 sf).
- Work through Examples 17 and 18.
- Ask students to complete Exercise 26M. Students who need more **support** should start the work on their own, and then work in pairs if necessary.

Plenary

- Students of a similar ability can work in pairs to make up an example that uses the sine rule.
- Let them swap their examples with a partner. Give them a few minutes to check their answers.

Homework and answers

Homework for this lesson is available in the homework folder. Consolidation and extension exercises are available in the further practice folder. Answers to the exercises can be found in the answers folder.

26.10b The cosine rule

Syllabus references

Statement	Extended
E6.5	Use the sine and cosine rules in calculations involving lengths and angles for any triangle. Use the formula area of triangle = $\frac{1}{2}ab \sin C$.

Collins references

- Student's Book pages 500–503

Learning aims

- Use the cosine rule to find sides and angles in any triangle.

Key words

- cosine rule, included angle

Prior knowledge

Students should know how to use sine, cosine, tangent and Pythagoras' theorem to calculate the angles and sides of right-angled triangles.

Common mistakes and remediation

Remind students that if they round off their answers in any intermediate working then the final answer could well be incorrect. Students who need more **support** may need to show intermediate working (to 4 sf). Show students who need more of a **challenge** how to enter a calculation directly into their calculators.

Useful tips

It would be useful to provide students with printed sheets of the various forms of the cosine rules. Students who need more **support** may need to refer to them constantly. Encourage students to draw separate diagrams to help them work out the different stages in a problem.

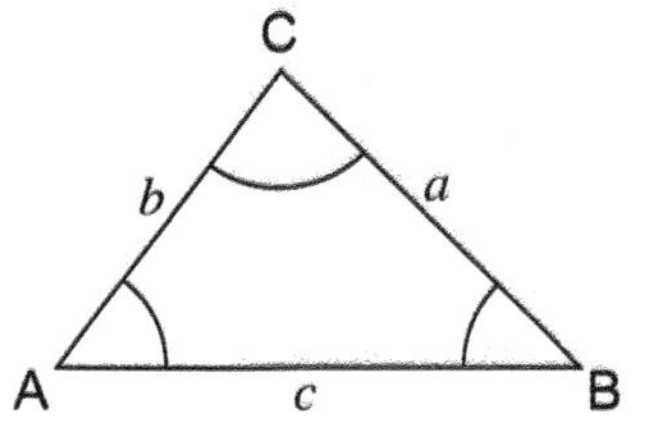

The cosine rule: $a^2 = b^2 + c^2 - 2bc \cos A$ and $\cos A = \frac{b^2 + c^2 - a^2}{2bc}$

Starter

- Recap the sine rule from Lesson 26.9a.
- Draw the triangle on the right on the board and remind students that $b \sin 50° = 5 \sin 65°$ and, therefore, that

$$\frac{b}{\sin 65°} = \frac{5}{\sin 65°}$$

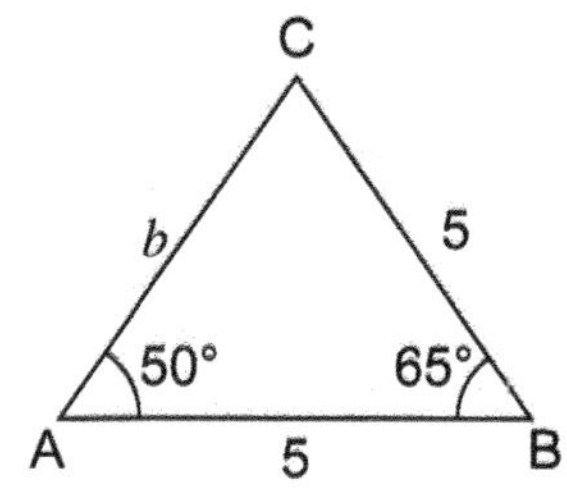

Main lesson activity

- Introduce the cosine rule. Draw the triangle on the board.
  The cosine rule is:

  $a^2 = b^2 + c^2 - 2bc \cos A$.

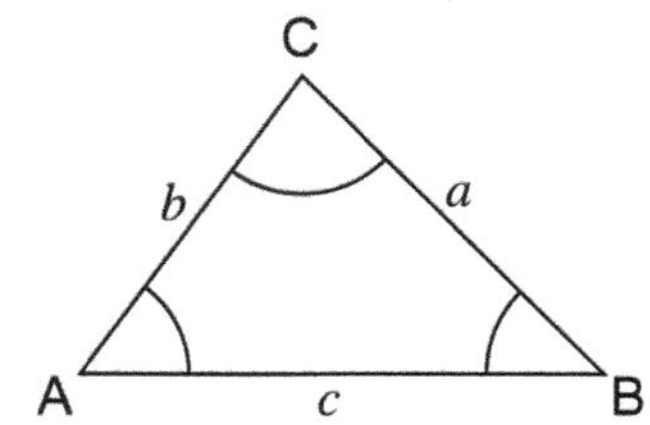

- This is used to find a if two sides and the included angle are known.
  To find b or c, rearrange the formula as:
  $b^2 = a^2 + c^2 - 2ac \cos B$ or $c^2 = a^2 + b^2 - 2ab \cos C$
- The formula can be rearranged to find an angle:

$$\cos A = \frac{b^2 + c^2 - a^2}{2bc}$$

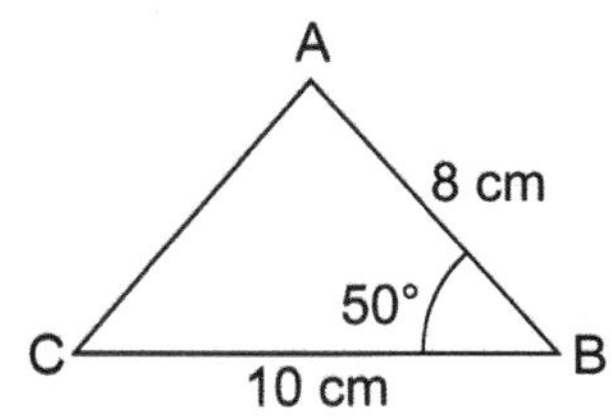

- Similar forms give $\cos B$ and $\cos C$.
- Go through an example to show how to apply the cosine rule.
  Calculate the length AC.
  - Let AC = b. So $b^2 = a^2 + c^2 - 2ac \cos B$.
  - Substitute the known values into the formula:

    $b^2 = 10^2 + 8^2 - 2 \times 10 \times 8 \times \cos 50°$

    (Demonstrate this on a calculator.)

    $b^2 = 61.15...$ So AC = = 7.82 cm (3 sf)
- Work through Examples 19 and 20.
- Ask students to complete Exercise 26N. Refer students who need more of a **challenge** to the Student's Book for an explanation of how to choose the correct rule, before they complete Exercise 26O.

Plenary

- Students of a similar ability can work in pairs to make up an example that uses the cosine rule.
- Let them swap their examples with a partner. Give them a few minutes to check their answers.

Homework and answers

Homework for this lesson is available in the homework folder. Consolidation and extension exercises are available in the further practice folder. Answers to the exercises can be found in the answers folder.

26.11 Using sine to find the area of a triangle

Syllabus references

Statement	Extended
E6.5	Use the sine and cosine rules in calculations involving lengths and angles for any triangle. Use the formula area of triangle = $\frac{1}{2}ab \sin C$.

Collins references

- Student's Book pages 504– 506

Learning aims

- Work out the area of a triangle if you know two sides and the included angle.

Key words

- area sine rule

Prior knowledge

Students should have covered the sine and cosine rules.

Common mistakes and remediation

Make students aware that if they round their answers in any intermediate working then the final answer might well be incorrect. Students who need more **support** may need to show intermediate working (to 4 sf). Show students who need more of a **challenge** how to enter a calculation directly into their calculators.

Useful tips

Display or give out a printed sheet with the following: Area of a triangle = $\frac{1}{2}ab \sin C$

Starter

- Illustrate a specific example of the formula for finding the area of a triangle, using two sides and the included angle, by working from first principles. Show students the triangle on the right.

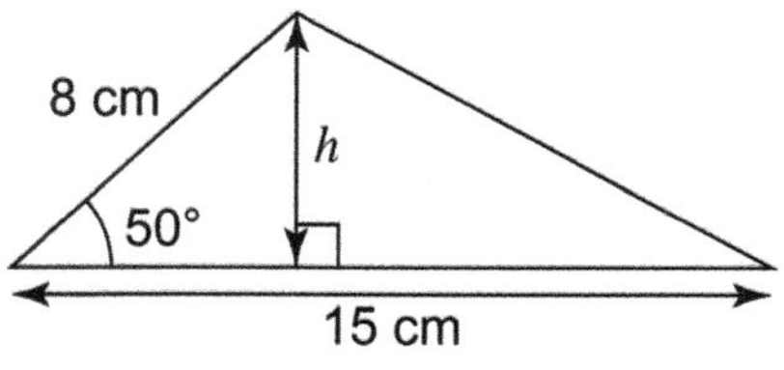

  - The area of the triangle is given by:
    Area = $\frac{1}{2} \times 15 \times h$
  - Now find h from: $\sin 50° = \frac{h}{8}$
  - So area = $\frac{1}{2} \times 15 \times 8 \times \sin 50° = 46.0 \text{ cm}^2$

Main lesson activity

- Explain that there is a formula for finding the area of a triangle that does not contain a right angle.
- Go through the proof, which is quite straightforward. Students who need more **support** could omit this.

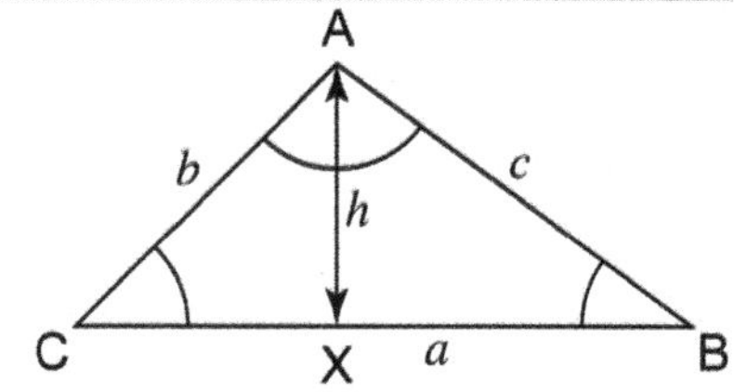

- Draw the general triangle on the right on the board and add the perpendicular height, h = AX.
- Then draw the smaller, right-angled triangle on the right, ACX, in which: $h = b \sin C$

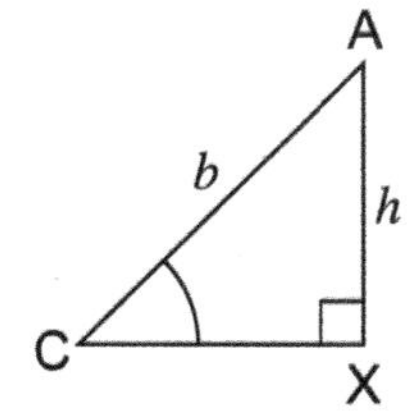

- To find the area of the general triangle ABC, use the formula:

  Area = $\frac{1}{2}$ (base × height). So the area = $\frac{1}{2} \times a \times b \sin C$

  Area = $\frac{1}{2} ab \sin C$
- This is the general formula to find the area of a triangle if two sides and the included angle are known. Using different perpendicular heights for the general triangle, the formula can also be written as:

  Area = $\frac{1}{2} ac \sin B$ or area = $\frac{1}{2} bc \sin A$
- Go through an example to show how to apply the formula.
  Work out the area of triangle XYZ.
  Two sides and an included angle are given, so substitute the values directly into the formula.

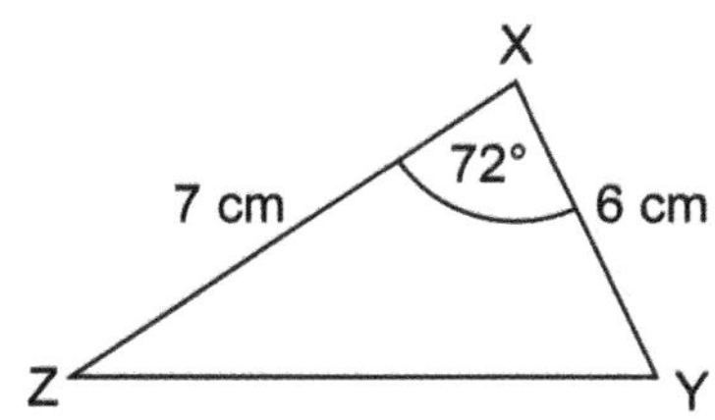

  Area = $\frac{1}{2} \times 7 \times 6 \times \sin 72°$
  So area = 20.0 cm^2 (3 sf)
- Questions sometimes need an application of the sine rule or cosine rule first. Show students who need more of a **challenge** an example. Work out the area of triangle ABC.
- Here three sides are given, so first find an angle.
  Any of the three angles would do.
- Find angle B, using the cosine rule.

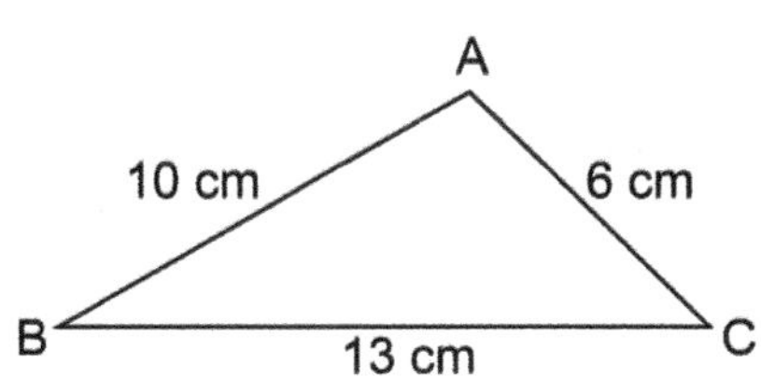

  $\cos B = \frac{13^2 + 10^2 - 6^2}{2 \times 13 \times 10}$

  so B = 26.34° (Leave the exact value on the calculator display.)
- Now apply the area formula: Area = $\frac{1}{2} \times 13 \times 10 \times \sin B$
- So the area is 28.8 cm^2 (3 sf). Use Examples 21 and 22 to consolidate this work.
- Students can now do Exercise 26P. The exercise is aimed at students who need more of a **challenge**. Students who need more **support** students can attempt the first few questions. Provide guidance as needed.

Plenary

- Draw the general triangle on the board.
- Invite students to sum up the three trigonometric formulae they have met.

Homework and answers

Homework for this lesson is available in the homework folder. Consolidation and extension exercises are available in the further practice folder. Answers to the exercises can be found in the answers folder.

26.12 Sine, cosine and tangent of any angle

Syllabus references

Statement	Extended
E6.4	Recognise, sketch and interpret the graphs of $y = \sin x$, $y = \cos x$ and $y = \tan x$ for $0° \leqslant x \leqslant 360°$. Solve trigonometric equations involving $\sin x$, $\cos x$ or $\tan x$, for $0° \leqslant x \leqslant 360°$.

Collins references

- Student's Book pages 506–513

Learning aims

- Find the sine, cosine and tangent of any angle.
- Solve simple trigonometric equations.

Prior knowledge

Students should have covered Section 26.9 on the sine and cosine of obtuse angles and be familiar with the idea of a rotating rod.

They should be able to solve simple equations.

Common mistakes and remediation

Students sometimes stop drawing a horizontal line when it meets the graph and miss one or more solutions. They should ensure that the line extends far enough to cover all solutions in the range required.

Sometimes a scientific calculator will give a negative answer when finding the inverse sine or inverse tangent of a negative number. Students need to be aware of the symmetries and periods of the trigonometric graphs to obtain answers in the required range.

Useful tips

Display or give out a printed sheet with the graphs of sine, cosine and tangent on them to help all, but particularly students who need more **support** when they start this chapter.

Starter

- Remind students about finding the sines and cosines of obtuse angles and extend this to the tangents of obtuse angles.
- Now ask students to find sin 210° on their calculator. (−0.5)
- Now ask them to enter $\sin^{-1}$ (−0.5) in their calculator, but DO NOT press the enter button – what do they think will appear?
- Explain why the answer given by calculators is not 210°, but −30°. (It gives the principal value, which for sine is defined as being between −90° and +90° inclusive). Students need to be aware of this and give the solutions between 0° and 360°.

Main lesson activity

- Go through the symmetries of the graphs of sine and cosine with your students. It is very useful for them to have a copy of the graphs beside them, but encourage them to sketch the graphs each time, marking in −1, 0 and 1 on the vertical axis and 0, 90, 180, 270 and 360 on the horizontal axis.
- Students should sketch the graphs each time to show their working in full.
- Go through the three parts of Example 23, emphasising the symmetry of each graph.
- They should check their solution satisfies the original equation each time.
- Encourage your students (especially those who need more of a **challenge**) to memorise the sines, cosines and tangents of 0°, 30°, 45°, 60° and 90° as these come up often in mathematics and are needed if a question asks for exact answers, in which case surd form may be appropriate. If they memorise these it allows them to make excellent sketches of the trigonometrical graphs. Here is a table of results:

	0°	30°	45°	60°	90°
sine	0	$\frac{1}{2}$	$\frac{1}{\sqrt{2}}$	$\frac{\sqrt{3}}{2}$	1
cosine	1	$\frac{\sqrt{3}}{2}$	$\frac{1}{\sqrt{2}}$	$\frac{1}{2}$	0
tangent	0	$\frac{1}{\sqrt{3}}$	1	$\sqrt{3}$	Undefined

- The above values can be found by working with an equilateral triangle or a right-angled isosceles triangle.
- Students can now do Exercise 26Q.
- Show students how the tangent graph differs from the sine and cosine graph and why it is undefined at 90°.
- Go through Example 24, mentioning in particular that the principal value given by their calculator is defined as being strictly between −90° and +90°, so they need to consider the tangent graph very carefully if the solutions to the equations need to be between 0° and 360°.
- Students can now do Exercise 26R.

Plenary

- Ask students to sketch the sine, cosine and tangent graphs between 0° and 360°. They should then exchange their sketches with a neighbour so they can be checked.

Homework and answers

Homework for this lesson is available in the homework folder. Consolidation and extension exercises are available in the further practice folder. Answers to the exercises can be found in the answers folder.

27.1a Perimeter and area of a rectangle

Syllabus references

Statement	Core
C5.2/E5.2	Carry out calculations involving the perimeter and area of a rectangle.

Collins references

- Student's Book pages 516–518

Learning aims

- Find the perimeter of a rectangle.
- Find the area of a rectangle.
- Use the formula for the area of a rectangle.

Key words

- length, width, perimeter, area

Prior knowledge

Students need to be aware of the idea of area as the amount of flat space covered by a 2D shape.

Common mistakes and remediation

Students commonly make three mistakes when calculating the area of a rectangle:

- confusing perimeter with area
- converting cm^2 into mm^2 by multiplying by 10
- converting m^2 into cm^2 by multiplying by 100.

Carefully explain the principle behind converting units in area questions, for example:

Area (in mm^2) = length × 10 (change to mm) × width × 10 (change to mm)

Before students start Exercise 27A, write the formulae for perimeter and area on the board, as a reminder.

Useful tips

It is important that students learn how to find the area of a simple shape such as a rectangle, as these form the basis for many compound shapes, which generally comprise rectangles, squares and triangles.

Starter

- Draw a 5 by 3 rectangle on the board. Ask students, 'How can I find the perimeter of this rectangle?' and then 'How can I find the area of this rectangle?' Repeat for two more rectangles.
- Encourage students who need more of a **challenge** to use the key formula $A = lw$ and understand that A = area, l = length and w = width.

Main lesson activity

- Draw a 5 by 4 rectangle on a square grid. Ask students to find its perimeter by counting the squares around its edges.
- Establish that perimeter can be found by adding width and length and multiplying by 2: $2(l + w)$.
- Now ask students to find the area by counting squares.
- Explain that the answer of 20 can also be found by multiplying 5 and 4.
- Walk students through Example 1 to reinforce these ideas.
- Now draw a 6 by 10 rectangle (not on a grid) and ask students to give its perimeter and area.
- Draw a rectangle with width and length labelled. Explain that: Area = length × width. This can also be written as a formula: $A = l \times w$ or $A = lw$.
- When students are calculating the area of each rectangle, ensure that they write down the formula before substituting values for length and width.
- Explain to students that questions on this topic will often require them to convert units.
- Conversion between area units is often a difficult idea for students who need more **support**. Even after being guided through an example, many students still think that changing square metres to square centimetres simply means multiplying by 100 rather than 100^2. This principle could be emphasised using this example:

  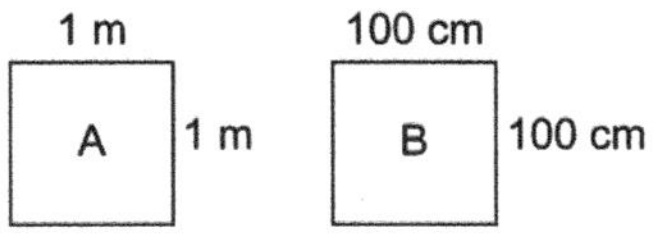

  Therefore, to change metres to centimetres, you multiply by 100. But to change square metres to square centimetres, you multiply by 10 000. For example, an area of 5 m^2 is the same as 50 000 cm^2 and an area of 8.2 m^2 is the same as 82 000 cm^2.
- Students can now do questions 1–7 of Exercise 27A. Students who need more **support** should be encouraged to draw diagrams for questions that do not supply them.

Plenary

- Summarise the work done in the lesson on area by ensuring students are aware of the rules, either in words or by using the formulae:

  $$A = lw \qquad l = \frac{A}{w} \qquad w = \frac{A}{l}$$

- Summarise the calculations needed for unit change of length and area, that is:
  - to change metres to centimetres, multiply by 100
  - to change centimetres to millimetres, multiply by 10
  - to change square metres to square centimetres, multiply by 10 000
  - to change square centimetres to square millimetres, multiply by 100.
- Compare the areas of these two rectangles:

  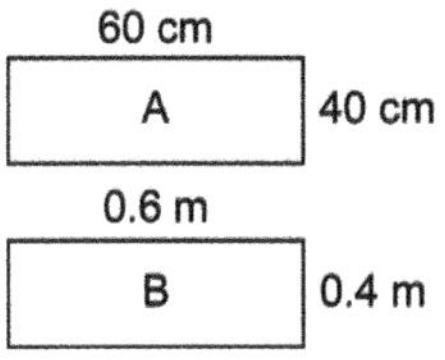

  Area of A = 40 × 60 = 2400 cm^2 Area of B = 0.4 × 0.6 = 0.24 m^2

Homework and answers

Homework for this lesson is available in the homework folder. Consolidation and extension exercises are available in the further practice folder. Answers to the exercises can be found in the answers folder.

27.1b Perimeter and area of a rectangle

Syllabus references

Statement	Core
C5.2/E5.2	Carry out calculations involving the perimeter and area of a rectangle.
C5.5/E5.5	Carry out calculations and solve problems involving perimeters and areas of compound shapes and parts of shapes.

Collins references

- Student's Book pages 516–519

Learning aims

- Find the area of a compound shape by splitting it into rectangles.

Key words

- length, width, perimeter, area

Prior knowledge

Students need to know how to calculate the perimeter and area of a rectangle as in Lesson 27.1a.

Common mistakes and remediation

Students often find it difficult to find the area of a compound shape.

Remind students that a compound shape is always made up of simpler shapes and finding the area is easier once they have split the shape up.

Remind them always to draw a diagram, showing how the shape is broken up and labelling all the lengths clearly.

Useful tips

Shapes can be cut up in any way and reassembled, and the area stays the same. You could emphasise this by starting with a rectangle of a known area and cutting and rearranging to get a shape of the same area. The cut lines do not have to be straight! Of course, the perimeter will change. Make sure the students realise this.

Starter

- Draw this shape on the board.

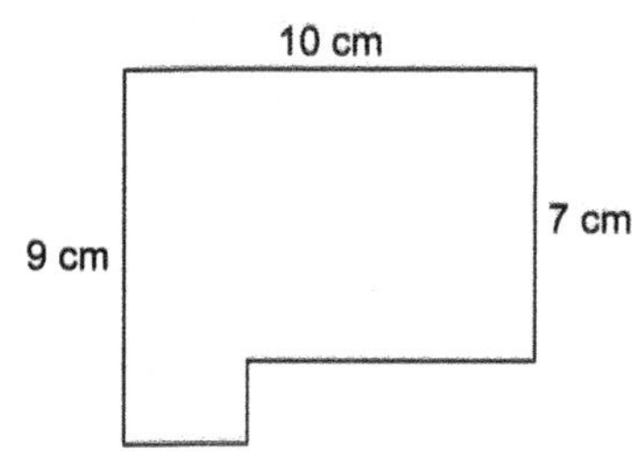

- Ask students: 'How might I find the area of this shape?' Split the shape into two rectangles and find the area of each before adding them.
- Ask: 'Is there another way to find the area of this shape?' Show that by splitting the shape into another two rectangles, the same answer is obtained.
- Emphasise to students that splitting a compound shape into rectangles, and clearly labelling those rectangles, will enable them to find the area of the shape.

Main lesson activity

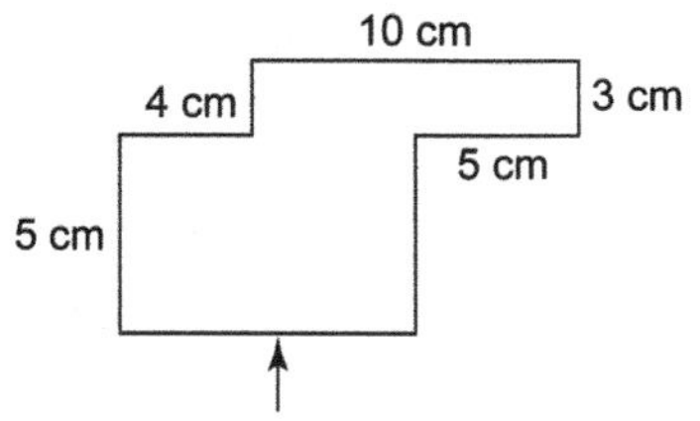

- Ask students to look at Example 2 and try to understand the method shown.
- Now, use the example shown here to talk students through this method as a class.
- Work out the area of this compound shape. Explain how to find the length marked with an arrow:

  $4 + (10 - 5) = 9$ cm

  So the area $= (10 \times 3) + (9 \times 5) = 30 + 45 = 75$ cm^2
- Students who need more of a **challenge** could draw their own compound shapes and ask other students to find the areas.
- Ensure students copy the diagram for each question and show precisely how they are splitting up the compound shape. Make sure they label each rectangle clearly.
- It is important that students set out their work clearly, as in Example 2, so they will still get marks for method even if they get a dimension wrong.
- Students can now do Exercise 27A questions 8–10.
- Students who need more **support** always find this a difficult topic. They could draw the first few examples in question 8 on centimetre-squared paper to reinforce the concept discussed in the lesson.

Plenary

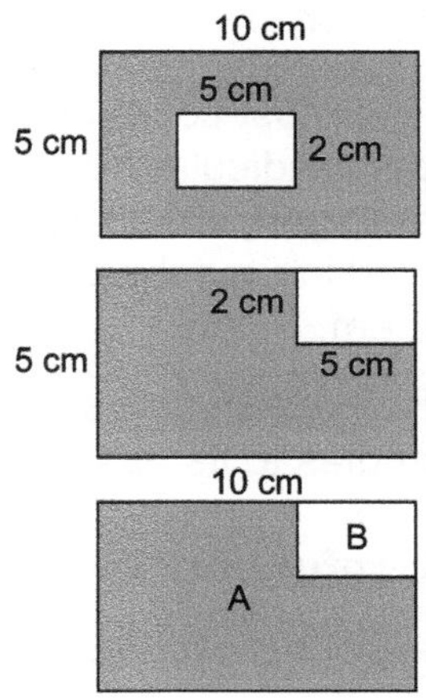

- Draw shape X on the board.
- Ask students if they can see a way to calculate the shaded area. They will probably suggest subtracting the two values.
- Now draw shape Y.
- Ask them if they can see a similar method of calculating the shaded area. Point out that sometimes the area of a compound shape may be worked out by subtracting areas rather than adding.
- Show students how to set out the working.
- Area of rectangle A $= 5 \times 10 = 50$ cm^2
- Area of rectangle B $= 2 \times 5 = 10$ cm^2
- Area shaded = Area of A – Area of B $= 50 - 10 = 40$ cm^2

Homework and answers

Homework for this lesson is available in the homework folder. Consolidation and extension exercises are available in the further practice folder. Answers to the exercises can be found in the answers folder.

27.2 Area of a triangle

Syllabus references

Statement	Core
C4.1/E4.1	Use and interpret the vocabulary of triangles.
C5.2/E5.2	Carry out calculations involving the perimeter and area of a triangle.
C5.5/E5.5	Carry out calculations and solve problems involving perimeters and areas of compound shapes and parts of shapes.

Collins references

- Student's Book pages 519–521

Learning aims

- Find the area of a triangle.
- Use the formula for the area of a triangle.

Key words

- base, perpendicular height

Prior knowledge

Students should know how to calculate the area of a rectangle, as covered in Lessons 27.1a and 27.1b.

Common mistakes and remediation

The most common error is to use a height other than the perpendicular height. However, there is unlikely to be any ambiguity at this level as students will have no way of working out the perpendicular height if it is not given to them.

Remind students that a perpendicular is always at right angles to its base, whichever side is taken as the base.

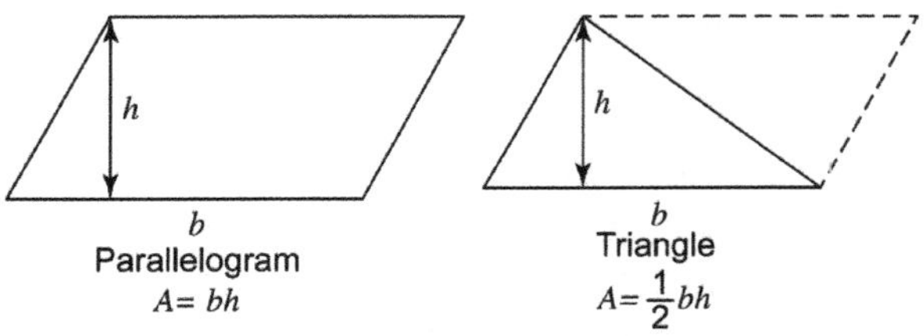

Useful tips

One way of finding the area of a triangle is to halve that of a parallelogram. Parallelograms are dealt with in Lesson 27.3, so you may find it useful to cover that first.

Starter

- Draw three triangles on the board, one right-angled, one isosceles and one equilateral. Ask students to describe each triangle, making sure they mention not only the shape but the correct names. Make sure that they can classify the three types.
- Ask students: 'What do we mean by 'perpendicular'? What do we mean by 'height'?' It might be useful to emphasise that other 'heights' exist, such as slant heights in cones.
- Ask students to label the base and the perpendicular height for each triangle. Ask other students to re-label the triangles, using a different side as the base and drawing in the corresponding perpendicular height.

Main lesson activity

- Show students that a right-angled triangle is half of a rectangle. This could be shown by cutting a rectangular piece of card into two right-angled triangles. If possible, give each student a piece of A4 paper and ask them to cut it into two right-angled triangles themselves. This will demonstrate the point more clearly.
- Discuss the terms *base* and *height* for a right-angled triangle.
- Lead students on to the idea that, in order to find the area of a right-angled triangle, they must multiply the base by the height and then divide by two. Put this into a formula for students:

  area of a triangle = $\frac{1}{2}$ base × height or $A = \frac{1}{2}bh$

- If the class needs more **support**, just use the formula in words initially, as they may not be able to follow an algebraic formula at this stage.
- Once students have mastered finding the area of a right-angled triangle, they can move on to finding the area of any triangle. If necessary, remind students that the area of any triangle is given by the formula:

  area = $\frac{1}{2}$ × base × perpendicular height or $A = \frac{1}{2}bh$

- If the class needs more **support**, continue to use the formula in words, until they can follow the algebraic formula.
- Use Example 3 to illustrate how this formula works in practice.
- This work can be reinforced by demonstrating that a rectangle can be drawn around the triangle.
- Students who need more **support** could be encouraged to draw the enclosing rectangle around the triangle, then to cut it out and show that the 'extra' parts can be combined to fit over the original triangle.
- Work through Example 4, which shows how the area of a compound shape can be calculated, using the fact that it can easily be split into a rectangle and a triangle. Encourage students to explore the ways that compound shapes can be split up, and to use the one that works best for them.
- Students can now do Exercise 27B.

Plenary

- Ask students to draw as many triangles as they can that have an area of 24 cm^2, but with bases of different lengths.
- Students who need more **support** could be asked to find two different triangles that both have an area of 12 cm^2.

Homework and answers

Homework for this lesson is available in the homework folder. Consolidation and extension exercises are available in the further practice folder. Answers to the exercises can be found in the answers folder.

27.3 Area of a parallelogram

Syllabus references

Statement	Core
C4.1/E4.1	Use and interpret the vocabulary of special quadrilaterals.
C5.2/E5.2	Carry out calculations involving the area of a parallelogram.
C5.5/E5.5	Carry out calculations and solve problems involving perimeters and areas of compound shapes and parts of shapes.

Collins references

- Student's Book pages 522–523

Learning aims

- Find the area of a parallelogram.
- Use the formula for the area of a parallelogram.

Key words

- parallelogram

Prior knowledge

Students should know how to calculate the area of a rectangle, as covered in Lesson 27.1a.

Common mistakes and remediation

When finding the area of a parallelogram, students may mistakenly halve the answer. They often use a height other than the perpendicular height if other measurements are given.

Ensure that each student has the formula for finding the area of a parallelogram written down clearly in their notes, with a diagram of the measurements that should be used.

Encourage students who need more **support** to refer to their notes before attempting each question.

Useful tips

Another way to find the area of a parallelogram is:

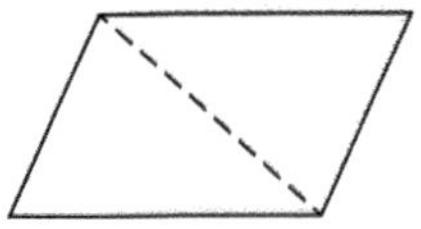

Area of parallelogram = area of two congruent triangles

$$= 2 \times \frac{1}{2} \times \text{base} \times \text{height} = \text{base} \times \text{height}$$

Starter

- Ask students: 'What is a parallelogram?' (A quadrilateral with two pairs of parallel sides.)
- Ask: 'Does this mean a rectangle is a parallelogram? Is a square a parallelogram?' (Yes, in both cases.)
- Ask students to point out parallelograms and rectangles in the classroom, explaining their decision each time.

Main lesson activity

- Draw a parallelogram on the board and show students that it can be split into a rectangle and two right-angled triangles that can be put together to form a rectangle. This can easily be shown on a whiteboard or by using a piece of card.
- Explain to the class that the area of a parallelogram can be found by multiplying the base by the height.

  $A = bh$

- The formula for the area of a parallelogram can be shown in two ways (see below and 'Useful tips').

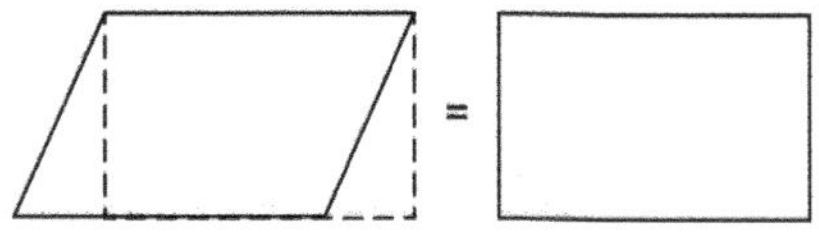

- Ensure final answers are given to a suitable degree of accuracy, normally 3 significant figures.
- Ask students to read Example 5. Check that they have understood the example and, in particular, that they understand that the perpendicular height is used in the calculation.
- Draw a parallelogram to explain the difference between a perpendicular height and a *slant height*.

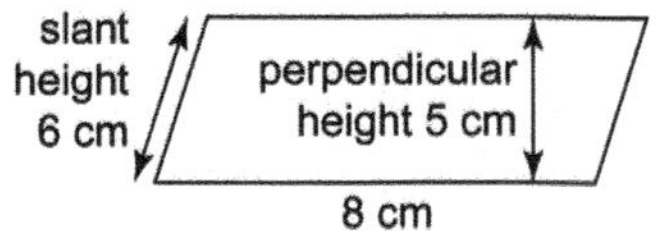

  So area = 8 × 5 = 40 cm²

- This example should clearly show students why they must always identify the base and perpendicular height first in questions on finding the area of a parallelogram.
- Students can now do Exercise 27C.

Plenary

- Ask students to draw three different parallelograms, each with an area of 12 cm².
- They should think how they could find the base and perpendicular height, given the area. They need to find different ways of giving an area of 12 cm². This should reinforce the concept of perpendicular height.
- Students who need more **support** could think about parallelograms with bases that are factors of 12.
- Students who need more of a **challenge** could be asked to find different parallelograms with areas of 24 cm² or 36 cm².

Homework and answers

Homework for this lesson is available in the homework folder. Consolidation and extension exercises are available in the further practice folder. Answers to the exercises can be found in the answers folder.

27.4 Area of a trapezium

Syllabus references

Statement	Core
C5.2/E5.2	Carry out calculations involving the area of a trapezium.

Collins references

- Student's Book pages 523–525

Learning aims

- Find the area of a trapezium.
- Use the formula for the area of a trapezium.

Key words

- trapezium

Prior knowledge

Students should know how to calculate the areas of rectangles, triangles and parallelograms, as covered in Lessons 27.1–27.3.

Common mistakes and remediation

The most common error is to calculate $(\frac{1}{2}a + b)$, which is generally down to lack of understanding in the use of the calculator. This can be avoided by encouraging students to label their trapezium with letters, a, b and h, before commencing the question and to set out their working appropriately as shown in the worked examples.

Students often try to split the trapezium into a rectangle and two right-angled triangles and then have difficulty finding the base of each triangle. Ensure that students have fully understood the method required before they attempt Exercise 27D.

As in Lesson 27.3 for the area of a parallelogram, it must always be emphasised in this context that height means perpendicular height and that we add the two *parallel* sides.

Useful tips

It may help students to recall that $\frac{1}{2}(a + b)$ is the mean of a and b. The formula therefore says that the area of a trapezium is the average length × the height.

Starter

- Ask students: 'What is a trapezium?' (A quadrilateral with just one pair of parallel sides.)
- Ask: 'Does a rectangle count as a trapezium? What about a square?' (No, in both cases, since they both have two pairs of parallel sides, although some mathematicians may define the trapezium differently.)

Main lesson activity

- The area of a trapezium is given by:

  A = half the sum of the parallel sides × height

  $A = \frac{1}{2}(a + b)h$

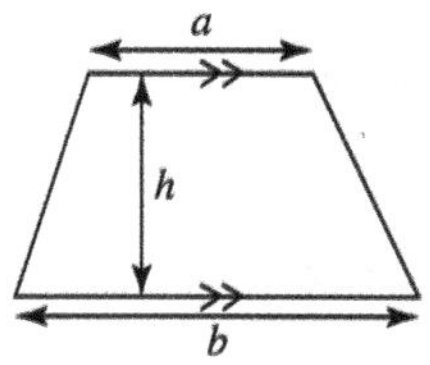

- This can be seen to be the case by splitting the trapezium into two triangles.

  Area of triangle 1 = $\frac{1}{2}$ × base × height = $\frac{1}{2}ah$

  Area of triangle 2 = $\frac{1}{2}$ × base × height = $\frac{1}{2}bh$

  Total area = $\frac{1}{2}ah + \frac{1}{2}bh = \frac{1}{2}h(a + b) = \frac{1}{2}(a + b)h$

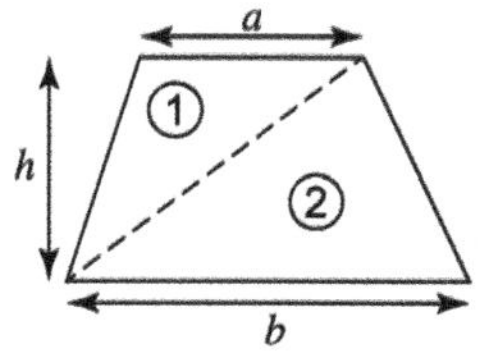

- Two identical trapeziums can be placed together (one upside down) to make a parallelogram.
- The area of the parallelogram is $(a + b)h$. So the area of one trapezium is half of this. This means that the area of a trapezium is

  $A = (a + b)h$

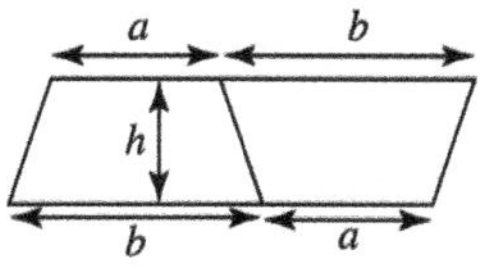

- Now take the class through an example, pointing out the best way to do the calculation.

  $A = \frac{1}{2}(a + b)h$ $\quad a + b = 16$

  $16 \times h = 16 \times 4 = 64$

  $64 \div 2 = 32$

  So $A = 32$ cm^2

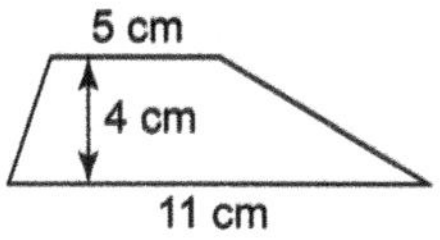

- Ask students to read Example 6 and match what they have just learnt to the calculation shown in it.
- Students can now do Exercise 27D. Students who need more **support** will find this section difficult and so they may find it useful first to draw and label the trapeziums in question 1, to help them understand how a trapezium can be formed from rectangles and triangles.

Plenary

- Draw the shapes below on the board:

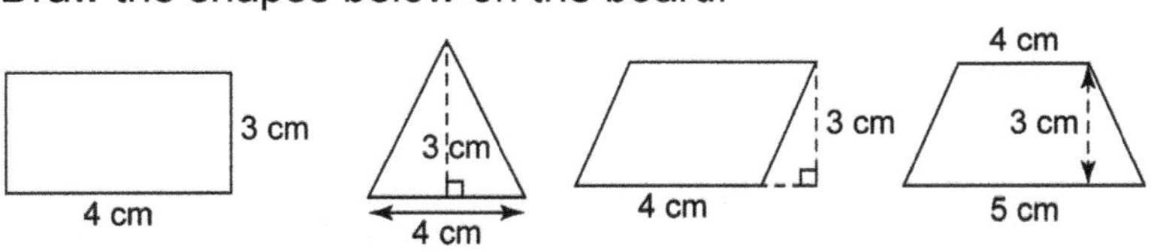

- Ask students: 'Which has the largest area?'

  Area of rectangle = 4 × 3 cm^2 = 12 cm^2

  Area of triangle = $\frac{1}{2}$ × 4 × 3 cm^2 = 6 cm^2

  Area of parallelogram = 4 × 3 cm^2 = 12 cm^2

  Area of trapezium = $\frac{1}{2}$ × (4 + 5) × 3 cm^2 = 13.5 cm^2

Homework and answers

Homework for this lesson is available in the homework folder. Consolidation and extension exercises are available in the further practice folder. Answers to the exercises can be found in the answers folder.

27.5a Circumference and area of a circle

Syllabus references

Statement	Core
C4.1/E4.1	Use and interpret the vocabulary of a circle.
C5.3/E5.3	Carry out calculations involving the circumference and area of a circle.

Collins references

- Student's Book page 526

Learning aims

- Calculate the circumference of a circle.

Key words

- circumference, diameter, radius, π (pronounced 'pi')

Prior knowledge

Students should be able to round to a given number of decimal places (usually one or two) to give sensible answers. They should also be familiar with substituting values into a formula.

Common mistakes and remediation

Students are often confused between the radius (r) and the diameter (d), and consequently may calculate the circumference as $C = \pi r$ or $C = 2\pi d$. Later, the problem may be confusing the formulae for circumference with that for the area, $A = \pi r^2$.

A simple poster, made by students, displayed on the classroom wall will help them to remember the formula: circumference, C, of a circle is given by $C = \pi d$ where d is the diameter and $\pi = 3.142$, or use the calculator button.

Students often forget to include the units with their answers. They should be continually reminded of this.

Useful tips

When using the π button on a calculator, students should remember to give the final answer to a sensible degree of accuracy, usually to 3 significant figures.

This lesson is designed to be used together with Lesson 27.5b on the area of a circle. At the end of both lessons students do Exercise 27E. For a class of students who need more of a **challenge** these two lessons can be delivered in the same classroom lesson.

Remind students that when the question asks for the answer to be given 'as a multiple of π' they should not be using $\pi = 3.142$ or replacing π by a numerical value. Their answers should include the symbol π instead as that gives an exact answer, not an approximate one.

Starter

- Provide students with six cylindrical objects – bottles, cans, tubes or piping – and about 2 m of string.
- Students measure the circumference of each one by wrapping string 10 times around the circumference of each one, measuring the string and dividing it by 10.
- They then measure the diameter of each one and divide the circumference by this value.

- Their results should all be about 3.14, rounded to 2 decimal places.
- Define this number as π and show students how they can calculate it accurately by pressing the π button on their calculator.

Main lesson activity

- Draw a circle on the board and introduce the formula $C = \pi d$.
- Point out that this can also be written as $C = 2\pi r$, since the diameter is twice the radius ($d = 2r$).
- If necessary, explain again that π is a number that is used in mathematics (and in lots of other scientific applications) when dealing with circles. Stress that π is not an exact number and can be written to millions of decimal places. In most cases π can be approximated to 3.142 when dealing with everyday examples. On most calculators, the π button gives π as 3.141 592 654. Students may need to be shown where the button is on their calculators. When checking answers to see if they are sensible, π can be taken to be 3.
- Go through a couple of examples to show students, particularly the ones who need more **support**, how to set out a solution to a problem.
- Calculate the circumference of this circle.

  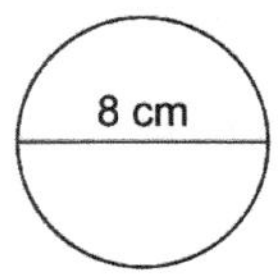

  $C = \pi d = \pi \times 8 = 25.1$ cm (to 1 decimal place)
- Point out that answers of 25 cm or 25.13 cm are also acceptable as well as 8π if the question asks for the answer as a multiple of π.
  Remind students always to put the correct units with the answer.
- Work through Example 7 with students, to reinforce the method and how to lay out the work.
- Calculate the radius of this circle, if it has a circumference of 4 m.

  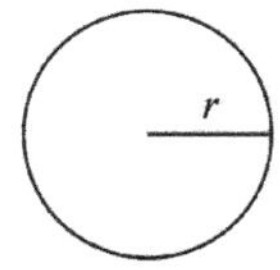

  $C = \pi d$

  $4 = \pi \times d$

  So $d = \dfrac{4}{\pi} = 1.2732...$

  So $r = 0.6$ m (to 1 decimal place)
- Point out to students that, on a calculator, this can be worked out as $4 \div \pi \div 2$.
- Use Example 8 to demonstrate again how to rearrange the formula and find the diameter of a circle, given the circumference.

Plenary

- Ask students to state the formula for the circumference of a circle. Describe some simple shapes and ask students to work out the circumference, mentally, using $\pi = 3$:

  head, $r = 9$ cm (54 cm)

  football, $r = 12$ cm (72 cm)

  gas storage tank, $r = 10$ m (60 m)

  cup, $r = 4$ cm (24 cm)

  oil drum, $r = 40$ cm (240 cm)

  pencil, $r = 8$ mm (24 mm).

Homework and answers

Homework for this lesson is available in the homework folder. Consolidation and extension exercises are available in the further practice folder. Answers to the exercises can be found in the answers folder.

27.5b Circumference and area of a circle

Syllabus references

Statement	Core
C5.3/E5.3	Carry out calculations involving the circumference and area of a circle.

Collins references

- Student's Book pages 527–529

Learning aims

- Calculate the area of a circle.

Key words

- circumference, diameter, radius, π

Prior knowledge

Students should be able to round to a given number of decimal places, to give sensible answers. They also need to be able to take squares and square roots, using calculators.

Common mistakes and remediation

It is common for students to confuse the radius (r) and the diameter (d). Some students may work out πr^2 as $(\pi r)^2$.

A simple poster, made by students, on the classroom wall will help them to remember the formula.

The area, A, of a circle is given by the formula $A = \pi r^2$, where r is the radius and $\pi = 3.142$, or use the calculator button.

Students often forget to include the units with their answers or use units of length. They should be continually reminded of this as they work through the exercise.

Remind students that when the question asks for the answer to be given 'as a multiple of π' they should not be using $\pi = 3.142$ or replacing π by a numerical value. Their answers should include the symbol π instead as that gives an exact answer, not an approximate one.

Useful tips

Students who need more **support** may need to be reminded of the circle formulae at the start of at least three or four lessons.

Students who need more **support** sometimes find it easier to remember the formula as $A = \pi \times r \times r$.

Starter

- Ask the class for the answers to 1^2, 2^2, 3^2 and 15^2, all calculated mentally. Then ask them to use a calculator to find 27.41^2.
- Discuss how students may take π to be 3 for checking answers and 3.142 on basic calculators, and use the π button on scientific calculators.

Main lesson activity

- Draw a circle on the board and introduce the formula for the area of a circle: $A = \pi r^2$.
- Explain to students that this can be written as $A = \pi \times r \times r$, but it is easier to use the x^2 button on a calculator.
- Students who need more of a **challenge** could be shown this diagram, to demonstrate that the area of the circle must be less than the area of the square that has an area of $4r^2$.

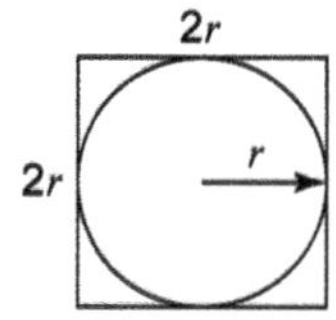

- Go through a couple of examples to show students who need more **support** how to set out a solution to a problem.
- Calculate the area of this circle.

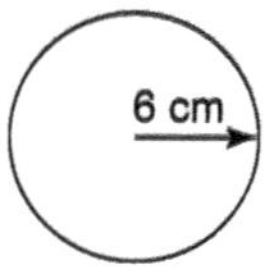

  $A = \pi r^2$

  $= \pi \times 36$

  $= 113.1\ \text{cm}^2$ (to 1 decimal place)

  $= 36\pi$ if the question asks for the answer as a multiple of π.

  Remind students that the unit for the area is square centimetres (cm^2).

- Go through Example 8 with the students, as extra practice.
- Calculate the area of this circular pond.

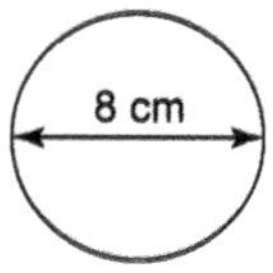

  $d = 4.8$ m, so $r = 2.4$ m

  $A = \pi r^2$

  $= \pi \times 2.4^2$

  $= 18.1\ \text{m}^2$ (to 1 decimal place)

  $= 5.76\pi$ if the question asks for the answer as a multiple of π.

  Remind students that the unit for the area is m^2.

- Work through Example 9, which is another problem in which the diameter is given. Use it as reinforcement.
- Students can now do Exercise 27E.
- Students should be encouraged to write down the formulae each time they do a question, as this will help them remember it.

Plenary

- Ask students to work out a few areas mentally using $\pi = 3$.
  For example, pond $r = 3$ m, window $d = 20$ cm, etc.
- Emphasise that radius must be used in the calculation, so if a diameter is given in the question it must be divided by two.

Homework and answers

Homework for this lesson is available in the homework folder. Consolidation and extension exercises are available in the further practice folder. Answers to the exercises can be found in the answers folder.

27.6 Surface area and volume of a cuboid

Syllabus references

Statement	Core
C4.1/E4.1	Use and interpret the vocabulary of simple solids.
C5.4/E5.4	Carry out calculations and solve problems involving the surface area and volume of a cuboid.
C5.5/E5.5	Carry out calculations and solve problems involving surface areas and volumes of compound solids and parts of solids.

Collins references

- Student's Book pages 529–531

Learning aims

- Calculate the surface area and volume of a cuboid.

Key words

- volume, cuboid, surface area, litre

Prior knowledge

Students need to understand the concept of volume and know the standard conversions for length: 1 cm = 10 mm, 1 m = 100 cm. They also need to be able to calculate the area of a rectangle.

Common mistakes and remediation

Students need to ensure they are working with consistent units. Difficulties also ensue whenever algebraic formulae need to be rearranged, for example, when students are required to find the length of a cuboid, given the width and height. Although students who need more of a **challenge** should be able to rearrange a formula such as $V = lbh$, students who need more **support** may struggle.

Useful tips

Allow students who need extra **support** to use multi-link cubes, draw diagrams and estimate answers.

Starter

- Write on the board the numbers 2, 3 and 5. Ask students to multiply them together, in the order they appear, 2 × 3 × 5 (to get 30). Then ask if they get the same answer if they multiply the same numbers in a different order, such as 5 × 2 × 3. A short discussion should lead to the answer that it makes no difference. Students who need more **support** will need to be shown that these answers, and maybe 3 × 5 × 2, are all the same.
- Now ask the class to multiply 2 × 17 × 5 (answer 170). Discuss with students how they did this. Draw out the fact that it is easiest to multiply 2 and 5 first, to get 10, then multiply 10 by 17 to get 170. This is a very useful fact that can save a lot of time and effort.

Main lesson activity

- Either draw a 2 × 3 × 4 cuboid on the board or, better still, use multi-cubes to show the shape. Ask the students what the volume is. (24 cm^3)
- Draw another cuboid on the board, say 2 × 4 × 5, and ask what the volume of this is. (40 cm^3)
- Ask the students how they are calculating the volumes. Elicit that they are multiplying all the numbers together. Show that this works because one layer is the length × the width, then the number of layers is the height. Lead students to see that they can calculate the volume of a cuboid as width × length × height.
- Work through Example 10 with the class, emphasising the rule for the volume and asking students to explain how to find the surface area, which involves calculating length × breadth for three pairs of faces. Get the students to explain why the formula for the total surface area needs two of each of the three different parts. Ask them how they would calculate with the 3.5 to get an answer without using a calculator. Help them to see that 4 × 3.5 = 14 is easier to work out than 24 × 3.5.
- Students who need more **support** will appreciate the explanation of the surface area calculation by means of multi-cubes, so that they can see why there are three pairs of faces and follow the reasoning.
- Explain to students who need more of a **challenge** that 1000 cm^3 = 1 litre, and work with them to deduce how many litres there are in a cubic metre.
- 1 m^3 = 100 × 100 × 100 cm^3 = 1 000 000 cm^3 (a million).

  So there will be 1 000 000 ÷ 1000 = 1000 litres.

  Bring this to the attention of the students who need more **support** when they need it in question 2.
- Students can now work through Exercise 27F.

Plenary

- Present this problem: 'There are two cuboids. One has dimensions 4 cm × 5 cm × 9 cm. The second has dimensions 3 cm × 6 cm × 10 cm. They both have an equal volume. Which cuboid has the larger surface area?'

Homework and answers

Homework for this lesson is available in the homework folder. Consolidation and extension exercises are available in the further practice folder. Answers to the exercises can be found in the answers folder.

27.7 Volume and surface area of a prism

Syllabus references

Statement	Core
C4.1/E4.1	Use and interpret the vocabulary of simple solids.
C5.4/E5.4	Carry out calculations and solve problems involving the surface area and volume of a prism.
C5.5/E5.5	Carry out calculations and solve problems involving surface areas and volumes of compound solids and parts of solids.

Collins references

- Student's Book pages 531–534

Learning aims

- Calculate the volume of a prism.
- Calculate the surface area of a prism.

Key words

- cross-section, prism

Prior knowledge

Students need to be able to calculate areas of rectangles, triangles, trapezia and compound shapes.

Common mistakes and remediation

Students will sometimes mistake the length of the prism for a length occurring in the common cross-section. They should be able to avoid this mistake by drawing a good diagram and solving the problem in two distinct stages.

Useful tips

Use the image of a loaf of bread, which gives similar slices wherever you cut it, as long as you cut it correctly.

Starter

- Ask students if they have heard the word 'prism' before and what they understand by it.
- Show students as many examples of prisms that you can find (for example, packaging boxes or pictures).
- Discuss what they have in common (i.e. a constant cross-section parallel to one plane). Include some examples that are not prisms to make the point.

Main lesson activity

- Draw a rectangle, triangle, circle and trapezium on the board. Ask students to recall the areas of these shapes. Extend the diagrams into 3D prisms by adding depth. Label them: rectangular prism, triangular prism, circular prism, trapezoidal prism.
- Ask what the rectangular prisms and circular prisms are more commonly called (cuboid, cylinder). Why is it that the other two prisms do not have special names?
- Explain that the volume of a prism is given by the product of the area of the cross-section of the prism and its length. This might more usefully be written as $V = A \times$ length.
- For most problems, it is therefore important to stress that students calculate the cross-sectional area of the prism as an essential first step.
- Show students that this is equivalent to the formula for the area of a cuboid. The formula for the volume of a prism is a generalisation of that.
- For most problems, emphasise that students should always begin by calculating the cross-sectional area of the prism.
- If you have some real boxes, you could ask students to make appropriate measurements so that they can calculate the volume. Also, work through Example 11.
- Remind students about finding the surface area of a solid and that sometimes they will need to calculate that as well as the volume of a prism. Exercise 27G question 5b is one such question. For further practice see Homework 27.7.
- Students can now complete Exercise 27G.

Plenary

- Ask students to sketch a prism with a volume of 60 cm^3. This could be done in pairs. Use mini-whiteboards if you have them. Ask one or two students who have drawn interesting prisms to show them to the class.

Homework and answers

Homework for this lesson is available in the homework folder. Consolidation and extension exercises are available in the further practice folder. Answers to the exercises can be found in the answers folder.

27.8 Volume and surface area of a cylinder

Syllabus references

Statement	Core
C4.1/E4.1	Use and interpret the vocabulary of simple solids.
C5.4/E5.4	Carry out calculations and solve problems involving the volume and surface area of a cylinder.

Collins references

- Student's Book pages 534–536

Learning aims

- Calculate the volume and surface area of a cylinder.

Key words

- curved surface, cylinder

Prior knowledge

Students should know how to find the volume of a prism, as covered in Lesson 27.7, and be able to calculate the area of a circle as covered in Lesson 27.5b. They should be prepared to leave their answer as a multiple of π in any question that involves circles.

Common mistakes and remediation

As in Lesson 27.5b, students who need more **support** may be confused between πr^2 and $(\pi r)^2$. They may find it helpful to use a simpler version of the formula in the form $V = \pi \times r \times r \times h$ until their confidence improves.

Useful tips

Students should be able to give numerical answers and answers in terms of π.

Starter

- Take a trundle wheel into the class or display a picture of one. Explain that a trundle wheel is used to measure distances. It has a wheel with a circumference of 1 metre.
- Ask for the diameter of the wheel (31.8 cm) and the area (796 cm^2). Students can work in pairs, as this is revision of formulae that they will need during the lesson.

Main lesson activity

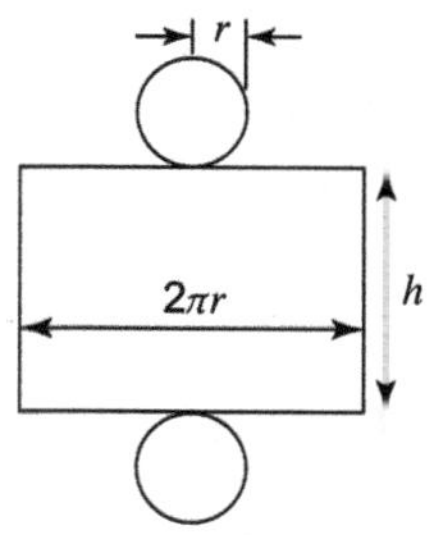

- Use an A4 sheet of paper to make a hollow cylinder by bending the long edges together. Remind students that a cylinder is a circular prism.
- Tell them that the sheet of paper is 29.7 cm long and 21.0 cm wide.
- Ask how they can find the area of the curved surface of the cylinder. This should be easy: $29.7 \times 21.0\ \text{cm}^2$.
- Now set the task of finding the volume of the cylinder. They could work in pairs to complete this. (The circumference of the base is 21.0 cm; radius = 3.34 cm; area = 35.09 cm^2; length = 29.7 cm; volume = 1040 cm^3 to 3 sf.)
- You could set an extension task of finding the volume of the cylinder formed by putting the shorter edges together. Is it the same? (No: volume = 1470 cm^3 to 3 sf.)
- Now consider a cylinder with two circular ends. The total surface area consists of a rectangle and two circles.
- Demonstrate that the curved surface area is πdh or $2\pi rh$, and the total surface area is $\pi dh + 2\pi r^2$ or $2\pi rh + 2\pi r^2$. Students can use either of these. You could show students who need more of a **challenge** that this is equivalent to $2\pi r(r + h)$.
- Work through Examples 12 and 13 with students.
- Now students can complete Exercise 27H. Students who need more **support** can complete questions they may find difficult with a partner.

Plenary

- Ask students (in pairs if you like) to find out the amount of metal that is used to make a coin.
- Discuss their answers as a class. For example, if they use 2 mm for the thickness of the coin and 26 mm for the diameter, they should get about 1060 mm^3 for the volume.
- Discuss the accuracy of the coin's 'thickness' measurement. Do the rim and the pattern make a difference?

Homework and answers

Homework for this lesson is available in the homework folder. Consolidation and extension exercises are available in the further practice folder. Answers to the exercises can be found in the answers folder.

27.9 Sectors and arcs: 1

Syllabus references

Statement	Core
C5.3/E5.3	Carry out calculations involving arc length and sector area as fractions of the circumference and area of a circle, where the sector angle is a factor of 360°.

Collins references

- Student's Book pages 536–538

Learning aims

- Calculate the length of an arc and the area of a sector.

Key words

- arc, sector, subtend

Prior knowledge

Students need to be able to calculate the area and circumference of a circle, as covered in Lesson 27.5 of the Student's Book.

Common mistakes and remediation

Students often forget that the perimeter of a sector will be the sum of the arc length plus two radii, not just the arc length alone. Emphasise this point.

Useful tips

Encourage students to use the correct terms 'arc' and 'sector'.

Remind students that when the question asks for the answer to be given 'as a multiple of π' they should not be using $\pi = 3.142$ or replacing π by a numerical value. Their answers should include the symbol π instead as that gives an exact answer, not an approximate one.

Starter

- Recap on students' knowledge of circles.
- Draw circles of radii 5 cm and 10 cm on the board. Ask students what they can tell you about the circumferences. (They may tell you how to work them out.) Try to elicit the fact that one circumference is double the other.
- Ask if you can generalise. (If the radius of a circle is doubled, so is the circumference.)
- Now ask about the areas. Is the same thing true? (This time one area is four times the other.)

Main lesson activity

- Following their work on the area and circumference of circles in Lesson 27.5, most students should be in a position to begin to discover the formulae for the arc perimeter and area of a sector.
- Ask students to write down the formula for the circumference of a circle. Then ask them to write a formula for the curved part of a semi-circle, and again for a quarter circle. Can they make the leap to finding the formula for an arc sector, given a sector angle?
- Assist if necessary by, for example, drawing a sector with a given sector angle and asking what fraction of the circle (or circumference) has been drawn. The area of a sector could be found in a similar way.
- Explain that if an arc ab subtends an angle $\theta°$ at the centre of a circle, then the arc length ab is given by:

  $$\text{arc length} = \frac{\theta}{360} \times 2\pi r$$

  The area of the sector is given by:

  $$\text{area} = \frac{\theta}{360} \times 2\pi r^2$$

- Emphasise that the angle tells you the fraction of a whole circle you are considering. The formula is less important.
- Work through Example 14 briefly.
- Students can now complete Exercise 27I. If students who need more **support** are struggling to complete the exercise, you could advise them to work with a partner.

Plenary

- Draw two sectors: one with an angle of 90° and radius 4 cm, the other with an angle of 45° and radius 8 cm.
- Ask: Which arc has the longer arc length? Which sector has the bigger area?
- Students should think about this in pairs for a couple of minutes. Then ask them to feed back their comments.
  (They have the same arc length. The second is double the area of the first.)
- Ask students for more examples of sectors with the same arc length as the first two.
  (Other examples are 180° and 2 cm; 60° and 6 cm; 120° and 3 cm; 22.5° and 16 cm.)

Homework and answers

Homework for this lesson is available in the homework folder. Consolidation and extension exercises are available in the further practice folder. Answers to the exercises can be found in the answers folder.

27.10 Sectors and arcs: 2

Syllabus references

Statement	Extended
E5.3	Carry out calculations involving the arc length and sector area as fractions of the circumference and area of a circle.

Collins references

- Student's Book pages 538–540

Learning aims

- Calculate the length of an arc and the area of a sector.

Key words

- arc, sector, subtend

Prior knowledge

Students need to be able to calculate the area and circumference of a circle, as covered in Lesson 27.5. They should also have worked through Lesson 27.9 where the sector angle was a factor of 360°.

Common mistakes and remediation

Students often forget that the perimeter of a sector will be the sum of the arc length plus two radii, not just the arc length alone. Emphasise this point.

Useful tips

Encourage students to use the correct terms 'arc' and 'sector'.

Remind students that when the question asks for the answer to be given 'as a multiple of π' they should not be using $\pi = 3.142$ or replacing π by a numerical value. Their answers should include the symbol π instead as that gives an exact answer, not an approximate one.

Starter

- Recap on students' knowledge of arcs and sectors

Main lesson activity

- Following their work on arcs and sectors, students should not need much time introducing this section and you might want to do this lesson at the same time as 27.9 with students who need more of a **challenge**.
- Remind students that if an arc ab subtends an angle $\theta°$ at the centre of a circle, then the arc length ab is given by:

  arc length $= \frac{\theta}{360} \times 2\pi r$

  The area of the sector is given by:

  area $= \frac{\theta}{360} \times 2\pi r^2$
- Emphasise that the angle tells you the fraction of a whole circle you are considering. The formula is less important.
- Work through Example 15 briefly.
- Students can now complete Exercise 27J. If students who need more **support** are struggling to complete the exercise, you could advise them to work with a partner.

Plenary

- Draw two sectors: one with an angle of 40° and radius 4 cm, the other with an angle of 20° and radius 8 cm.
- Ask: Which arc has the longer arc length? Which arc has the longer perimeter? Which sector has the bigger area?
- Students should think about this in pairs for a couple of minutes and then ask them to feed back their comments.
  (They have the same arc length. The second has a longer perimeter. The second is double the area of the first.)
- Ask students for more examples of sectors with the same perimeter as the first one.
  (There are an infinite number of examples – students can work in pairs to check each other's answers.)

Homework and answers

Homework for this lesson is available in the homework folder. Consolidation and extension exercises are available in the further practice folder. Answers to the exercises can be found in the answers folder.

27.11 Volume of a pyramid

Syllabus references

Statement	Core
C5.4/E5.4	Carry out calculations and solve problems involving the surface area and volume of a pyramid.
C5.5/E5.5	Carry out calculations and solve problems involving surface areas and volumes of compound solids and parts of solids.

Collins references

- Student's Book pages 540–543

Learning aims

- Calculate the volume of a pyramid.

Key words

- pyramid, vertical height, vertex

Prior knowledge

Students should know how to calculate the areas of simple shapes such as circles, rectangles and triangles.

Common mistakes and remediation

Students sometimes assume that the height is given by the length of a sloping edge rather than the vertical height. Point out this error whenever it occurs in order to minimise it.

Useful tips

Usually the apex of a pyramid is vertically above the centre of the base, but this does not have to be the case. You will need some real examples of pyramids to show students. Sets of such models can be purchased. Also look out for packaging and paperweights in the shape of pyramids.

Starter

- Show students a picture of the Great Pyramid of Giza. Tell them that the base was originally a square of side 230.4 m and the height was 146.6 m. Its weight is estimated as 5.9 million tonnes.
- Ask students to estimate the volume. You are not looking for accuracy, just a reasonable guess, so suggest that they compare it with a cuboid of a similar size (about 2.5 million m^3).
- Finally, ask them to calculate: side divided by height and double it. Where have they seen this number before? It is extremely close to π. So the ratio of side to height is $\frac{\pi}{2}$. Is this a coincidence? Nobody knows.

Main lesson activity

- Explain to students that there is a formula for the volume of a pyramid that involves the base area and the height.
- Write on the board: Volume = ? × area of base × height
- Explain that the question mark stands for a number. What number? The plenary should help to reinforce that it is $\frac{1}{3}$.
- You could compare this to the formula for the area of a triangle:

  area = $\frac{1}{2}$ × length of base × height
- Use the formula to find the volume of the Great Pyramid.
- If you have any models of pyramids, ask students what measurements you need to take to find the volumes. Get them to work out the measurements and the calculations.
- Explain that pyramids do not necessarily have a square base, although this is the most common type.
- Show any examples you have. The formula for the volume is the same, whatever the shape of the base.
- Work through Examples 16 and 17 with students.
- Now students can complete Exercise 27K. Students who need more **support** can complete questions 1–3 on their own, and then work with a partner to complete the rest. Provide support as needed.

Plenary

- Ask students to sketch a pyramid with a volume of 64 cm^3. Collect suggestions.

  Three possibilities are: base 4 × 4, height 12; base 8 × 8, height 3; base 2 × 2, height 48.

Homework and answers

Homework for this lesson is available in the homework folder. Consolidation and extension exercises are available in the further practice folder. Answers to the exercises can be found in the answers folder.

27.12 Volume and surface area of a cone

Syllabus references

Statement	Core
C5.4/E5.4	Carry out calculations and solve problems involving the surface area and volume of a cone.
C5.5/E5.5	Carry out calculations and solve problems involving surface areas and volumes of compound solids and parts of solids.

Collins references

- Student's Book pages 543–545

Learning aims

- Calculate the volume and surface area of a cone.

Key words

- cone, vertical height, slant height

Prior knowledge

Students should know how to find the volume of a pyramid, as covered in Lesson 27.11.

They should also be able to give their answers in terms of π where requested.

Common mistakes and remediation

Students often confuse the height and the slant height. Make sure they understand the difference between them.

Useful tips

Some real examples of cones to show students would be useful. Try to find examples with different angles at the apex.

Starter

- Give students a circular piece of paper (or a circle printed on paper for them to cut out). Ask them to find the centre. They can do this by folding the circle in half in two different ways and finding where the diameters cross, but give students a chance to work out a method by themselves.
- Ask them to draw in any two radii and cut along them. They now have two sectors.
- Show them how to tape or glue the straight edges to make a cone. Do this with both sectors.
- Ask students to describe the similarities and differences. Main points to bring out:
- All have the same slant height – introduce this terminology and distinguish it from the (actual) height.
- They have different heights and bases. Make it clear that the height is measured perpendicular to the base.
- The greater the height, the smaller the base area/diameter, and vice versa.
- Ask them to measure the height and base diameter of each cone.

Main lesson activity

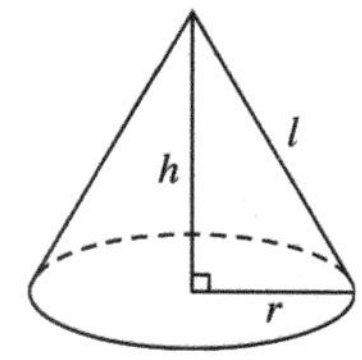

- Ask students how they can find the volume of a cone.
  They should realise that a cone is a pyramid with a circular base, so the volume is $\frac{1}{3}$ × area of base × height.
  For a cone, volume $= \frac{1}{3} \times \pi r^2 \times h$ or $V = \frac{1}{3}\pi r^2 h h$
- Ask them to calculate the volume of each cone they made in the starter. Now ask them how they can find the area of the curved surface of the cone.
- In discussion bring out these points:

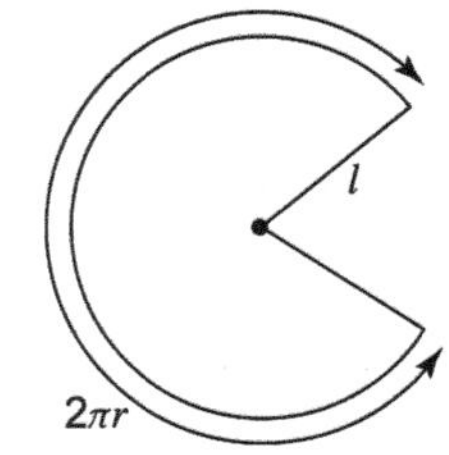

- The curved surface area is the same as the area of the original sector.
- The radius of the original sector is the slant height of the cone. Call it l.
- The area of the original circle is πl^2.
- The circumference of the original circle is $2\pi l$.
- The length of the arc of the sector is $2\pi r$, where r is the radius of the circular base of the cone.
- The sector is a fraction of the original circle; it is $\frac{2\pi r}{2\pi l} = \frac{r}{l}$.
- The area of the curved surface is $\frac{r}{l} \times \pi l^2 = \pi r l$.
- Some students will find the details of this demanding. Emphasise the final result.
- Ask students to find the area of the curved surface of each of their cones.
- Finally, ask students to imagine that they have added a circular base to their cone. What is the total surface area of each one (curved surface + base)? Show only students who need more of a **challenge** that a formula for the total surface area of a cone is $A = \pi rl + \pi r^2 = \pi r(l + r)$.
- Look at Example 18 with students.
- Now students can complete Exercise 27L. Allow students who need more **support** to complete the questions in pairs, if necessary. Provide support as needed.

Plenary

- Look at students' answers to question 4.
- Ask if there is a connection between the angle of the sector and the diameter of the base.
- In fact $d = \frac{\text{angle}}{180} \times l$, but students may express this in different ways.

Homework and answers

Homework for this lesson is available in the homework folder. Consolidation and extension exercises are available in the further practice folder. Answers to the exercises can be found in the answers folder.

27.13 Volume and surface area of a sphere

Syllabus references

Statement	Core
C5.4/E5.4	Carry out calculations and solve problems involving the surface area and volume of a sphere.

Collins references

- Student's Book pages 545–547

Learning aims

- Calculate the volume and surface area of a sphere.

Key words

- sphere

Prior knowledge

Students should have covered Lessons 27.11 and 27.12.

They should also be able to give their answers in terms of π where requested.

Common mistakes and remediation

Students may make errors keying $\frac{4}{3}$ in the volume formula into their calculators. Make sure they can use their own calculators efficiently and correctly. Tell students that using an approximation such as 1.33 can lead to errors and should be avoided.

Useful tips

The main activity requires an orange, which should be roughly spherical; and, if possible to obtain, some centicubes.

Starter

- Revise the formulae covered so far.
- Draw a shape and ask for formulae for area or volume as appropriate.

Main lesson activity

- Explain that the final shape students will look at is the sphere. Give them the following formulae:
  - volume = $\frac{4}{3}\pi r^3$
  - surface area = $4\pi r^2$
- Show students the orange. Say that you will assume it is spherical. Ask what measurements they need to find the volume and surface area of the orange. The radius is difficult to measure, but the diameter is possible if you have callipers or put a book on top, a table below, and then measure the gap.
- Ask students to calculate the volume and surface area, remembering to include units. Ask if their results seem reasonable. (A cube made of centicubes will help them see if the volume seems reasonable.)
- A sphere cannot be flattened like a cone so that one can look at the area. Demonstrate this by peeling the orange. The peel cannot be flattened completely, but if the pieces are small they will be fairly flat.
- To estimate the surface area of the orange, place the pieces on a sheet of centimetre square paper, draw around the outline and count the squares.
- Check that it is close to the calculated value. You could have a good discussion of what 'close' means in this context.
- Work through Example 19 with students.
- Now students can complete Exercise 27M, in which the last question is for students who need more of a **challenge** only.

Plenary

- Set this question: The Earth is approximately spherical. The length of the equator is 40 000 km. Two-thirds of the Earth's surface is covered with water. What area of the Earth's surface is covered with water? (About 340 million km^2.)
- Let students do the question in pairs. Discuss how they worked out the answer and the appropriate degree of accuracy for the area. ($\frac{2}{3}$ is an approximation, so the answer should also be an approximation.)

Homework and answers

Homework for this lesson is available in the homework folder. Consolidation and extension exercises are available in the further practice folder. Answers to the exercises can be found in the answers folder.

28.1 Lines of symmetry

Syllabus references

Statement	Core
C4.5/E4.5	Recognise line symmetry in two dimensions.

Collins references

- Student's Book pages 550–551

Learning aims

- Draw the lines of symmetry on a 2D shape.
- Recognise shapes with reflective symmetry.

Key words

- line of symmetry, mirror line

Prior knowledge

Knowing the names of common 2D shapes is an advantage.

Common mistakes and remediation

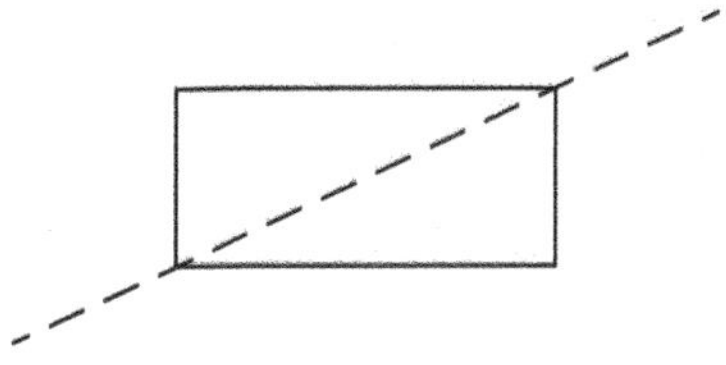

Drawing this as a line of symmetry is by far the most common error, closely followed by not putting in all the lines when dealing with polygons – for example, putting only three lines of symmetry on a regular hexagon.

Students who need more **support** benefit from using a mirror initially, especially with rectangles, parallelograms, pentagons and hexagons.

Useful tips

A mirror or tracing paper can be used to check for lines of symmetry.

Starter

- Have some large shapes made out of thin paper, such as a square, a rectangle and a parallelogram. Ask students where the lines of symmetry are for each. Fold the shapes along these.
- For the square, this will give each half exactly covering the other. It is likely that the students will designate a diagonal of the rectangle as a line of symmetry. Make sure that they see this does not work, similarly with the parallelogram.

Main lesson activity

- Students should copy the following into their books. Students who need more **support** could be given copies of the shapes.

Lines of symmetry

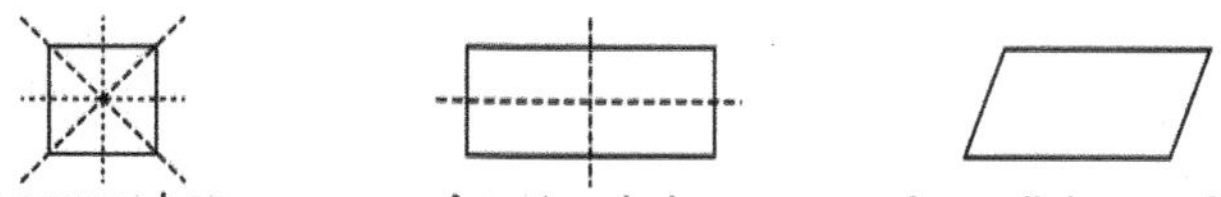

A square has 4 lines of symmetry | A rectangle has 2 lines of symmetry | A parallelogram has 0 lines of symmetry

- Encourage students to use a mirror or tracing paper to check their lines of symmetry, especially the parallelogram.
- Now let them do the same to check capital letters for symmetry. For example:

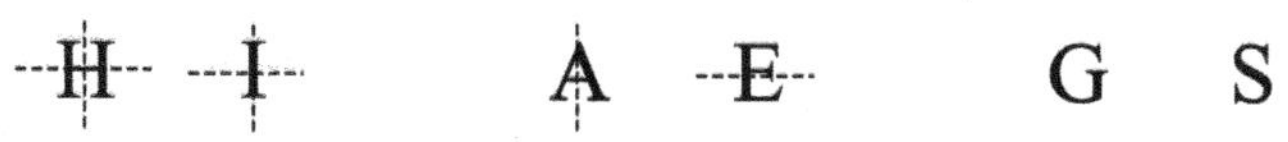

These letters have 2 lines of symmetry. | These letters have 1 lines of symmetry. | These letters have 0 lines of symmetry.

- Students who need more of a **challenge** could discuss the number of lines of symmetry for a circle or more complex shapes. For example:

A circle has an infinite number of line of symmetry | This star shape has 5 lines of symmetry

- Students may find it difficult to see lines of symmetry that are not horizontal or vertical. Encourage students who need more **support** to turn the book, so that the lines of symmetry appear horizontal or vertical.
- Initially, encourage students who need more **support** to use a mirror to check for lines of symmetry. Encourage them to do without the mirrors, as they gain confidence.
- Work through Example 1 with students if necessary before they do Exercise 28A.

Plenary

Ask students questions about line symmetry.

- How many lines of symmetry are there in this shape?

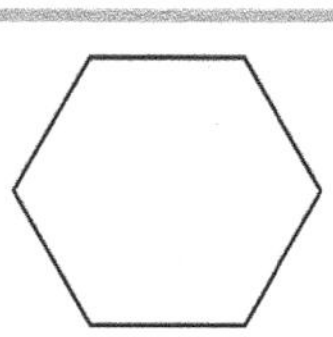

- How many lines of symmetry are there in each of these letters?

O T I S

- How many lines of symmetry are there in these regular shapes?

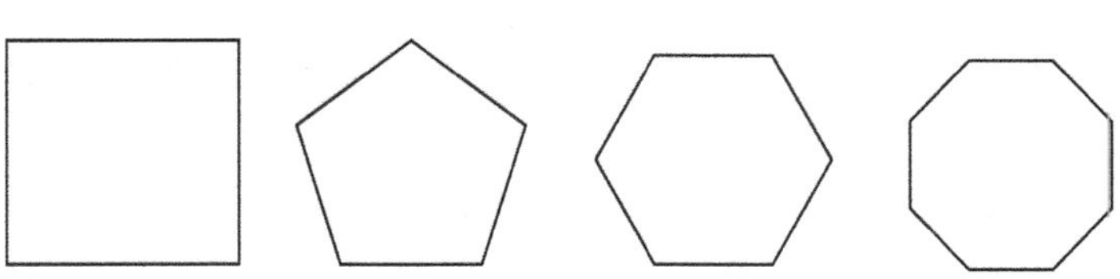

- Discuss the connection between the number of sides of a regular polygon and the number of lines of symmetry.

Homework and answers

Homework for this lesson is available in the homework folder. Consolidation and extension exercises are available in the further practice folder. Answers to the exercises can be found in the answers folder.

28.2 Rotational symmetry

Syllabus references

Statement	Core
C4.5/E4.5	Recognise order of rotational symmetry in two dimensions.

Collins references

- Student's Book pages 552–553

Learning aims

- Find the order of rotational symmetry for a 2D shape.
- Recognise shapes with rotational symmetry.

Key words

- rotational symmetry, order of rotational symmetry

Prior knowledge

Students need to know the names of the commonly used 2D shapes.

Common mistakes and remediation

Not using tracing paper usually leads to inaccurate answers on the more complicated shapes. Students often say a shape has order 0 instead of order 1 for 'no' rotational symmetry.

Students often confuse line symmetry and rotational symmetry.

Students who need more **support** should always have access to mirrors and tracing paper and practice using them. A line of symmetry can also be referred to as a mirror line, so remember to use a mirror for lines of symmetry.

Useful tips

Students can use tracing paper when finding the order of rotational symmetry.

Starter

- Draw a square with a dot in the top right-hand corner: 
- In your head, turn it once clockwise and draw it: 
- Do it again: 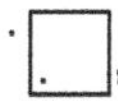, and again: 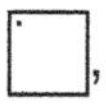, and again: 
- How many different squares are there? (4)
- Do the same with a rectangle (2), an equilateral triangle (3) and a hexagon (6).
- What about a circle? (infinity) Or a right-angled triangle? (0)

Main lesson activity

- After the starter, discuss with students how a square can be turned four different ways to fit exactly on top of itself. Explain that this is called rotational symmetry. Because a square can be fitted exactly over itself in four ways, we say that a square has rotational symmetry of order 4.
- Use the whiteboard, or tracing paper, to demonstrate this to the class. Work through Example 2, going through it step by step, to ensure that students recognise that the method works for any shape with rotational symmetry.
- Ask the class to tell you the order of rotational symmetry for a rectangle and for a parallelogram. Again, demonstrate how to use tracing paper or the whiteboard to check.
- Students should copy the following diagrams, with the notes, into their books. Students who need more **support** could be given copies of the shapes.

Rotational symmetry

A square has rotational symmetry of order 4

A rectangle has rotational symmetry of order 2

A parallelogram has rotational symmetry of order 2

- Encourage students, particularly those who need more **support**, to use tracing paper to check the order of rotational symmetry for shapes.
- Now follow through the same procedure for an equilateral triangle and an isosceles triangle.
- Explain that the isosceles triangle has no rotational symmetry. Since it fits exactly on top of itself only once, we say that it has rotational symmetry of order 1.
- An equilateral triangle has rotational symmetry of order 3.

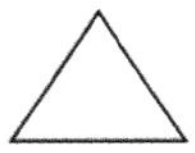
An equilateral triangle has rotational symmetry of order 3

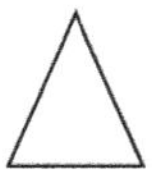
An isosceles triangle has rotational symmetry of order 1

- Students who need more of a **challenge** could discuss the order of rotational symmetry that a circle has, and then consider more complex shapes. For example:

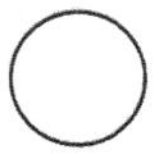
A circle has rotational symmetry of order infinity

This shape has rotational symmetry of order 8

- Students can now do Exercise 28B.

Plenary

- Ask students to name any capital letters with line symmetry. List them on the board. (A, B, C, D, E, H, I, M, O, Q, T, U, V, W, X, Y)
- Ask students to name any letters with rotational symmetry. (H, I, N, O, S, X, Z)
- Ask the class what is special about the words: BED? (line symmetry); NOON? (rotational symmetry). Are there any others? (CHOICE, HEX, COD etc.)

Homework and answers

Homework for this lesson is available in the homework folder. Consolidation and extension exercises are available in the further practice folder. Answers to the exercises can be found in the answers folder.

28.3 Symmetry of special shapes in two dimensions

Syllabus references

Statement	Core
C4.5/E4.5	Recognise line symmetry and order of rotational symmetry in two dimensions.

Collins references

- Student's Book pages 553–554

Learning aims

- To recognise the symmetry properties of particular triangles and quadrilaterals.

Key words

- rectangle, square, parallelogram, kite, rhombus, trapezium, isosceles triangle, equilateral triangle

Prior knowledge

Understand lines of symmetry and the order of rotational symmetry of a two dimensional shape from Lessons 28.1 and 28.2.

Common mistakes and remediation

Students may need to be reminded of the special names of different three- and four-sided shapes. A set of flash cards can be a useful way of revising them.

Students often say a shape has order 0 instead of order 1 for 'no' rotational symmetry.

Students often confuse line symmetry and rotational symmetry.

Useful tips

Students should understand that many combinations of rotational and reflection symmetry are not possible and they may be asked to find a shape which is impossible.

Starter

- Draw a three-by-three grid of nine squares. Ask students to shade <u>four</u> squares to give:
  - a design with one line of symmetry
  - a design with two lines of symmetry
  - a design with four lines of symmetry
  - a design with no lines of symmetry and rotational symmetry of order 2
  - a design with no lines of symmetry and rotational symmetry of order 4.
- Ask students to give their answers in each case. It is useful to discuss whether designs are different or essentially the same (just rotated or reflected). Note that the second and fifth cases are impossible if exactly four squares are to be shaded.

Main lesson activity

- Ask students to draw triangles that have no, one, two or three lines of symmetry.
- Compare their answers and bring out the following points.
  - A wide variety of triangles have no lines of symmetry.
  - All triangles with one line of symmetry are isosceles.
  - Two lines of symmetry is impossible.
  - All triangles with three lines of symmetry are isosceles and all the angles are 60 degrees.
- Remind students that we have special names for some types of quadrilaterals. See the key words above. Ask students to name a four-sided shape with exactly one line of symmetry. Answers: rectangle, kite and, in special cases, trapezium.
- Ask students to name a four-sided shape with exactly two lines of symmetry. Answers: rectangle and rhombus.
- Students can now do Exercise 28C.

Plenary

- Ask students to draw a four-sided shape to fit in each cell of this table:

	Two lines of symmetry	**Four lines of symmetry**
Rotational symmetry of order 2		
Rotational symmetry of order 4		

- Here are possible answers:

	Two lines of symmetry	**Four lines of symmetry**
Rotational symmetry of order 2	Rectangle or rhombus	Impossible
Rotational symmetry of order 4	Impossible	Square

- Discuss the students' answers and agree that all their drawings fit these descriptions.

Homework and answers

Homework for this lesson is available in the homework folder. Consolidation and extension exercises are available in the further practice folder. Answers to the exercises can be found in the answers folder.

28.4 Symmetry in three dimensions

Syllabus references

Statement	Extended
E4.1	Use and interpret geometrical terms.
E4.5	Recognise the symmetry properties of the prisms, cylinders, pyramids and cones.

Collins references

- Student's Book pages 555–557

Learning aims

- To find the planes and axes of symmetry of simple three-dimensional shapes.

Key words

- plane of symmetry, axis of symmetry, cuboid, prism, pyramid, cone, cylinder, reflection

Prior knowledge

Reflection and rotation symmetry of two-dimensional shapes.

Common mistakes and remediation

Students may think that a plane through a cuboid (rectangular box) joining one edge to an opposite edge is a plane of symmetry. Discuss this at an appropriate point and clarify why it is incorrect. It is similar to thinking that the diagonal of a rectangle is a line of symmetry.

Useful tips

Make a collection of boxes and containers of different shapes to use with students. Look out for packaging of chocolates, gifts and food items in unusual shapes.

Starter

- Draw a coordinate grid and mark the point (4, 2). Say that this is one point on a two-dimensional shape. Say that both the x-axis and the y-axis are lines of symmetry. Also, the shape has rotational symmetry of order 4 about the origin. Ask the students to add as many more points on the shape as they can.
- They should find seven more points, giving eight all together. They should have (± 4, ± 2) and (± 2, ± 4). They make the corners of an octagon, centre the origin. They should be able to justify the positioning of each point – for example, a 90 degree clockwise rotation about the origin (for symmetry of order 4) takes (4, 2) onto (–2, 4).

Main lesson activity

- Hold up a rectangular box (cuboid) such as a food box or a shoe box. Explain how a mirror through the centre of the box parallel to two faces would reflect one half of the box exactly onto the other. In how many ways can we do this? (There are three ways, corresponding to the three different opposite pairs of parallel faces.)
- Say that the mirror corresponds to a plane of symmetry. This corresponds to a line of symmetry for a two-dimensional shape. Discuss how this could be drawn. The diagram at the start of 28.4 in the Student's Book shows a way to do this.
- For rotational symmetry there needs to be an axis of symmetry. The shape can rotate about this. This corresponds to a centre of rotation for a two-dimensional shape and in fact you can imagine an axis of symmetry passing through the centre of rotation for a two-dimensional shape. Demonstrate this with an axis passing through the centres of opposite faces. If possible show this practically with a stick passing through the box. Show that the shape has rotational symmetry of order 2 about this axis. Ask students to find other axes of symmetry. (There are two more. There is one through the centres of each pair of opposite faces.)
- Ask the students to draw an axis of symmetry on a diagram of the box. Again, there is an example in the Student's Book. Using dashed lines where the axis is inside the shape can be useful.
- Students can now start Exercise 28D.

Plenary

- Hold up a box in the shape of a cube and, if possible, give each student a cube to look at. A dice can be used for this purpose or construction cubes. Ask students how many axes of symmetry does a cube have? What is the order of rotational symmetry about each one? This is a good exercise for students to work on in pairs.
- Answer: There are 13.
    - Three pass through the mid points of opposite faces and the order of rotational symmetry is 4. This is similar to the rectangular box but the order is 4 this time instead of 2.
    - Six pass through the mid points of opposite edges and the order of rotational symmetry is 2.
    - Four pass through diagonally opposite corners and the order of rotational symmetry is 3. This one is the hardest to visualise.
- Geologists classify crystals of minerals by their symmetry properties and so 3D symmetry is very important to them in identifying minerals.

Homework and answers

Homework for this lesson is available in the homework folder. Consolidation and extension exercises are available in the further practice folder. Answers to the exercises can be found in the answers folder.

28.5 Symmetry in circles

Syllabus references

Statement	Extended
E4.8	Use the following symmetry properties of circles: equal chords are equidistant from the centre, the perpendicular bisector of a chord passes through the centre, tangents from an external point are equal in length.

Collins references

- Student's Book pages 557–559

Learning aims

- Chords of the same length are the same distance from the centre of a circle.
- The perpendicular bisector of a chord passes through the centre of the circle.
- The two tangents to a circle from an external point are equal in length.

Key words

- centre, chord, tangent, perpendicular bisector

Prior knowledge

It is helpful if students know how to construct the perpendicular bisector of a line using compasses and a straight edge, even though this construction is not officially part of the syllabus.

They should be familiar with chords and tangents and know that a tangent and radius at a point on a circle are perpendicular.

Common mistakes and remediation

Students do not always give reasons for their answers when asked to do so. The exercise gives them a chance to practise this skill and it should be emphasised when discussing the answers.

Useful tips

Encourage students to use the correct vocabulary throughout this section.

Starter

- Students will need a pair of compasses.
- Ask students to:
  - Draw a triangle; any shape or size but not too small.
  - Construct the perpendicular bisector of each side in turn. A check on accuracy is that each of these three lines should meet at one point.
  - Using the meeting point as the centre, draw a circle that passes through each corner of the triangle.
- Check that students know that each of the sides of the triangle is a chord of the circle.
- Ask students to measure the distance from the chord to the centre. This is done by measuring along the perpendicular bisector.

Main lesson activity

- This is a visualisation exercise. It should be carried out with eyes closed and students should remain silent throughout.
  - Imagine a chord passing through the centre of the circle. In your mind construct the perpendicular bisector of the chord. This will pass through the centre of the circle.
  - Imagine the ends of the chord moving round the circle without changing the length of the chord. The perpendicular bisector moves too.
  - What is the path of the centre of the chord?
  - How does the perpendicular bisector move?
  - What happens to the distance of the chord from the centre of the circle?
- Now ask students to open their eyes and discuss what happened and establish the following:
  - the centre of the chord follows a circular path
  - the bisector rotates about the centre of the circle, always passing through it
  - the distance of the chord from the centre does not change.
- State the following results:
  - The perpendicular bisector of a chord passes through the centre of the circle.
  - For a given circle, chords of the same length are the same distance from the centre.
- Finally ask students to draw a circle. Choose any point outside the circle and draw two tangents to the circle, one on each side. Remind them if necessary that each tangent will just touch the circle at one point and will be perpendicular to the radius at the point of contact. Measure the length of each tangent, from external point to point of contact. These should be the same for each particular example.
- This is not a proof that tangents from an external point are equal in length but students should agree that it seems reasonable. The fact that the diagram looks symmetrical will help to convince them.
- Students can now work through Exercise 28E.

Plenary

- Ask students to imagine an inverted hollow cone with its axis vertical and with a sphere inside it. What can they say about the points of contact between the circle and the sphere? Can they justify their answer?
- It is a circle. One way to see this is to imagine cutting the cone and sphere down a plane of symmetry. This will be any vertical plane through the centre of the sphere. The cut surface will look like a circle with two tangents from an external point and the points where the tangents touch the circle are the points where the cone meets the sphere.

Homework and answers

Homework for this lesson is available in the homework folder. Consolidation and extension exercises are available in the further practice folder. Answers to the exercises can be found in the answers folder.

29.1 Translations

Syllabus references

Statement	Core
C7.1/E7.1	Translate a shape by a vector $\begin{pmatrix} x \\ y \end{pmatrix}$.

Collins references

- Student's Book pages 562–564

Learning aims

- Translate a 2D shape.

Key words

- transformation, translation, vector

Prior knowledge

Students need very little prior knowledge to begin basic work on transformations. Recent experience of work involving Cartesian coordinates might be useful, but is not essential.

Common mistakes and remediation

Students who need more **support** may be confused by the terms 'transformation' and 'translation'. Ensure that they understand that a translation is a specific type of transformation, as are a rotation and a reflection.

Sometimes students mark vertices on the shapes that do not correspond and hence make mistakes when giving the vector translation. Make sure that students know the vertices they mark should be corresponding vertices on the object and image.

Remind students that column vectors are not fractions and should not have a line between the x and y numbers.

Useful tips

To build confidence, encourage students to use the terms 'transformation' and 'translation' in the correct context.

Remind them that when a triangle has simply been translated, it has not been reflected or rotated.

Starter

- Begin by discussing the general idea of the transformation of an object, and what it means. Many students will have come across different types of transformations before, so ask them to name specific types.
- Emphasise that students will study each of the four basic transformations – translation, reflection, rotation and enlargement – in turn. They need to be able to recognise, describe and perform these transformations.

Main lesson activity

- Explain that a translation is a transformation that moves a shape from one position to another.
- Triangle A has been translated to triangle B. Point out that under the translation, the triangle keeps its orientation: it has not been reflected or rotated. Also, triangle A is congruent to triangle B. Say: 'Triangle A has been mapped onto triangle B by a translation.' Explain that this just means one triangle has been changed by being slid into another position.

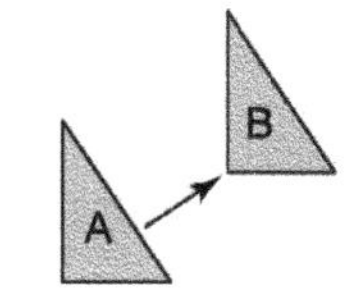

- Discuss how to describe different translations on a coordinate grid.

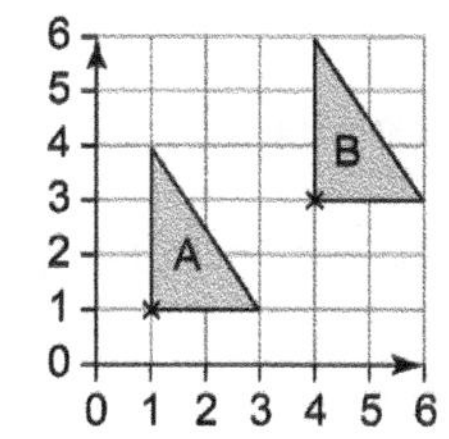

- Put a cross on a vertex on both triangles. Explain that to move from its position in triangle A to triangle B the cross moves three squares to the right and two squares up. Show students that it is the same movement for the other two vertices.
- This movement or translation is represented by the vector $\begin{pmatrix} 3 \\ 2 \end{pmatrix}$.
- A vector has two distinct components:

  $\begin{pmatrix} x \\ y \end{pmatrix}$ x → number of squares moved horizontally; y → number of squares moved vertically

  If x is positive, it describes a movement to the right.

  If x is negative, it describes a movement to the left.

  If y is positive, it describes a movement up.

  If y is negative, it describes a movement down.
- Take students through Example 1, explaining how this works. Reinforce with students who need more **support** that a vector is not a coordinate pair.
- Now students can complete Exercise 29A. Suggest that students who need more **support** complete questions 1 to 3 in pairs before moving on to complete the exercise. Provide support as needed.

Plenary

- Draw on the board any two congruent shapes, for example:

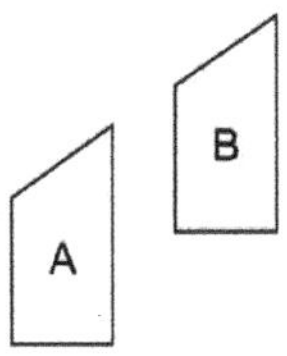

- Tell students that shape A is mapped onto shape B by the vector $\begin{pmatrix} x \\ y \end{pmatrix}$.
- Ask students to give the vector that would map shape B onto shape A. Answer: $\begin{pmatrix} -x \\ -y \end{pmatrix}$.

Homework and answers

Homework for this lesson is available in the homework folder. Consolidation and extension exercises are available in the further practice folder. Answers to the exercises can be found in the answers folder.

29.2 Reflections: 1

Syllabus references

Statement	Core
C7.1/E7.1	Reflect a shape in a vertical or horizontal line.

Collins references

- Student's Book pages 564–565

Learning aims

- Reflect a 2D shape in a mirror line.

Key words

- reflection, object, image, mirror line

Prior knowledge

Students should have completed Lesson 29.1 on translations and be familiar with equations of lines of the type:

$y = x$, $y = -x$, $y = a$ and $x = b$

Common mistakes and remediation

Students often draw the image and the object at different distances from the mirror line.

Starter

- Draw a set of axes, each labelled from –5 to 5. Draw some horizontal and vertical lines and ask students for their equations.
- Make sure they understand that lines of the form $x = b$ are vertical and those of the form $y = a$ are horizontal. Finish with $y = x$ and $y = -x$.

Main lesson activity

- Explain that a reflection is a transformation that changes a shape into a mirror image of itself. The original shape is the object and the reflected shape is the image.
- In this diagram triangle X has been reflected in the mirror line to give triangle Y.

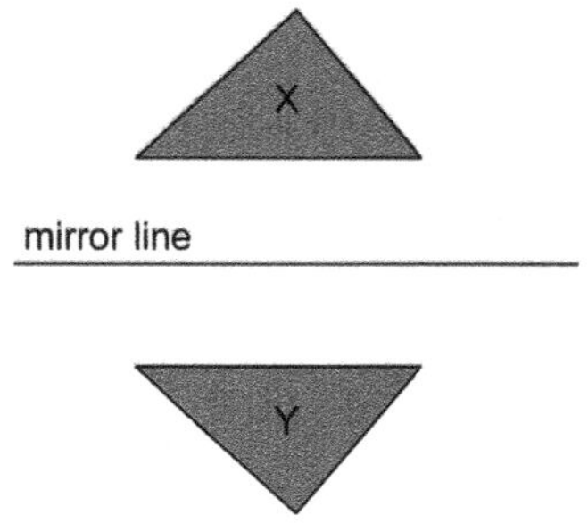

- Point out that, under the reflection, each vertex on the object is the same distance from the mirror line as each corresponding vertex on the image.
- Point out that triangle X is congruent to triangle Y. Tell students that Triangle X has been mapped onto triangle Y by a reflection in the mirror line. Look at Example 2 in the Student's Book.
- Discuss how to describe different reflections on a coordinate grid.
- Establish that the best way to do this is to give the equation of the mirror line. Demonstrate, with an example.

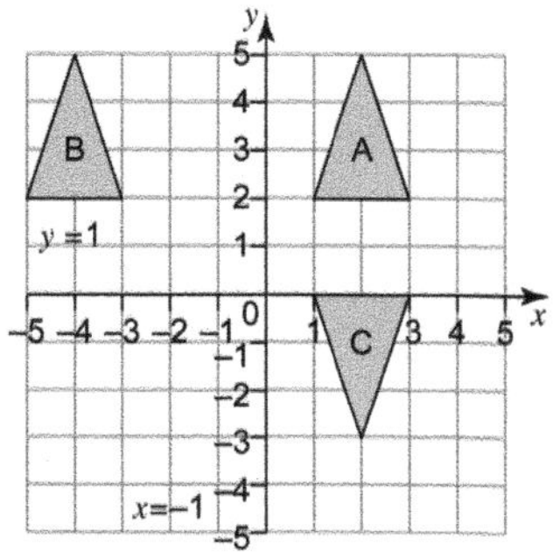

- Triangle B is the reflection of triangle A in the line $x = -1$.
- Triangle C is the reflection of triangle A in the line $y = 1$.
- Students can now complete Exercise 29B.

Plenary

- Ask students to reflect some points in the line $y = x$ and write down the coordinates of the object points with their image points. Ask them what they notice.

Homework and answers

Homework for this lesson is available in the homework folder. Consolidation and extension exercises are available in the further practice folder. Answers to the exercises can be found in the answers folder.

29.3 Reflections: 2

Syllabus references

Statement	Extended
E7.1	Reflect a shape in a straight line.

Collins references

- Student's Book pages 566–568

Learning aims

- Reflect a 2D shape in a diagonal mirror line.

Prior knowledge

Students should have completed Lesson 29.1 on translations and be familiar with equations of lines of the type:

$y = x$, $y = -x$, $y = a$ and $x = b$

They should also have completed Lesson 29.2 on simple reflections and be familiar with finding the equation of the mirror line.

Common mistakes and remediation

Students who need more **support** find it more difficult to reflect shapes in a diagonal mirror line.

Students often draw the image and the object at different distances from the mirror line. When reflecting shapes in diagonal lines, they may also draw them like this.

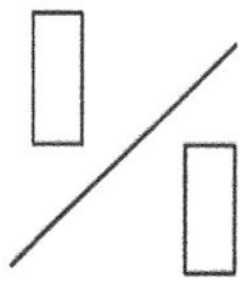

Demonstrate that if they rotate the page so that the diagonal line becomes a horizontal line it is much easier to see that the reflection should be like this.

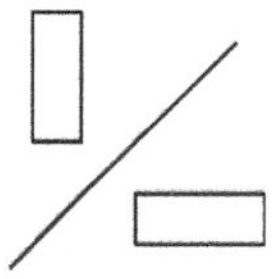

Useful tips

When reflecting shapes in diagonal lines, it is easier if students turn the page so that the mirror line is horizontal or vertical.

Starter

- Recap how to reflect a shape in horizontal and vertical mirror lines.
- Now draw on the board the first diagram in Example 3 and ask students to work out how they will reflect the shape t in a diagonal line.

Main lesson activity

- Demonstrate using the starter diagram how to draw lines from the vertices perpendicular to the mirror line to form the reflection.
- Point out that the reflection can be termed M(t), that is, the mirror of t.
- Students who need more **support** find it more difficult to reflect shapes in a diagonal mirror line. Demonstrate with another example.

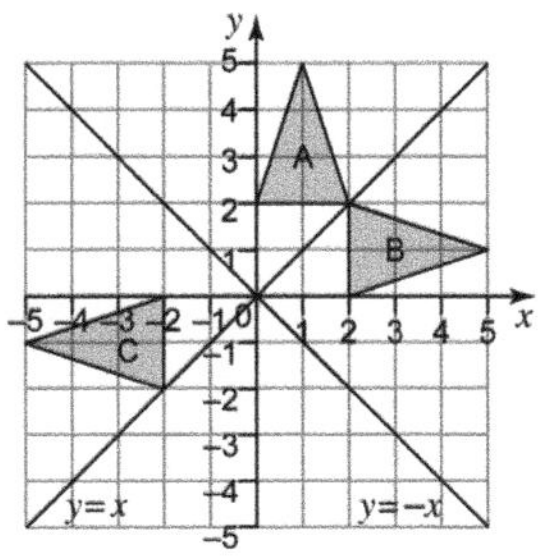

- Triangle B is the reflection of triangle A in the line $y = x$.
- Triangle C is the reflection of triangle A in the line $y = -x$.
- Explain how to count across the squares from a vertex to the mirror line and then count across the same number of squares to find the position of the reflection of that vertex.
- Now students can complete Exercise 29C. Suggest that students who need more **support** complete questions 1 to 4 in pairs before moving on to complete the exercise. Provide support as needed.

Plenary

- Ask students who need more of a **challenge** to explain to the class that if the point (a, b) is reflected in the line $y = x$, then the image point is (b, a).

Homework and answers

Homework for this lesson is available in the homework folder. Consolidation and extension exercises are available in the further practice folder. Answers to the exercises can be found in the answers folder.

29.4 Rotations: 1

Syllabus references

Statement	Core
C7.1	Rotate a shape about the origin, vertices or midpoints of edges of the shape, through multiples of 90°.

Collins references

- Student's Book pages 568–571

Learning aims

- Rotate a shape about a vertex, the midpoint of an edge, or the origin on a coordinate grid.

Key words

- rotation, centre of rotation, angle of rotation, clockwise, anticlockwise

Prior knowledge

Students should understand the word transformation and have covered Lessons 29.1 to 29.3 on translations and reflections.

Common mistakes and remediation

When describing a rotation, students often forget that they need to provide three pieces of information: the centre of rotation, the angle of rotation and the direction of the rotation. Emphasise this point during the lesson.

Useful tips

Encourage students to use tracing paper for any problems involving a rotation.

Starter

- Ask students to draw a T-shape, like this.

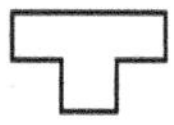

- Ask them to draw another T-shape to add to the first, to make a shape with rotational symmetry of order 2.

  Here are some examples of what students may produce.

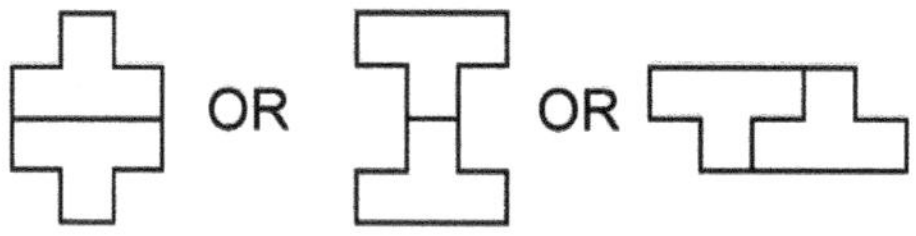

- Ask students for any similarity between the diagrams. In each case the second T-shape is a 180° rotation of the first T-shape.

Main lesson activity

- Explain that a rotation is a transformation that changes a shape by turning it about a fixed point. The new shape is the image of the original

- Ask students to draw a triangle on square paper **[1]**. Ask them to imagine it is rotated 180° about vertex A. This means that vertex A stays in the same place. Ask them to draw the image **[2]**. Now ask them to rotate the original triangle 90° clockwise about vertex A **[3]**.
- Point out that a rotation of 90° clockwise has the same result as a rotation of 270° anticlockwise. You can use either description. For a rotation of 180°, clockwise and anticlockwise have the same result.
- Now ask them to rotate the original triangle 90° anticlockwise (or 270° clockwise) about vertex A **[4]**.
- Ask students to draw the original triangle again **[1]**. Repeat the exercise of three different rotations, but this time about vertex B **[5]**.
- Now ask the students the draw the original triangle again and mark in the midpoint of AC. Label it M. Ask them to rotate the triangle 180° about M. They should have a rectangle **[6]**.
- Now ask them to draw the triangle on a coordinate grid **[7]**. Ask them to rotate it 180° about the origin. Tell them to look at the coordinates to help them do this **[8]**. Point out how the coordinates have changed: A (6, 1) → (−6, −1), and so on. Ask them to rotate the original triangle 90° clockwise about the origin **[9]**. This time, A (6, 1) → (1, −6), and so on. Finally, rotate the original triangle 90° anticlockwise about the origin **[10]**.
- Students who have difficulty drawing the image in the correct place can use tracing paper to help them. Put the point of a pencil on the origin and rotate the tracing paper. This can be less accurate so use the coordinates as well.
- Students can now do Exercise 29D.

1
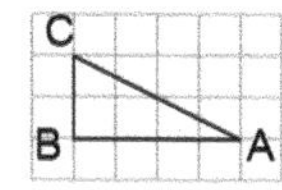

2
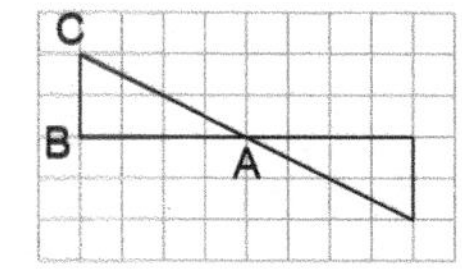

3
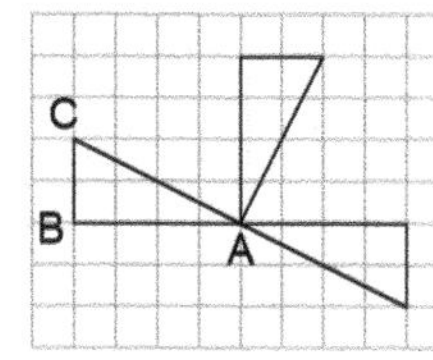

4
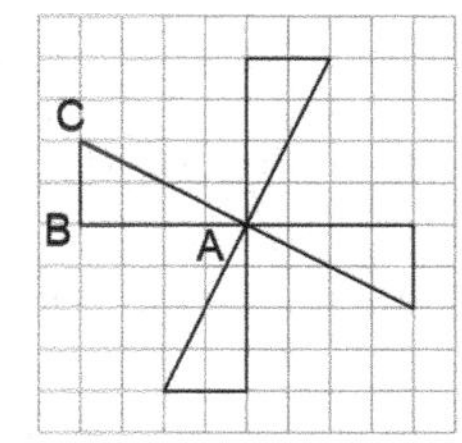

5
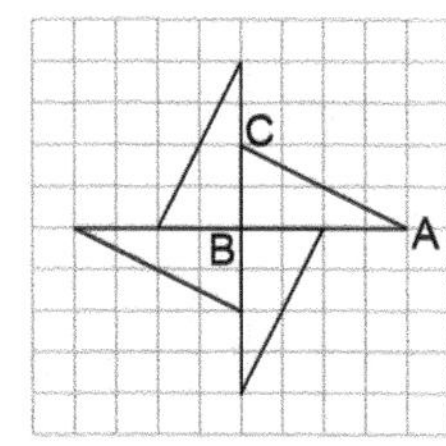

6
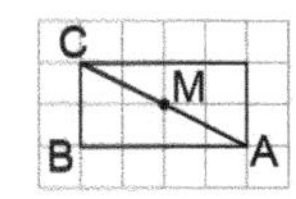

7
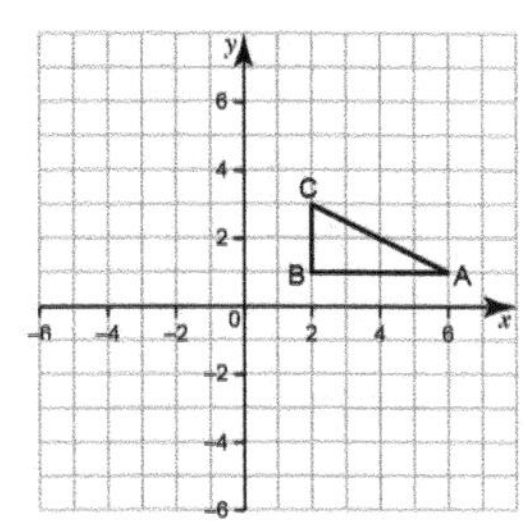

8
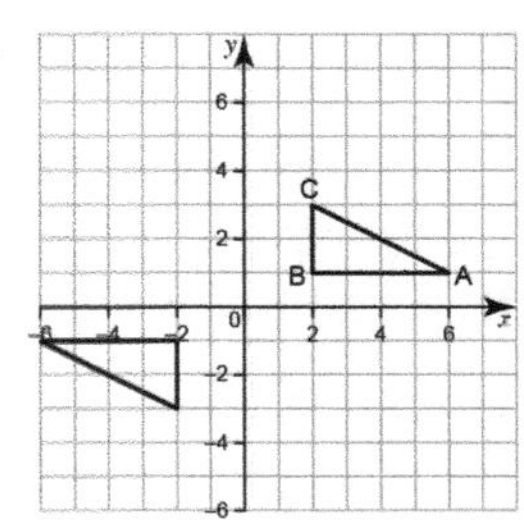

9
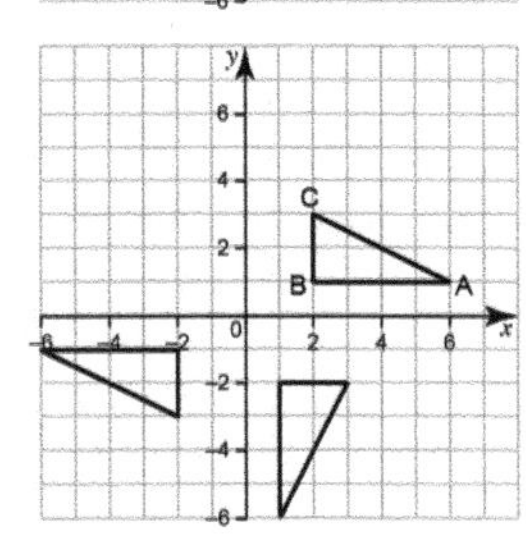

10
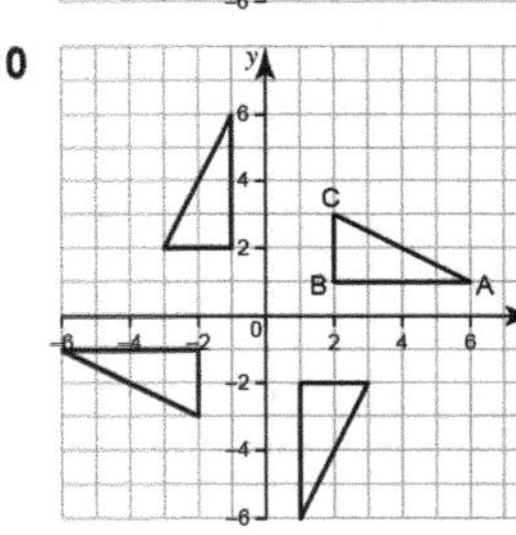

Plenary

- On the drawing they did in the Starter, ask students to mark the centre of the rotation from the original to the new T-shape. Was it at a vertex or at the midpoint of an edge?
- For the three examples given in the Starter, all three centres were in the midpoint of an edge. Students can look at the diagram of a neighbour to check that it is correct.

Homework and answers

Homework for this lesson is available in the homework folder. Consolidation and extension exercises are available in the further practice folder. Answers to the exercises can be found in the answers folder.

29.5 Rotations: 2

Syllabus references

Statement	Extended
E7.1	Rotate a shape about a centre through multiples of 90°.

Collins references

- Student's Book pages 571–574

Learning aims

- Recognise and draw rotations of a shape about any point.

Prior knowledge

Students should have covered Lessons 29.1 to 29.4 on translations, reflections and simple rotations.

Common mistakes and remediation

When describing a rotation, students often forget that they need to provide three pieces of information: the centre of rotation, the angle of rotation and the direction of the rotation (unless it is 180°). Emphasise this point again during the lesson.

Students who need more **support** find it more difficult to rotate a shape about point that is not the origin and not on the shape.

Useful tips

Encourage students to use tracing paper for any problems involving a rotation if they are finding it difficult.

Starter

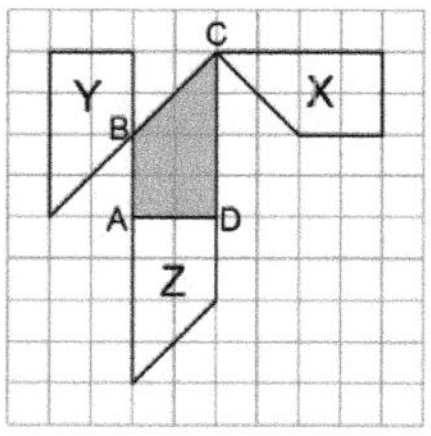

- Show this diagram. Ask students to describe the rotation of shape ABCD onto X. Then the rotation of ABCD onto Y. Then the rotation of ABCD onto Z.
- The answer are X: 90° anticlockwise (or 270° clockwise) about C. Y: 180° about B. Z: 180° about the midpoint of AD.
- Remind students that for 90° they must say clockwise or anticlockwise but this is not necessary for 360°.

Main lesson activity

Say that students need to be able to draw the rotation of a shape about any point. Show them diagram **[1]**. It is a triangle that was used in lesson 29.4.

1
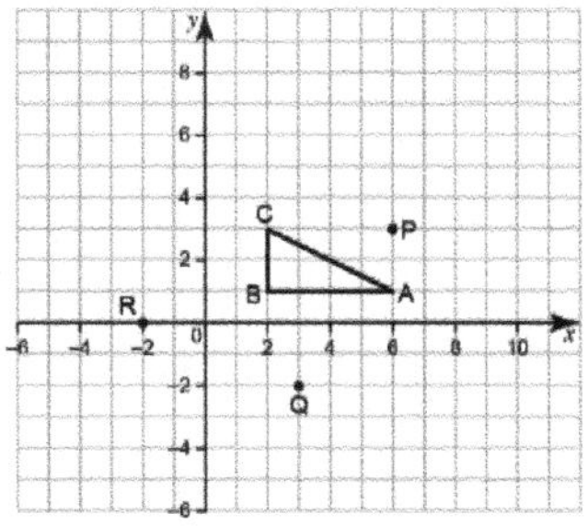

2
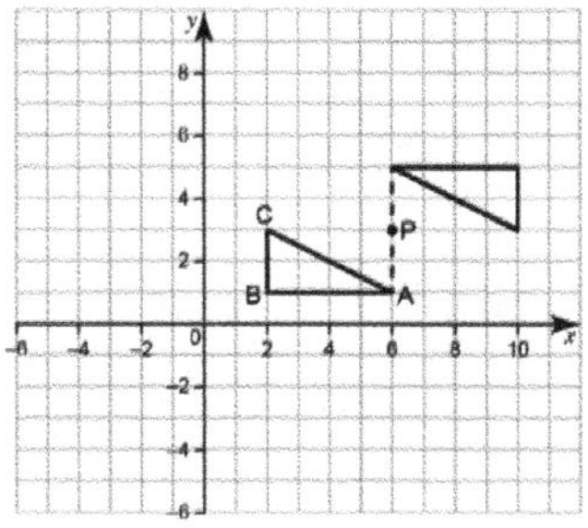

3
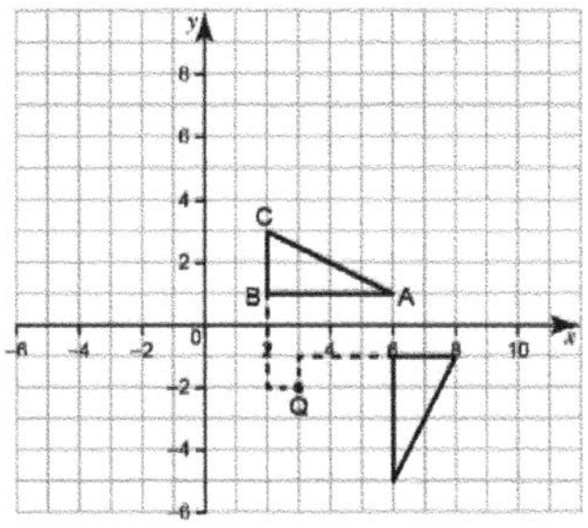

4
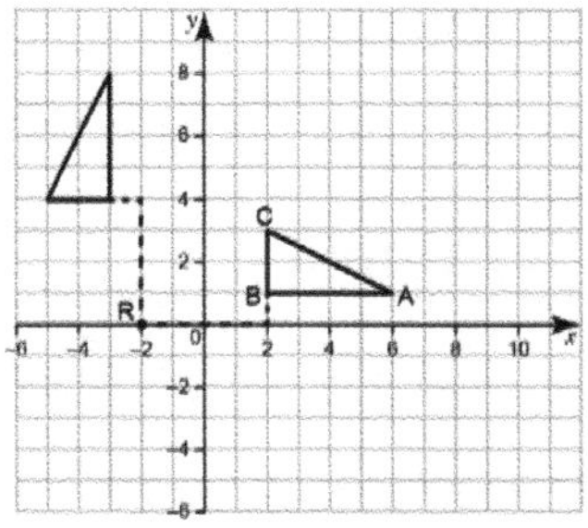

- Ask students to copy it and then draw a 180° rotation of ABC about P (6, 3). See diagram **[2]**.
- Explain how drawing the broken line between P and A and rotating it 180° will help to get the image of vertex A in the correct position.
- Now ask the students to draw a 90°clockwise rotation of ABC about point Q (3, −2). See diagram **[3]**
- Explain how drawing the broken line between Q and B and then rotating it 90° clockwise will get the image of B in the correct position. They could do the same for A and C.
- Now ask for a 90°anticlockwise rotation of ABC about point R (−2, 0). See diagram **[4]**.
- Ask students to explain how they drew it. How did they find the image of each point?
- Now look at Example 4 at the start of section 29.5. Ask students how they would have drawn the image? They would probably start by finding the image of the point on the shape closest to the centre of rotation.
- Say that they can use tracing paper to give an idea of where the image will be or to check their drawing when they have made it.
- Students can now start Exercise 19E.

Plenary

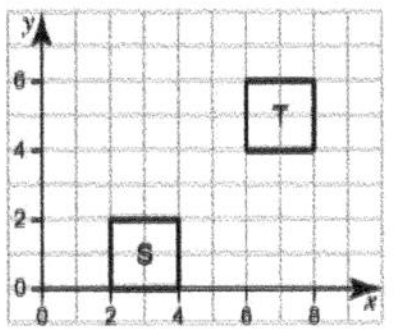

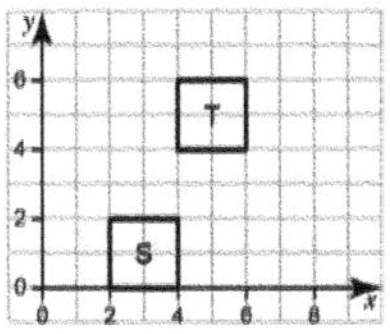

- Show the top diagram. Ask for the centre of a 180° rotation that will take S onto T. It is (5, 3).
- Now ask for the centre of a 90° rotation that will take S onto T. Is there more than one? This has two possible answers: 90° clockwise about (6, 1) and 90° anticlockwise about (3, 5)
- Now ask the same questions about the second diagram, where T is in a different position. The rotations are 180° about (4, 3); 90° clockwise about (6,2); 90° anticlockwise about (2, 4)

Homework and answers

Homework for this lesson is available in the homework folder. Consolidation and extension exercises are available in the further practice folder. Answers to the exercises can be found in the answers folder.

29.6 Enlargements: 1

Syllabus references

Statement	Core
C7.1/E7.1	Enlarge a shape from a given centre by a scale factor.

Collins references

- Student's Book pages 574–580.

Learning aims

- Enlarge a shape by a positive or fractional scale factor.

Key words

- scale factor, enlargement, centre of enlargement, ray method, coordinate method

Prior knowledge

Students should be familiar with translations, reflections and rotations of shapes

Common mistakes and remediation

In mathematics an enlargement always needs a centre. Encourage students to mark the given centre clearly each time and to use it.

Useful tips

Suggest that when students use coordinates to draw an enlargement, they could add the rays to check that they pass through the centre of enlargement. The ray method can be inaccurate if the rays are not drawn with great care.

Starter

- Ask students what they think it means to enlarge a shape by a scale factor of 2. Most are likely to have the (correct) idea that the sides of the shapes will double in length.
- Then ask what they think it means to enlarge a shape by a scale factor of 3, 4, 6 … Check that students have understood that the scale factor affects the lengths of the sides.

Main lesson activity

- Explain that an enlargement is a transformation that changes the size of a shape. The original shape is the object and the enlarged shape is the image.
- Draw triangle ABC (it does not need to be equilateral) and mark point O. Ask students to draw their own version. The exact shape and position of O do not matter in the student's version.
- Say that you will talk through how to draw an enlargement of ABC with a centre at O and a scale factor of 2. Ask students to copy each step.
- Draw a line from O through A and extend it.
- Measure the length of OA = x cm
- Multiply by 2. Mark A' on the line where OA' = $2x$ cm
- Repeat with B to find B'.
- Repeat with C to find C'.

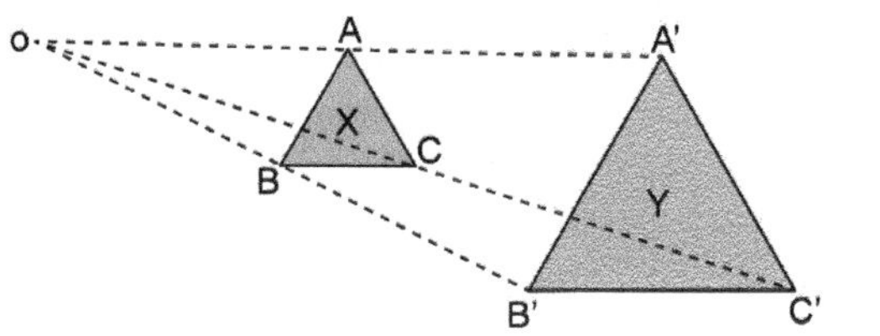

- Draw triangle A'B'C'.
- Ask students to check the accuracy of their enlargement. Is A'B' parallel to AB and twice as long?
- Check B'C' and C'A' in the same way.
- Point out that the two triangles should be similar. They should have the same angles.
- Explain that you can also draw an enlargement using coordinates.
- Draw this pair of axes and triangle
- Say that you are going to draw an enlargement of triangle X with centre at the origin and scale factor 3.
- From the origin the top of triangle X is 2 units to the right and 3 units up. Multiply these numbers by 3 to get 6 and 9. These are the coordinates of the top of triangle Y.
- Find the other two vertices of Y in the same way.
- Check that Y is similar to X: the same angles and sided three times longer.

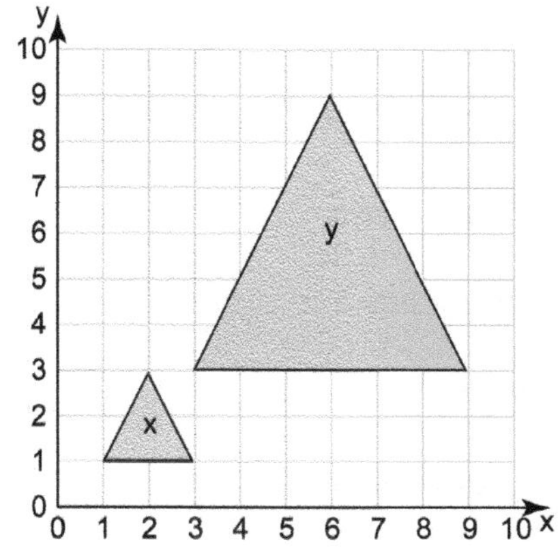

- Example 7 in section 29.6 shows how to use point other than the origin as the centre of enlargement. Talk through this with the students.
- now say that the scale factor can be a fraction. In the first example , if you started with A'B'C' and halved the distances you would find ABC – an enlargement with a scale factor of ½ .
- Finally talk through Example 8 with the students. This shows how to do an enlargement on a grid when there are no coordinates.
- Now students can do Exercise 29F.

Plenary

- Ask students to look at the questions they have answered. What happens to the area of a shape in an enlargement?
- In general if the scale factor of the enlargement is k then the area is multiplied by k^2.
- Students may give particular examples of this. For a scale factor of 2 the area is multiplied by 4. For a scale factor of 3 the area is multiplied by 9. For a scale factor of $\frac{1}{2}$ the area is multiplied by $\frac{1}{4}$.

Homework and answers

Homework for this lesson is available in the homework folder. Consolidation and extension exercises are available in the further practice folder. Answers to the exercises can be found in the answers folder.

29.7 Enlargements: 2

Syllabus references

Statement	Extended
E7.1	Enlarge a shape from a given centre by a negative scale factor.

Collins references

- Student's Book pages 580–585

Learning aims

- Recognise and draw the enlargement of a shape by a negative scale factor.

Key words

- negative scale factor

Prior knowledge

Students understand enlargements with a positive or fractional scale factor.

Common mistakes and remediation

Students can make errors in getting an enlargement with a negative scale factor in the correct position.

It is useful to rotate the paper 180° and check that the image looks similar to the original shape.

Useful tips

Point out that an enlargement with a negative scale factor can be thought of as a positive enlargement followed by a rotation of 180° about the centre of enlargement.

Starter

- Draw this shape and ask students to copy it on square paper.

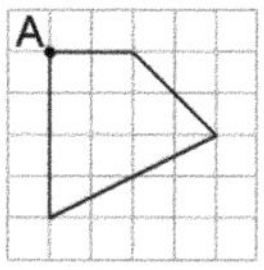

- Ask them to draw an enlargement, centre A, scale factor 1.5. The result should look like this.

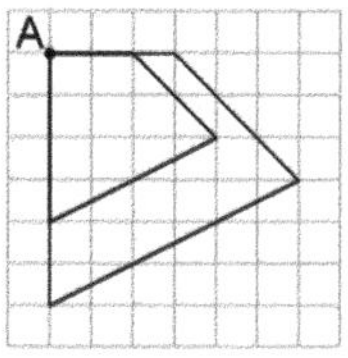

- Check their diagrams are correct. Are corresponding lines parallel? Ask each student to compare theirs with a neighbour's and check that they look the same. A common error is to measure the new length from the vertex and not from the centre of enlargement.

Main lesson activity

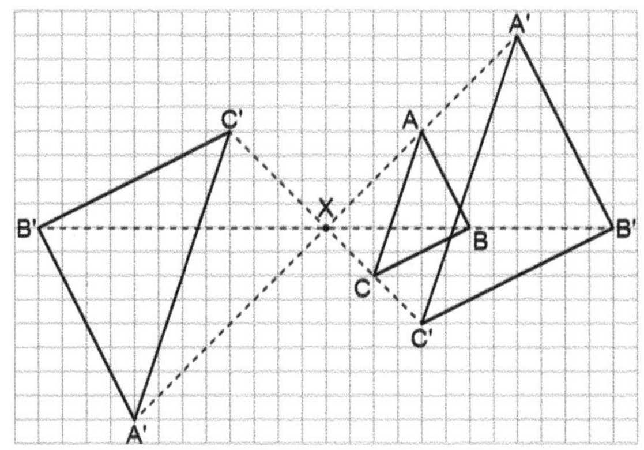

- Say that you are going to show students how to enlarge a shape with a negative scale factor.
- Draw triangle ABC and point X from this diagram. Do not draw anything else yet.
- Ask students to put point X in the middle of a sheet of square paper and then draw triangle ABC. Make sure it is correctly positioned. For example, A is 4 to the right and 4 up from X.
- Ask students to enlarge ABC with centre X and scale factor X to draw triangle A'B'C'.
- Students should check that corresponding lines are parallel. Draw triangle A'B'C' on your diagram too.
- Now say that for an enlargement with a scale factor of -2 the distances from X are doubled but they are measured in the opposite direction. Demonstrate by marking A" on your diagram, 8 squares left of X and 8 squares down.
- Ask students to mark B" and C" on their diagrams and draw triangle A"B"C". Draw it on your diagram too.
- Ask students what they notice about triangles A'B'C' and A"B"C". They should see that one is a rotation of the other through 180° about X. This is generally true: an enlargement with a negative scale factor is equivalent to n enlargement with a positive scale factor followed by a rotation of 180° about the centre of enlargement.
- Finally draw in the broken lines. If the diagram is drawn accurately the sets of 3 points will be on straight lines passing through X.
- Now talk through Example 9 with the students. This shows an enlargement with a scale factor of −2 on a coordinate grid where the centre of enlargement is at (1, 0)
- Ask the students to describe a transformation that will map A'B'C' onto ABC in Example 9. It is an enlargement, centre (1, 0), scale factor $-\frac{1}{2}$.
- Students can now complete Exercise 29G.

Plenary

- At this point ask students to summarise the information needed for the transformations they have learned so far before going on to combined transformations in the next lesson.

Transformation	**Information needed**
Translation	column vector
Reflection	equation of mirror line
Rotation	angle, direction, centre of rotation
Enlargement	scale factor, centre of enlargement

Homework and answers

Homework for this lesson is available in the homework folder. Consolidation and extension exercises are available in the further practice folder. Answers to the exercises can be found in the answers folder.

29.8 Combined transformations

Syllabus references

Statement	Extended
E7.1	Reflect a shape in a straight line. Rotate a shape about a given centre through multiples of 90°. Enlarge a shape from a given centre by a negative scale factor.

Collins references

- Student's Book pages 585–587

Learning aims

- Combine transformations.

Prior knowledge

Students should have covered Lessons 29.1 to 29.7 on translations, reflections, rotations and enlargements.

Common mistakes and remediation

Students often miss out some of the information required to explain rotations and enlargements fully. Throughout the lesson, reinforce the information in the table below.

Useful tips

Display the summary of all the transformations covered so far. Always have tracing paper available.

Transformation	Information needed
Translation	column vector
Reflection	equation of mirror line
Rotation	angle, direction, centre of rotation
Enlargement	scale factor, centre of enlargement

Starter

- Display a coordinate grid, with x-axis and y-axis labelled appropriately.
- Add two squares: A with vertices at (–2, 0), (–2, 2), (0, 0) and (0, 2) and B with vertices (0, 0), (0, 2), (2, 2) and (2, 0).
- Ask students if B is a transformation of A and, if so, if they can describe the transformation.
- Ask: 'Is it possible to describe it as a translation? Similarly, can it be described as a reflection, a rotation or an enlargement?' Expected answers include:
  - translation by the vector $\begin{pmatrix} 2 \\ 0 \end{pmatrix}$
  - reflection in the line $x = 0$ or the y-axis
  - rotation of 90° clockwise about the origin, or 90° anticlockwise about (0, 2)
  - no enlargement, as B is not enlarged.

Main lesson activity

- Explain that this lesson covers all the transformations they have met and combinations of them.
- Consider enlargements first, as they are the only transformations that change the size of the object and, hence, they are easily identified. Having ruled out enlargement, a translation is the next easiest to spot.
- Sometimes a shape may be transformed more than once, by combined transformations. Work through two examples.
- **Example 1**: Display this diagram. Give students who need more **support** a copy or let them draw it themselves on squared paper.

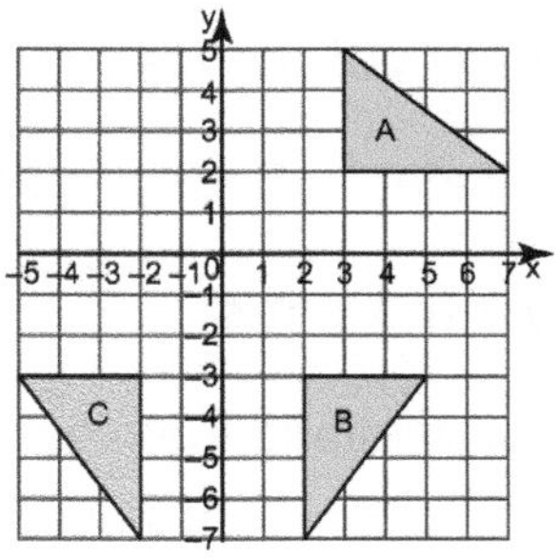

- Ask students to describe fully the single transformation that maps triangle A onto triangle B: a rotation of 90° clockwise about the origin or (0, 0).
- Now ask students to describe fully the single transformation that maps triangle B onto triangle C: a reflection in the line x = 0 or the y-axis.
- Explain that the combined transformation that maps triangle A onto triangle C is: a rotation of 90° clockwise about the origin, followed by a reflection in the line x = 0. Now ask students to give the single transformation that maps triangle C onto triangle A. Make sure they give a full description of the reflection: a reflection in the line $y = -x$.
- **Example 2**: Display this diagram. Give students who need more **support** a copy or let them draw it themselves on squared paper.

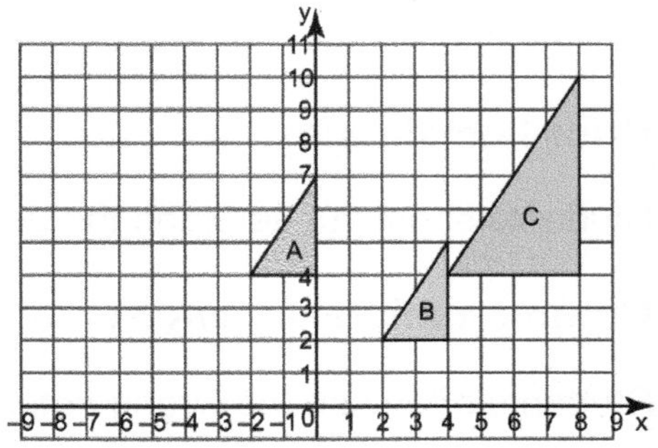

- Ask students to describe fully the single transformation that maps triangle A onto triangle B; a translation by the vector $\begin{pmatrix} 4 \\ -2 \end{pmatrix}$.
- Now ask students to describe fully the single transformation that maps triangle B onto triangle C: an enlargement by scale factor 2 about the origin or (0, 0).
- Explain that the combined transformation that maps triangle A onto triangle C is: a translation by the vector $\begin{pmatrix} 4 \\ -2 \end{pmatrix}$ followed by an enlargement by scale factor 2 about the origin. Now ask students to give the single transformation that maps triangle C onto triangle A.
- Make sure students give a full description of the enlargement: an enlargement by scale factor $\frac{1}{2}$ about (–8, 4).
- Students can complete Exercise 29H. Students who need more **support** can work in pairs.

Plenary

- If they need help, show this thought-process diagram and ask students to complete it.

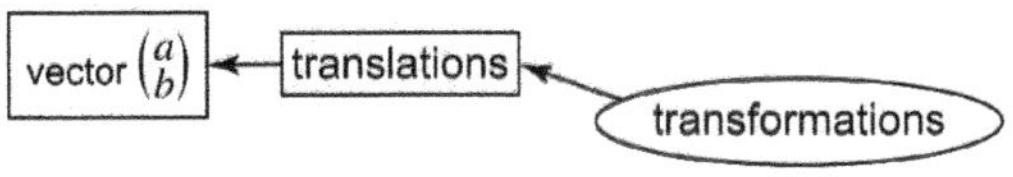

Homework and answers

Homework for this lesson is available in the homework folder. Consolidation and extension exercises are available in the further practice folder. Answers to the exercises can be found in the answers folder.

30.1 Introduction to vectors

Syllabus references

Statement	Extended
E7.2	Describe a translation by using a vector represented by e.g. $\binom{x}{y}$, $\overrightarrow{AB}$ or $\mathbf{a}$. Add and subtract vectors. Multiply a vector by a scalar.

Collins references

- Student's Book pages 590–592

Learning aims

- Understand the meaning of a vector.
- Know different ways of representing vectors.
- Add, subtract and multiply vectors.

Key words

- magnitude, direction, vector, column vector, scalar, coordinate grid

Prior knowledge

Students should understand the use of coordinates on a coordinate grid.

Common mistakes and remediation

Students often use the wrong notation for vectors. Point out that when using a letter (e.g. $\mathbf{a}$) in handwritten form this is shown as $\underline{a}$, to distinguish it from a, which is an algebraic variable that represents a number. The form $\mathbf{a}$ or $\underline{a}$ represents a journey or movement. Similarly, $\overrightarrow{AB}$ represents the length of a line from A to B.

Students often put a line between two elements of a column vector so it looks like a fraction. That is incorrect and should be brought to their attention so they do not do it. They must put brackets around a column vector.

Useful tips

Encourage students to think of a vector as a journey or a movement.

Starter

- Draw these points on a square grid.
- Ask students to give the column vectors for $\overrightarrow{AB}$, $\overrightarrow{BA}$, $\overrightarrow{CE}$, $\overrightarrow{EC}$, etc.
- Using the same grid, explain that one unit across = $\mathbf{a}$ and one unit up = $\mathbf{b}$.
- Ask students to express the column vectors for $\overrightarrow{AB}$, $\overrightarrow{BA}$, $\overrightarrow{CE}$, $\overrightarrow{EC}$, etc. in terms of $\mathbf{a}$ and $\mathbf{b}$. For example,

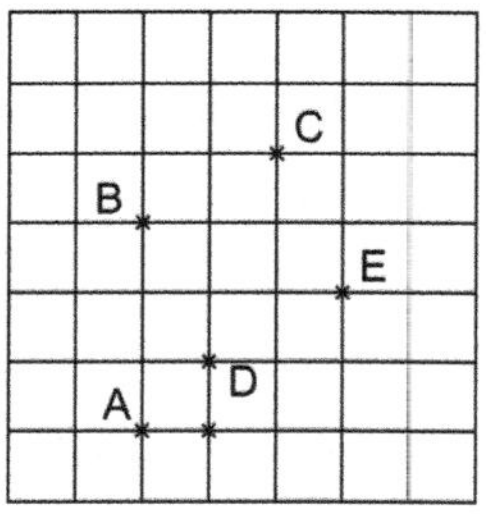

$$\overrightarrow{AB} = \begin{pmatrix}0\\3\end{pmatrix} = 3\mathbf{b},\ \overrightarrow{BA} = \begin{pmatrix}0\\-3\end{pmatrix} = -3\mathbf{b},\ \overrightarrow{CE} = \begin{pmatrix}1\\-2\end{pmatrix} = \mathbf{a} - 2\mathbf{b},\ \overrightarrow{EC}\begin{pmatrix}-1\\2\end{pmatrix} = -\mathbf{a} + 2\mathbf{b}$$

Main lesson activity

- Make sure students are familiar with the representation and the notation used for vectors. They should be clear that a vector has both magnitude (length or distance) and direction. Refer to the vector in the Student's Book, which represents a translation that can be considered as a journey or movement from A to B.
- On a diagram, the direction of the vector is shown by an arrow. This vector can be represented by $\mathbf{a}$ or $\underline{a}$ or $\overrightarrow{AB}$. (The bold $\mathbf{a}$ is the printed notation in books and $\underline{a}$ is the written notation.)
  The magnitude of the vector is represented by the length of the line from A to B.
- A vector can also be negative. For example, vector $-\mathbf{c}$ has the same length (magnitude) as vector $\mathbf{c}$, but acts in the opposite direction.

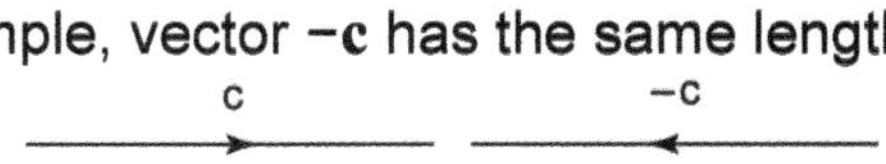

Addition, subtraction and multiplication of vectors

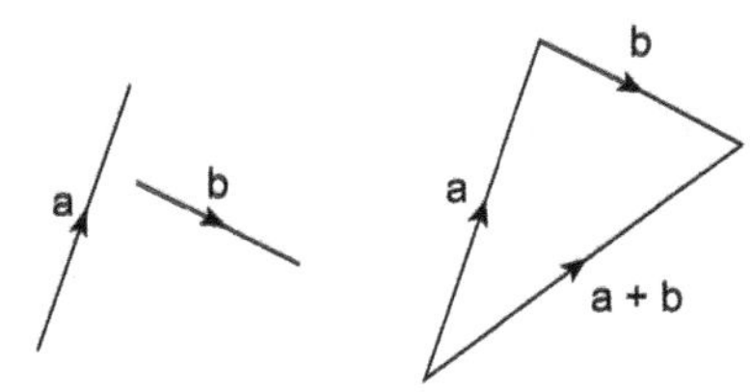

- Explain how to add and subtract vectors. For example, for two vectors $\mathbf{a}$ and $\mathbf{b}$, $\mathbf{a} + \mathbf{b}$ is defined as the translation of $\mathbf{a}$ followed by the translation of $\mathbf{b}$.
- Suggest that students think of this vector addition as a journey from A to B followed by a journey from B to C. $\overrightarrow{AB} + \overrightarrow{BC} = \overrightarrow{AC}$
- Similarly, for two vectors $\mathbf{a}$ and $\mathbf{b}$, $\mathbf{a} - \mathbf{b}$ is the translation of $\mathbf{a}$ followed by the translation of $-\mathbf{b}$.

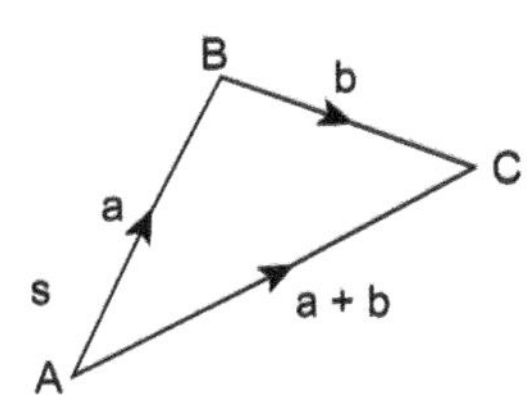

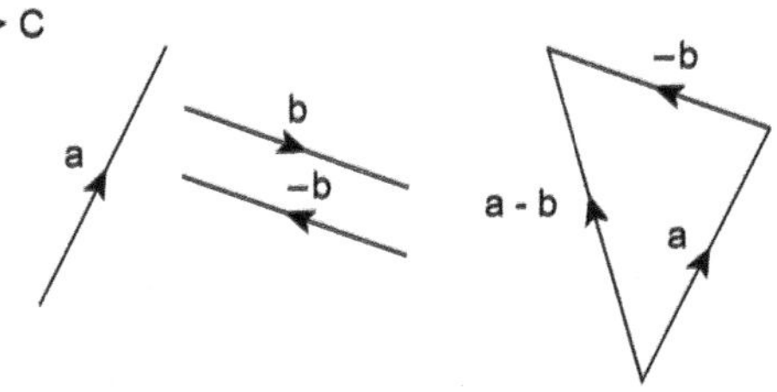

- Draw this grid on the board.

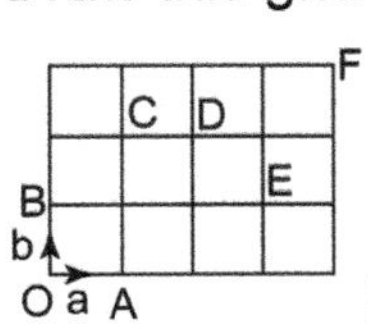

$\overrightarrow{OA} = \mathbf{a}$, $\overrightarrow{OB} = \mathbf{b}$

- Show students how to find the vectors $\overrightarrow{OC}$, $\overrightarrow{BF}$, $\overrightarrow{CE}$, $\overrightarrow{FA}$ and $\overrightarrow{DE}$ ensuring they use the correct notation. ($\overrightarrow{OC} = \mathbf{a} + 2\mathbf{b}$, $\overrightarrow{BF} = 4\mathbf{a} + 2\mathbf{b}$, $\overrightarrow{CE} = 2\mathbf{a} - \mathbf{b}$, $\overrightarrow{FA} = -3\mathbf{a} - 3\mathbf{b}$, $\overrightarrow{DE} = \mathbf{a} - \mathbf{b}$)
- Explain that a vector can be multiplied by a number (or a scalar quantity). For example, vector $2\mathbf{a}$ is twice as long as vector $\mathbf{a}$, acting in the same direction.
- Students can now do Exercise 30A. Check that they use the correct notation.

Plenary

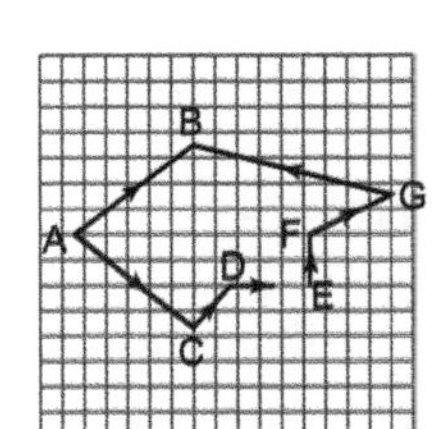

- On a grid on the board, draw a vector $\overrightarrow{AB}$. Ask someone to give the column vector for $\overrightarrow{AB}$.
- Now add point C, drawing the vectors $\overrightarrow{AC}$ and $\overrightarrow{CB}$. Ask what $\overrightarrow{AC} + \overrightarrow{BC}$ equals, and why.
- Returning to the original vector diagram $\overrightarrow{AB}$, add new points C, D, E, F and G. Again, ask for the sum of all the individual vectors $\overrightarrow{AC}$, $\overrightarrow{CD}$, $\overrightarrow{DE}$, $\overrightarrow{EF}$, $\overrightarrow{FG}$, $\overrightarrow{GB}$.

Homework and answers

Homework for this lesson is available in the homework folder. Consolidation and extension exercises are available in the further practice folder. Answers to the exercises can be found in the answers folder.

30.2 Using vectors

Syllabus references

Statement	Extended
E7.4	Represent vectors by directed line segments. Use position vectors. Use the sum and difference of two vectors to express given vectors in terms of two coplanar vectors. Use vectors to reason and to solve geometric problems.

Collins references

- Student's Book pages 593–597

Learning aims

- Understand and use position vectors.

Key words

- position vector

Prior knowledge

Students should be confident with finding and using vectors such as
$\frac{3}{2}\mathbf{a} - \frac{3}{2}\mathbf{b}$ and $2(2\mathbf{a} - \mathbf{b}) - 3(\mathbf{a} - \mathbf{b})$.

Common mistakes and remediation

Students find it difficult to write down vectors from a line segment given in a ratio form.

Before they start Exercise 30B, it could be useful to show students the following example.

B is a point on AC such that AB : BC = 2 : 1. 

If $\overrightarrow{AC} = \mathbf{c}$, then $\overrightarrow{AB} = \frac{2}{3}\mathbf{c}$ and $\overrightarrow{BC} = \frac{1}{3}\mathbf{c}$.

Useful tips

In algebra it is usual, if possible, to write an expression so that it starts with a positive variable, so $-2x + 3y$ would be written as $3y - 2x$. The same idea applies to vectors, so $-2\mathbf{a} + 3\mathbf{b}$ can be written as $3\mathbf{b} - 2\mathbf{a}$. Also, as in algebra, $3(\mathbf{a} + \mathbf{b}) = 3\mathbf{a} + 3\mathbf{b}$.

Starter

- Draw the parallelogram PQRS. Ask students to express $\overrightarrow{PS}$, $\overrightarrow{PQ}$, $\overrightarrow{PR}$ and $\overrightarrow{QS}$ in terms of $\mathbf{a}$ and $\mathbf{b}$.

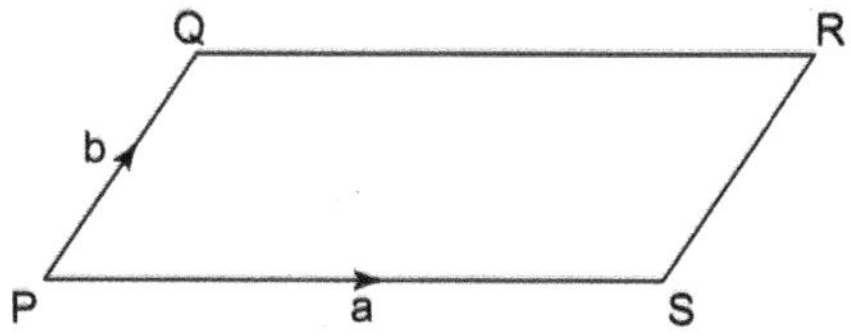

($\overrightarrow{PS} = \mathbf{a}$, $\overrightarrow{PQ} = \mathbf{b}$, $\overrightarrow{PR} = \mathbf{a} + \mathbf{b}$, $\overrightarrow{QS} = \mathbf{a} - \mathbf{b}$)

- Draw the triangle PQS. Ask: 'What is $\overrightarrow{QS}$ in terms of $\mathbf{a}$ and $\mathbf{b}$?'
  (To get from Q to S, go from Q to P, then P to S, so $\overrightarrow{QS} = -\mathbf{b} + \mathbf{a}$ or $\mathbf{a} - \mathbf{b}$.)

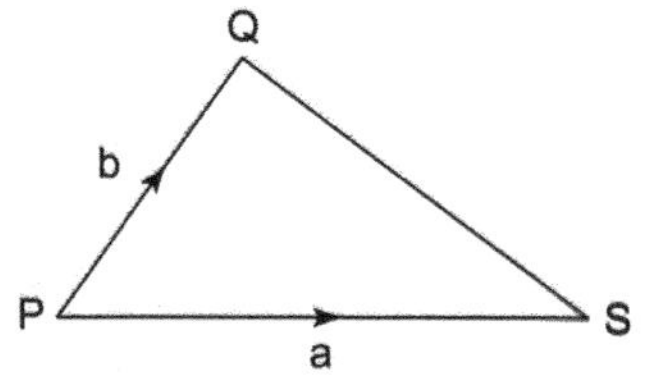

Main lesson activity

- Ensure that students have understood the two examples in the starter, then go through the following example, showing them how to think of vectors as journeys. It is vital that they use the correct notation in their work.
- OAB is a triangle and M is the midpoint of OB. C and D are points on AB with AC = CD = DB.

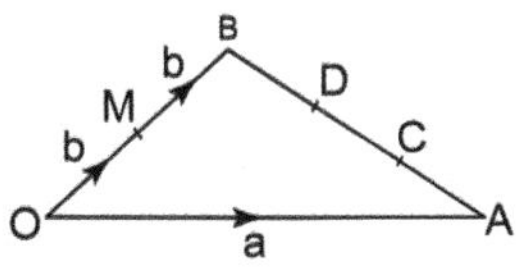

  $\overrightarrow{OA} = \mathbf{a}$ and $\overrightarrow{OB} = 2\mathbf{b}$

- Explain how to find the three vectors $\overrightarrow{BA}$, $\overrightarrow{MD}$ and $\overrightarrow{OC}$ in terms of $\mathbf{a}$ and $\mathbf{b}$, giving answers in their simplest form.

  $\overrightarrow{BA} = \overrightarrow{BO} + \overrightarrow{OA} = -2\mathbf{b} + \mathbf{a} = \mathbf{a} - 2\mathbf{b}$

  $\overrightarrow{MD} = \overrightarrow{MB} + \overrightarrow{BD} = \overrightarrow{MB} + \frac{1}{3}\overrightarrow{BA} = \mathbf{b} + \frac{1}{3}\mathbf{a} - \frac{2}{3}\mathbf{b} = \frac{1}{3}\mathbf{a} + \frac{1}{3}\mathbf{b}$

  $\overrightarrow{OC} = \overrightarrow{OA} + \overrightarrow{AC} = \overrightarrow{OA} + \frac{1}{3}\overrightarrow{AB} = \overrightarrow{OA} - \frac{1}{3}\overrightarrow{BA} = \mathbf{a} - \frac{1}{3}\mathbf{a} + \frac{2}{3}\mathbf{b} = \frac{2}{3}\mathbf{a} + \frac{2}{3}\mathbf{b}$

- Now let students work through Example 1.
- Students can complete Exercise 30B.

Plenary

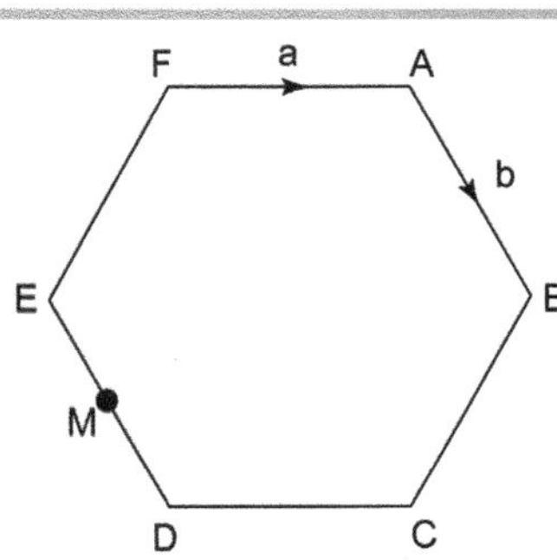

- Draw this diagram on the board.
- Explain how to find the vector $\overrightarrow{AM}$ where ABCDEF is a regular hexagon and M is the midpoint of DE.

Homework and answers

Homework for this lesson is available in the homework folder. Consolidation and extension exercises are available in the further practice folder. Answers to the exercises can be found in the answers folder.

30.3 The magnitude of a vector

Syllabus references

Statement	Extended
E7.3	Calculate the magnitude of a vector $\binom{x}{y}$ as $\sqrt{x^2 + y^2}$.

Collins references

- Student's Book pages 598–599

Learning aims

- Find the magnitude of a vector using Pythagoras' theorem.

Key words

- magnitude, Pythagoras' theorem

Prior knowledge

Students should have covered Lessons 30.1 and 30.2. They also need to be familiar with Pythagoras' theorem.

Common mistakes and remediation

Some students may think that the magnitude of a vector can be calculated as a linear measurement, e.g. the number of squares between its coordinate points. Make it clear at the start of the lesson and throughout that the magnitude of a vector must be calculated and this can be done using trigonometry.

Useful tips

Ask students in the plenary to make a chart to guide them through all the terminology used for vectors, including the modulus signs which indicate magnitude.

Starter

- Remind students that vectors have magnitude as well as direction.
- Introduce them to the use of the modulus signs for magnitude:
  $|\boldsymbol{a}|$ or $|\overrightarrow{\mathrm{AB}}|$
- Draw a simple $|\overrightarrow{\mathrm{AB}}|$ vector on the board with coordinates $\binom{4}{-3}$ like the one shown on Student's Book page 546.
- Make this into the hypotenuse of a triangle.
- Check that students understand why the x value is positive and the y value is negative. (The vector moves to the right along x and downwards along y.)
- Now take students through the calculation of the vector using Pythagoras' theorem as in the Student's Book.

Main lesson activity

- Project or draw the vector and grid from Example 2 on the board:

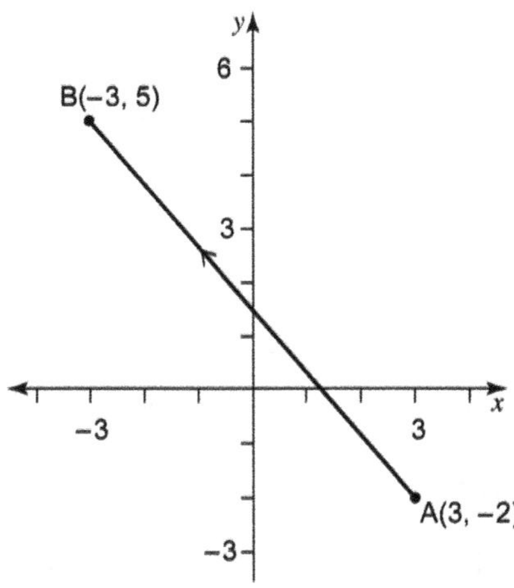

- Ask students to identify the coordinates of vector $|\overrightarrow{AB}|$.
- Some students may find it helpful to have the other 'sides of the triangle' drawn in like this:

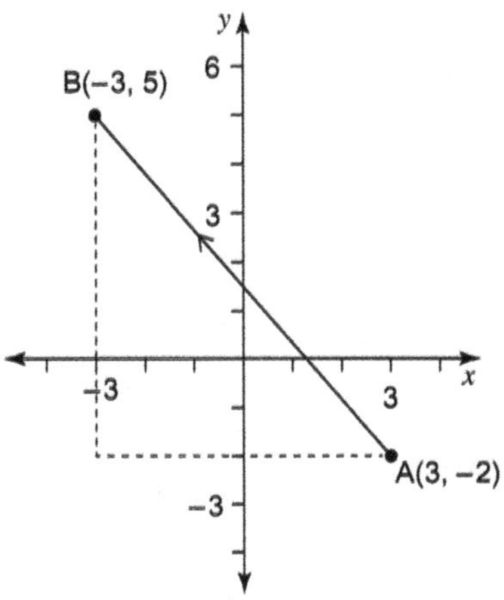

- Ask students to calculate the magnitude of the vector and then to look at Example 2 to check their working.
- Emphasise that a diagram is not necessary if you have the coordinates of the vector and establish that:

  If $\mathbf{a} = \binom{x}{y}$ then $|a| = \sqrt{x^2 + y^2}$
- Students can now do Exercise 30C.

Plenary

- Students draw up their own guides to the terminology and symbols used in vectors.
- Ask students to contribute one definition each to a class guide written up on the board. This also allows you to check that all definitions are correct.

Homework and answers

Homework for this lesson is available in the homework folder. Consolidation and extension exercises are available in the further practice folder. Answers to the exercises can be found in the answers folder.

31.1 Frequency tables

Syllabus references

Statement	Core
C9.1/E9.1 **C9.2/E9.2**	Collect, classify and tabulate statistical data. Read, interpret and draw simple inferences from tables and statistical diagrams. Appreciate restrictions on drawing conclusions from given data.

Collins references

- Student's Book pages 620–623

Learning aims

- Collect and represent discrete and grouped data using tally charts and frequency tables.
- Appreciate the fact that conclusions from given data might not apply in different situations.

Key words

- tally chart, frequency, frequency table, classes, class interval, grouped frequency table

Prior knowledge

Students must know how to use a tallying process accurately (with both discrete and grouped data) and must be able to understand information that has been presented in tabular form.

Common mistakes and remediation

Students often find it difficult to distinguish between the types of situations which suit the different methods of data collection. Plenty of examples and clear explanations are helpful here.

It is also very common for students to try to find shortcuts in the tallying process, as the correct method is often seen as very time consuming. It is worth comparing results of tallying, as you are likely to find differences in what should be a straightforward activity. If students can see that differences occur so easily, it gives them more reason to follow the correct procedure and to be more careful than they might think is necessary.

Useful tips

It is essential to be very methodical when collecting and using data.

Starter

- Tell the class that you are going to collect information about how many siblings they have (include half brothers and sisters, etc.). Ask each student for his or her answer. Write the results on the board as a list.
- Once all of the data has been collected, discuss how it could be analysed. Mention that this is raw data at the moment. Ask if there are better ways to show the data. Putting it in order may be suggested. This would give ordered data. A table may be mentioned. Organise the data into a table, using tallies to record each value.
- Compare the advantages of raw data, ordered data and a table.
- Now ask them if they could use their conclusions to make predictions from: a) another mathematics class in i) the same year group in their school and ii) the same year group in a school in a different

country; b) a group of people in the 30–40 age range; c) a group of people in the 80–90 age range. This should help them realise that there are restrictions in the use of their conclusions.

Main lesson activity

- Introduce a data collection task, e.g. where did students in the school go on holiday last year?
- Talk about this initially and elicit from the students the fact that they need to tie this down to regions rather than every single place! Such regions could be: Europe, America, Africa or Asia.
- Discuss how you would get this data and whether you would ask everyone in the school or simply take a sample. For students who need more of a **challenge** you can talk about the difference between a random sample – people are selected completely randomly – and stratified samples – similar numbers from each age group and each sex are selected to get a balanced view.
- Ask students to imagine they are asking a random set of people. Tell them that they have a clipboard with a tally chart ready for completion. Put this up on the board with your regions and space for tallies. Emphasise the need for a methodical approach to collecting and organising data.
- Go round the class quickly asking students individually which region they went to and put a stroke in the correct space in the tally chart, remembering to show them how each fifth entry is gated to give sets of 5 for easier counting. You can emphasise this with the students who need more **support** so that they appreciate the reason for gating, as counting in fives is easier than just adding up a lot of single sticks.
- Ask different students to give you the total frequency for each region. Make a comment about which was the most popular.
- Go through Examples 1 and 2 in the Student's Book.
- Introduce the idea of a situation where we need a grouped frequency table, e.g. the house number where people live. Ask students for their house number – you should get a range of small and large numbers.
- Ask students how they could collect this data sensibly without noting every single number. Lead students to suggest another tally chart split up with numbers such as 1–30, 31–60, 61–90, over 90.
- Put another tally chart up on the board showing these groups and again go round the class asking for their house numbers, filling the table in as you go. Again total this up at the end. Make some comment about which group appears to be the most common.
- Talk about the ways data can be collected: taking a sample (as in the examples above), making an observation (such as the number of people in passing cars), doing an experiment (such as finding the total when two dice are rolled).
- The class can now do Exercise 31A in the Student's Book.

Plenary

- Ask for examples of occasions when sampling, observation and experiment should be used.

Homework and answers

Homework for this lesson is available in the homework folder. Consolidation and extension exercises are available in the further practice folder. Answers to the exercises can be found in the answers folder.

31.2 Pictograms

Syllabus references

Statement	Core
C9.4/E9.4	Draw and interpret pictograms.

Collins references

- Student's Book pages 623–625

Learning aims

- Show collected data as pictograms.
- Interpret pictograms.

Key words

- pictogram, symbol, key

Prior knowledge

Students must be able to understand data presented in tabular form. It is also helpful to have a good command of the multiplication tables for this lesson.

Common mistakes and remediation

Students who need more **support** often use a symbol to represent 1 unit of data. Encourage them to avoid this wherever it is more sensible to choose another scale.

A common mistake is to use different symbols in the different classes of the pictogram. Remind students, whenever this occurs, that the symbols are used to compare the frequencies, so they must be the same size throughout.

Useful tips

Using squared paper will help students to keep the symbols in their pictograms the same size.

Starter

- Draw a number line on the board, divided into 10. Write 0 at one end and any multiple of 10 (e.g. 30) at the other.
- Point at different marks on the line and ask students to give the value.
- Point at some positions between marks and ask for the value (all values are integers). Repeat with other multiples of 10, but end with 40.
- Now draw the following on the board: ◇◇◇◇◇◇◇◇◇◇
- Tell students that each diamond is worth four and we are counting from left to right. Point at a diamond and ask what the value is. The link between this and the previous activity should be clear.
- Ask how we could show 10 with this notation. A half diamond may be suggested, but make the point that using symbols is not always as useful as a continuous line.

Main lesson activity

- Put on the board a very simple pictogram such as: Walk, Bus, Car, and some simple stick people after each one (e.g. four people walk, three people by bus and one person by car).
- Ask the class what this information might be representing. From the discussion, a suggestion might be: Ways of getting to school.
- Ask the students: 'How many walked to school?' They may give the answer 4 or they may say that they cannot tell, which is correct.
- Go through Example 3 in the Student's Book, which shows a basic pictogram with one symbol representing one phone call.
- Talk about the need for a key to the figures, as the whole point of a pictogram is that one little drawing can represent a number of people. For the travelling to school example, suggest that each stick person represents six people. Then ask how many people are in each category. Students who need more **support** need this stage to be done carefully so that they can visualise one stick man = six people, as it is a concept they find difficult.
- Ask the question: 'How could you represent four people travelling by car rather than six?' Lead students to the fact that the symbol can be divided up – it is convenient that the stick man has six parts to it (head, body, two arms and two legs). So in this case a person could be drawn with two bits missing!
- Go through Example 4 in the Student's Book.
- Talk about how important it is to decide on how many each symbol will represent, as often it is not necessary to divide this symbol up. Care needs to be taken when deciding on a symbol and also on deciding how many that symbol will represent. Ask students who need more of a **challenge** for examples of a good choice of numbers that could be represented by one symbol and examples of numbers that would be a poor choice. (Good numbers tend to be 2 and 4 because you can easily subdivide the symbol in half or quarters, poor numbers would be ones like 3 or 7 as they are difficult to subdivide.)
- Talk about the relevance of the symbol, that it should be a simple diagram or shape that has something to do with the data. Encourage students who need more **support** always to use the same symbol in a pictogram, whereas students who need more of a **challenge** could cope with different ones provided they are all the same size.
- The class can now do Exercise 31B in the Student's Book.

Plenary

- Recap with small groups of numbers and a symbol unit, and ask for the quantity of symbols to represent each number. For example: 'How many symbols are required for the numbers 18, 21, 36 and 42, with each symbol representing 6?' Repeat with variations.

Homework and answers

Homework for this lesson is available in the homework folder. Consolidation and extension exercises are available in the further practice folder. Answers to the exercises can be found in the answers folder.

31.3 Bar charts

Syllabus references

Statement	Core
C9.4/E9.4	Draw and interpret bar charts and simple frequency distributions.

Collins references

- Student's Book pages 625–629

Learning aims

- Draw and interpret bar charts that represent statistical data.

Key words

- bar chart, axis

Prior knowledge

Students must be able to understand data presented in tabular form. One question in this exercise requires students to be able to express one number as a fraction of another.

Common mistakes and remediation

Students who need more **support** sometimes confuse the two axes and consequently draw the bars the wrong way round, possibly without a correctly spaced frequency axis. This is most likely to happen if the data is numerical. Go through examples carefully with the students in order to minimise these errors.

Useful tips

Students may find it easier to include as many divisions on the frequency axis as possible, even if only some of them are numbered.

Starter

- Select some newspaper clippings with bar charts. Distribute these among the class (this works best in pairs or small groups).
- Ask for a summary of what each bar chart is depicting.
- Discuss why the journalists have decided to use bar charts.

Main lesson activity

- Either ask the class to look at Example 5 in the Student's Book or display it on the board. Then talk about the familiar bar chart. Ask the students to talk about the special aspects of a bar chart that they must bear in mind when drawing one (same width bars and labelled axes).
- Talk about the fact that they are usually vertical, but they can be drawn horizontally. You can mention to students who need more of a **challenge** that a bar chart can even be drawn in a flower shape where they are drawn outside a regular polygon (these are often called bar chart roses).
- Using Example 5, ask a few questions relating to the data on the bar chart.
- Then ask students to look at Example 6 or display it on the board. Ask the class what is so different about this one (dual bar chart, showing more than one set of data). In this case a key is essential to identify which data set is which. Students who need more **support** will need to be reminded that this is essentially two bar charts just drawn on the same diagram and that they must keep reminding themselves of the key – which bar chart is for which set of data.
- Again, ask questions about the chart, e.g. 'What was the temperature in Turkey in May?' (80 degrees.) 'How much cooler was England than Turkey in August?' (30 degrees.)
- Once you are happy that the class are confident about looking at a dual bar chart, ask them to do Exercise 31C in the Student's Book.

Plenary

- Ask students to summarise the main points of the lesson. Focus on the fact that the bars are drawn from the data axis and extend to the length of the frequency value.

Homework and answers

Homework for this lesson is available in the homework folder. Consolidation and extension exercises are available in the further practice folder. Answers to the exercises can be found in the answers folder.

31.4 Pie charts

Syllabus references

Statement	Core
C9.4/E9.4	Draw and interpret pie charts.

Collins references

- Student's Book pages 629–633

Learning aims

- Draw pie charts.
- Interpret pie charts.

Key words

- pie chart, angle, sector

Prior knowledge

Students must know how to express one number as a fraction of another and how to calculate a fraction of a number.

Students must know how to draw and measure angles.

Common mistakes and remediation

The most common mistake when drawing a pie chart is to measure each angle from the same starting point.

Demonstrate the correct method of drawing pie charts to the class, with particular emphasis on starting the measurement of each angle at the end of the previous sector.

Useful tips

The sum of the angles of the sectors of any pie chart must always be 360°. It can be useful to check, before starting to draw, that this is the case. Sometimes, through approximating, the sum may be one degree higher or lower, and students will need to understand this.

Starter

- Ask students to find all the factor pairs of 360. They will need these in the exercise, so it may be worthwhile to ask them to write them down in their notes.

Main lesson activity

- Use Example 8 to illustrate what a pie chart is. Explain that it is a circular diagram (shaped like a pie), split into slices or sectors. Each sector represents a category of the data, and its size represents the frequency. Stress that each sector should be labelled clearly and the pie chart should also have a title.
- Explain that, in general, they will need to do some calculations when they draw pie charts themselves. Use Example 8 to illustrate this. Point out the table of results and explain that the circle will be split into sectors. The size of each sector indicates the frequency of the data it represents.

- Ask: 'What fraction of the people surveyed used a train?' Students who need more **support** will need help here to see that this is $\frac{24}{120}$. Show the class how to reach this fraction. First, ask how many people are involved (120), then how many used the train. (The answer is 24. This is 24 out of 120, or $\frac{24}{120}$). Write this fraction on the board, then go through each category, to give practice in finding these fractions.
- Say: 'We know that $\frac{24}{120}$ of the population asked use the train, but how do we show this fraction on the chart?'
- Students who need more **support** may not see how this fraction links to the circle. Lead the class to understand that they need to find this fraction of 360° (the angle at the centre of the circle), which will give the angle of the sector. Students who need more of a **challenge** may see that 120 divides into 360 three times, so one person will be represented by 3°, so they can just multiply the frequency by 3° each time to get the angle of the sector.
- Ask: 'How do we find $\frac{24}{120}$ of 360°?' Look for understanding that they have to multiply 360 by this fraction. If students recognise it as 'divide 360 by 120 and then multiply by 24', that is excellent. Ensure that students have calculators, and that they know how to do this calculation.
- Go through each calculation to ensure that all students see how to find the angles in the right-hand column.
- Discuss how to draw this pie chart, using a protractor for the sector angles. Students who need more **support** could benefit from practice at drawing angles before they attempt to draw pie charts. When drawing the pie chart, it is good practice drawing the smallest angles first so that any cumulative errors will be made on the largest angle.
- Ask the students who need more of a **challenge** to suggest why this is the case. If necessary, explain that it is because the fractional error is small on the large angle, but the same-sized error on a small angle could be as great as half that angle.
- Students can now work through Exercise 31D.

Plenary

- Draw a pie chart on the board, with sectors of 120°, 90°, 45° (the other angle is 105°). Label the sectors, e.g. 'red', 'blue', 'white' and 'yellow'. Give it a suitable title.
- Say that this represents 420 people. Ask students how many each sector represents. Lead them to note that 90° is a quarter of 360°, so that sector represents 105 people. Repeat for the sectors of 120° and 45°. Ignore the 105° unless asked about it.
- Now say that the pie chart represents a different set of people and that the 90° sector represents 60 people. Ask: 'How many do the sectors of 120° and 45° represent?' Again, ignore the 105° sector.
- For students who need more of a **challenge**, change the numbers so that the 120° sector represents 300 people. Ask how many the other sectors represent. Work out what 30° represents, then 90°, then halve this to get to 45°.
- Ask: 'Does the pie chart on its own tell you how many people it represents altogether?' Make sure that students realise that it does not. This is an important idea that is vital when trying to compare one pie chart with another. Pie charts show proportions, not actual numbers.

Homework and answers

Homework for this lesson is available in the homework folder. Consolidation and extension exercises are available in the further practice folder. Answers to the exercises can be found in the answers folder.

31.5 Scatter diagrams

Syllabus references

Statement	Core
C9.5/E9.5	Draw and interpret scatter diagrams. Understand what is meant by positive, negative and zero correlation. Draw by eye, interpret and use a straight line of best fit.

Collins references

- Student's Book pages 634–638

Learning aims

- Draw, interpret and use scatter diagrams.

Key words

- scatter diagram, variables, correlation, positive correlation, negative correlation, zero correlation, line of best fit

Prior knowledge

Students must be able to plot points, using coordinates.

Students must be able to use appropriate scales on axes and interpret intermediate positions on them.

Common mistakes and remediation

The greatest difficulty that students have is drawing the line of best fit appropriately. Students often try to join all the plotted points, as in a line graph. They also try to make the line of best fit pass through the origin. Ensure that students have adequate practice to overcome this.

Useful tips

Make sure that students know that they need to have as many points above the line as below it, and that the line does not have to pass through any of the points. Best fit does not mean exact fit.

Starter

- Draw a coordinate grid. Give students coordinates of points and ask them to plot them on the grid.
- Show students sections of number lines, marked in different scales, and ask for intermediate values. First, indicate random points and ask students to identify them, then tell students values and ask them to indicate where they are on the lines.
- Draw a coordinate grid and add a straight-line graph to it. Give students the value of one variable on the line and ask them to state the corresponding value of the other.

Main lesson activity

- Ask: 'If you get good marks in your mathematics test are you likely also to get a good mark in your science test?' This should lead to some useful discussion about trends and people that buck the trend.

- Look at the scatter diagram showing mathematics and science test results in the Student's Book. Say that the points have been plotted from data and the pattern they show clearly demonstrates that, for this group of students, there is a clear link between a good science mark and a good mathematics mark – the higher the science mark, the higher the mathematics mark is likely to be.
- Introduce the word *correlation*. The scatter diagram shows *positive correlation*, since as one variable increases in value, so does the other.
- This is not always the case. Refer students to the three diagrams in the section on correlation, which show positive, negative and no correlation. Ensure that students appreciate the clear difference between positive and negative correlation. Explain that the classification of no correlation still says something. It indicates that there is no clear link between the two sets of data, which is often very useful to know. Use Example 10 to reinforce this work.
- With students who need more of a **challenge**, extend this discussion to include *strong positive* and *weak positive* correlation, then *strong negative* and *weak negative*. Patterns showing a slight correlation can be described as weak, those showing a very clear link are strong. Sketch some examples of these on the board.
- Talk about *lines of best fit*. Discuss the diagrams in the Student's Book and ask the class to describe the lines.
- Draw out the fact that a line of best fit is just a straight line that shows the trend. It does not need to go through all of the points, or even through any of them, but must simply be as close to them all as possible. There should be as many points on one side of the line as on the other.
- Explain to students who need more of a **challenge** that a line of best fit can be a curve but, again, must just be one smooth curve, not an attempt to join all the points together.
- Moving on from lines of best fit, ask about marks that are missing from the set. Ask: 'What science mark might you expect from Kim, who gained 70 marks in the mathematics test?' Ensure that students know they can use the line of best fit to answer, by following the line from 70 on the mathematics axis to the line of best fit, then back along to the science axis to find the mark of about 68. Ensure that the students who need more of a **challenge** see this as an approximation.
- Make sure students are aware there are restrictions on using the line of best fit. It should not be used to extrapolate data (i.e. make predictions for data outside the range of the data given). It is generally OK to use it to interpolate data (i.e. make predictions for data inside the range of the data given).
- The class can now work through Exercise 31E. Plotting scatter diagrams can be time-consuming, so students could do selected questions.

Plenary

- Ask: 'What correlation is there between the distance you live from school and your marks in your last mathematics test?' Students should recognise that there is no correlation.
- Ask students for any correlations they can think about relating to themselves. Possible responses include distance of home from school and time taken to get to school (positive).
- Now ask about any correlations relating to cars. Possible responses include:
    - value of car and its age (negative)
    - speed and fuel consumption (positive)
    - size and seats (no correlation or weak positive).

Homework and answers

Homework for this lesson is available in the homework folder. Consolidation and extension exercises are available in the further practice folder. Answers to the exercises can be found in the answers folder.

31.6 Histograms

Syllabus references

Statement	Extended
E9.7	Draw and interpret histograms. Calculate with frequency density.

Collins references

- Student's Book pages 639–645

Learning aims

- Draw histograms.
- Calculate frequencies from a histogram.

Key words

- class width, frequency density

Prior knowledge

Students should understand how to draw a chart from grouped data.

Common mistakes and remediation

Students may confuse bar charts and histograms. Check frequently that they remember and understand the differences.

Useful tips

The most important point is that in a histogram the area represents the frequency

Starter

- Recap what a histogram is. Some students may think it is a bar chart with no gaps, but this is not accurate. Explain that it is a bar chart in which the area of the bars, not the height, represents the frequency and the vertical axis is the frequency density.
- Referring to the text under 31.7 on page 604, recap with students the meaning of class interval and introduce the idea of class frequency.

Main lesson activity

- Show this table of times a group of 50 people take to get to work.

Time, t minutes	$0 < t \leq 20$	$20 < t \leq 30$	$30 < t \leq 40$	$40 < t \leq 50$	$50 < t \leq 90$
Frequency	6	12	7	15	10

- Ask questions to check students understanding of the table. For example: How many people took between 40 and 50 minutes? (15) A person took 30 minutes. Which class are they in? ($20 < t \leq 30$) What can you say about the longest time anyone took? (it is not more than 90 minutes).
- Point out that times are continuous data, unlike the marks in the starter which are discrete.

- Say you are going to draw a chart to show this continuous data. It is called a **histogram**. To do this you need to calculate the **frequency density** for each class. The frequency density is the frequency divided by the class width.
- Draw this table to calculate the frequency densities:

Time, *t* minutes	Frequency	Class width	Frequency density
$0 < t \leq 20$	6	20	0.3
$20 < t \leq 30$	12	10	1.2
$30 < t \leq 40$	7	10	0.7
$40 < t \leq 50$	15	10	1.5
$50 < t \leq 90$	10	40	0.25

- Explain as you fill in the last column that 6 ÷ 20 = 0.3, 12 ÷ 10 = 1.2 and so on.
  Each student needs graph paper. Take them through the steps to draw this histogram:

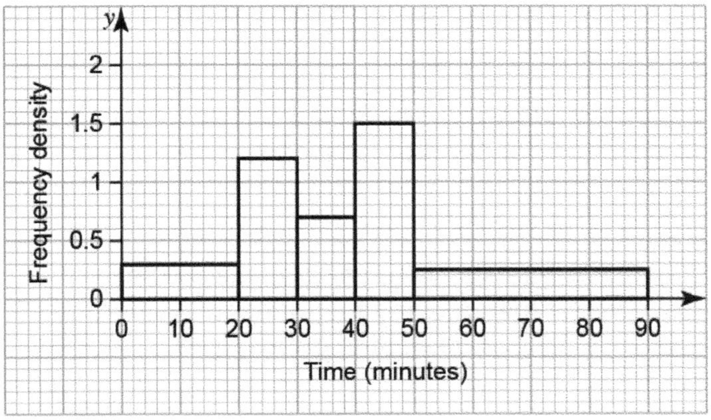

- Point out the continuous scale on the horizontal (time) axis. The boundaries of the classes are the edges of the columns There is no gap between the columns.
- Ask students to check that the area of each column is equal to the frequency. 20 × 0.3 = 6, 10 × 1.2 = 12 and so on. This is the essential feature of a histogram. Emphasise that the height of the column is not the frequency but the frequency density.
- Talk through Examples 11 and 12 in section 31.6; these explain how to interpret and draw a histogram.
- Now students can do Exercise 31F. If necessary, students can support each other by working in pairs.

Plenary

- Can students recognise whether a chart is a bar chart or a histogram? Ask students to explain the difference between a bar chart and a histogram.

Homework and answers

Homework for this lesson is available in the homework folder. Consolidation and extension exercises are available in the further practice folder. Answers to the exercises can be found in the answers folder.

32.1 The mode

Syllabus references

Statement	Core
C9.3/E9.3	Calculate the mode for individual data and distinguish between the purposes for which it is used.

Collins references

- Student's Book pages 648–650

Learning aims

- Find the mode from lists of data and from frequency tables.

Key words

- average, mode, frequency, modal value

Prior knowledge

Students should be able to extract information from a simple bar chart or frequency table.

Common mistakes and remediation

Students might easily confuse highest *value* with highest *frequency*. When finding the mode from a frequency table, a common mistake is to look for the number that occurs most often in the frequency column. Remind students that the mode (if there is one) is always one of the data values.

Useful tips

Continually remind students that mode means most, in order to help them avoid muddling the meanings of the different types of average.

Starter

- Go round the class and ask students for their shoe sizes. Record these on the board. If the class has boys and girls keep their results separate.
- Ask students to suggest the best value to represent the average shoe size of the class. This may lead to discussions of the most common, calculating a mean, etc.
- If appropriate discuss whether data should be split into boys and girls.

Main lesson activity

- Ask for the mode of the shoe sizes of the whole class. If appropriate ask the students for the shoe size that is occurring most often for the girls. Tell them that this is the mode. Ask for the mode of the boys' shoe size.
- Now ask who has the shoe size which is the mode for the whole class and introduce the word 'modal', by telling them this is the modal shoe size of the whole class.
- Use expressions such as the 'most common', 'most popular' or 'most frequent' to describe the mode and explain that finding the mode is quick and easy, whereas the other averages that students will look at later involve a little more work.
- Ask: 'What could go wrong with this idea of mode being the most common? When might it cause a problem?' (Where there is more than one mode, and where there is no mode at all.)
- Illustrate each of the above by a simple list on the board:
    - More than one mode: 5, 5, 5, 6, 6, 6, 8, 9
    - No mode: 4, 5, 6, 8, 9, 10
    - Ask students who need more of a **challenge** whether this last example actually has no mode or whether they are all in fact the mode. Both answers, paradoxically, are correct.
- Explain that if there is not a single mode then this particular average is not so useful, although the simple fact that there is no mode still says something about the data.
- Work through Example 1 with the class.
- Ask: 'What is the modal age of the class? What is the modal number of brothers and sisters of the members of the class? What is the modal pet of students in the class? Is there a modal name for students in the class?'
- Explain that the mode is the only average that can be used for qualitative data. They will need to understand this point in order to answer question 6 of Exercise 32A.
- The students can now do Exercise 32A.

Plenary

- Ask the class for some common everyday modes, e.g. modal gender in class, modal record in charts, modal name in class, modal way of getting to school, modal pet of class.

Homework and answers

Homework for this lesson is available in the homework folder. Consolidation and extension exercises are available in the further practice folder. Answers to the exercises can be found in the answers folder.

32.2 The median

Syllabus references

Statement	Core
C9.3/E9.3	Calculate the median for individual data and distinguish between the purposes for which it is used.

Collins references

- Student's Book pages 650–652

Learning aims

- Find the median from a list of data and a table of data.

Key words

- median, middle value

Prior knowledge

Students must be able to order numbers, including decimals and directed numbers. They must be able to read information from tables and bar charts.

Common mistakes and remediation

Students often forget to put lists of data into numerical order before locating the median. Students often forget to add one to the number of data items before halving to find the central one.

When using a frequency table, a common error is to use the central class from the table, instead of the central data value.

Make students aware of these errors so that they can avoid them.

Useful tips

Continually remind students that *median* means middle (when data items are arranged in order) in order to help students avoid muddling the meanings of the different types of average.

Starter

- State pairs of numbers and ask students which number is higher and which lower. Use various types of number, including some more difficult decimals and some negative numbers. Move on to three numbers, asking students to arrange them in order, and so on.
- State pairs of numbers and ask students for the number that is halfway between them. Start with simple pairs (2 and 4, 9 and 11) and then use increasingly difficult pairs (3 and 6, 22 and 50, 29 and 5, 2.1 and 8.7). Encourage students to develop a method for finding the midpoint.

Main lesson activity

- Ask: 'What is the average height in this class? Whose height is average? How can you find out?' Establish that the average height could be that of the 'middle' student, arranged by height. Demonstrate, by calling five students to the front and asking the rest of the class to put them in order of height. Explain that this average is the *median*. The median value separates the ordered data into two sections, each with the same number of values. Repeat with seven different students, again finding the median height.
- Now call four students to the front and ask the class to find their median height. When the students are standing in height order, let the class discuss what to do next. Lead them to suggest that they need to find the value halfway between the two middle heights. Demonstrate how to do this on the board. Add the two heights and halve the sum, rounding the result to the nearest centimetre or half-centimetre, if possible.
- Repeat with six more students, again asking the class to find the median height. Put the students in order of height, take the two middle values, add them together and find the halfway value.
- Ask students who need more of a **challenge**: 'When might it not be possible to find the median of a set of data?' The problem arises if the data cannot be sorted into order, as is the case for non-numerical or qualitative data. Numerical or quantitative data can be arranged in order. The median is the middle value (or one that is halfway between the two middle values, for an even number of data items).
- Work through Example 2 with the students before they do questions 1–4 in Exercise 32B.
- Once the class are confident at putting the data into order and finding the middle value, discuss with them how to find the middle value in longer lists of numbers. Ask which number is the middle out of 5 (3rd), out of 9 (5th) and out of 13 (7th). Ask if they can see a pattern or describe a way of identifying which would be the middle out of a set of 27 numbers. (Add 1 to the number of numbers and halve it.) Show that the middle of 27 numbers is the (27 – 1) halved, which is the 14th.
- Ask students what would happen if they had an even number of data values, such as 28. Lead students to the answer of (28 + 1) halved or 29 halved, which is 14.5. This is halfway between the 14th and the 15th number.
- Explain that they will now find the median height of the whole class. Count the students in the class, determine which number will give the middle student and then line up the students in order and count up to the median position. Introduce the 'even and odd' aspect by including yourself where necessary.
- Work through Example 3 with the class, making sure they understand how to find the median when using frequency tables.
- The class can now complete Exercise 32B.

Plenary

- Give a median and ask for examples of lists of three numbers that would give this median.
- Give another median and ask for examples of lists of four numbers that would give this median.
- Give a number and ask for two numbers that are the same distance from it.
- Ask for a definition of the median and a method for remembering it.

Homework and answers

Homework for this lesson is available in the homework folder. Consolidation and extension exercises are available in the further practice folder. Answers to the exercises can be found in the answers folder.

32.3 The mean

Syllabus references

Statement	Core
C9.3/E9.3	Calculate the mean for individual data and distinguish between the purposes for which it is used.

Collins references

- Student's Book pages 652–655

Learning aims

- Calculate the mean of a set of data.

Key words

- average, mean

Prior knowledge

Some questions in Exercise 32C require the students to be able to change the units of measurements, e.g. from hours and minutes or fractions of hours into minutes.

Common mistakes and remediation

The most common mistake when using a calculator to find the mean is to try to do the calculation in one step, without finding the total first. Encourage students to do the two stages separately, using whatever method is appropriate for the calculators they are using.

Useful tips

Students who need more **support** should not try to use the statistical functionality of the calculators to calculate the mean. It is easier for them simply to find the total of the numbers and then divide. It is only useful for students who need more of a **challenge** if they are likely to be doing a lot more statistics, taking them to the higher level.

Starter

- Pick some pairs of consecutive even numbers. Ask students to add them and divide the answer by 2, and then to comment on the result. Do the same for pairs of consecutive odd numbers. Pick pairs of non-consecutive even numbers and non-consecutive odd numbers and repeat.
- Pick some sets of three consecutive numbers. Ask students to add them and divide the answer by 3, and then to comment on the result. Repeat the process with four consecutive numbers and dividing by 4, then 5, and so on as far as possible.
- Select a group of four or five students and give each of them some cards (or similar), ensuring that they are not distributed equally. Ask the students to share the cards fairly or equally. Ask them how they could work out how many each person should get.

Main lesson activity

- Put on the board the numbers 2, 2, 3, 5, 88. Ask: 'What is the mode?' (2, the most common) Ask: 'What is the median?' (3, the middle value)
- Ask: 'What is wrong with these two averages in this case?' Lead students to see that these averages do not take into account the much larger final number, 88, and that a big number should make a difference to the average. Point out that the mode and median would be the same if 88 were replaced by 8.
- Explain that there is a final average, the *mean*, which is the only average which takes into account *all* the possible data. It is the most commonly used average because of this, but it is the most difficult to calculate (hence its name – mean!).
- Explain to students that all the data values are added together, so in the example this gives 100, then the total is divided by the number of numbers, here 5, so the mean would be 100 ÷ 5 which is 20.
- Go through Example 4. Emphasise the importance of the two-stage process (addition and then division) and encourage students to show their working at each stage, even when using a calculator.
- Explain that calculating the mean rarely gives a whole-number answer. For example, change the list of numbers on the board by adding the number 7. This will change the total to 107 and the number of numbers to 6, so the mean is 107 ÷ 6 = 17.8333333. Such an answer is usually rounded to one more decimal place than the data given. Here the mean should be rounded to 1 decimal place, giving 17.8. If necessary, remind students how to round numbers to a given number of decimal places.
- Show students who need more of a **challenge** possible shortcuts to calculating the mean if the numbers are all a similar size. For example, the mean of 92, 94, 96, 96 and 97 can be seen as 'ninety something', just requiring the mean of the unit digits. The mean of 2, 4, 6, 6 and 7 is 25 ÷ 5 = 5, so the mean is 95.
- Students can now do Exercise 32C.

Plenary

- Ask for summaries of the three different types of average, and for ways of remembering which is which.
- Give a number that is the mean of two numbers and ask for suggestions for what the two original numbers might have been. Do the same for sets of three or more numbers.

Homework and answers

Homework for this lesson is available in the homework folder. Consolidation and extension exercises are available in the further practice folder. Answers to the exercises can be found in the answers folder.

32.4 The range

Syllabus references

Statement	Core
C9.2/E9.2 C9.3/E9.3	Compare sets of data using tables, graphs and statistical measures. Calculate the range for individual data and distinguish between the purposes for which it is used.

Collins references

- Student's Book pages 655–657

Learning aims

- Find the range of a set of data and compare different sets of data, using the mean and the range.

Key words

- range, spread, consistency

Prior knowledge

Students must be able to carry out subtraction with decimal numbers and with negative numbers.

Common mistakes and remediation

Students sometimes use the first and last data values in a list, instead of checking to see if the list is ordered first. Stress that they should be using the *highest* and *lowest*.

Students often give the answer as, for example, 9 to 13, instead of 13 – 9 = 4. Stress that the range is the answer to the subtraction.

Useful tips

If students have difficulty subtracting decimal numbers or negative numbers, suggest they use a number line.

Starter

- Pick some pairs of numbers and ask for the difference between them. Pick some differences and ask for possible pairs of numbers.
- Pick some pairs of numbers that have the same mean and ask for the mean of each pair. Ask for suggestions about how the different pairs of numbers could be compared with each other if their means are the same.

Main lesson activity

- If the class has boys and girls ask: 'In this class, whose heights are more variable, the boys' or the girls'?' Discuss what students understand by 'variable'. Introduce the idea of differences between greatest and least values, and how this can indicate how much variation there is in the data. Explain that the difference between the highest and lowest values is the range.
- Find the range of the girls' heights and the range of the boys' heights and compare them. Establish which set of heights has the greater variability, which will be the one with the greater range.
- Emphasise that the range is just a number, the difference between highest and lowest values, and does not represent the actual heights.
- Write a few simple lists of numbers on the board and ask: 'What is the range?' Include decimal numbers and negative numbers in some of the lists. Illustrate each time the simple process: find the highest, find the lowest and subtract.
- Students who need more **support** may express the range as, for example, 5 to 12, which is a reasonable interpretation of the word. Explain that in a mathematical sense, range has a very specific meaning and a subtraction must always be carried out.
- Students who need more of a **challenge** may start to recognise that two sets of data may have the same number of values, but that those in one set may be closer together than those in the other, which may include extreme values. Encourage them to examine several data sets and compare the ranges.
- Explain how to compare two sets of data using the mean to give an idea of the size of the numbers, and the range to give an idea of how spread out they are. Point out that if a question asks for a comparison, there is not necessarily a right or wrong answer as long as they give a good explanation for their decision.
- Work through Example 5 with the class.
- The class can now do Exercise 32D.

Plenary

- Ask for a definition of the range.
- Discuss situations in which consistency might be an important element. Try to elicit real situations which the students have experienced, such as sporting situations, preparation for exams, part-time jobs they may have experience of, etc.

Homework and answers

Homework for this lesson is available in the homework folder. Consolidation and extension exercises are available in the further practice folder. Answers to the exercises can be found in the answers folder.

32.5 Which average to use

Syllabus references

Statement	Core
C9.3/E9.3	Calculate the mean, median and mode for individual data and distinguish between the purposes for which these are used.

Collins references

- Student's Book pages 658–660

Learning aims

- Understand the advantages and disadvantages of each type of average and decide which one to use in different situations.

Key words

- representative, appropriate, extreme values

Prior knowledge

Students should know how to calculate all three types of average and how to find the range of a set of data.

Common mistakes and remediation

Mistakes here generally come from a lack of thought about the situation, or a lack of clarity about the different uses of the three averages. Go through the table in the Student's Book carefully in order to strengthen students' understanding and to rule out as many errors as possible.

Useful tips

Go round the class asking students for the definitions of mode, median and mean, in a random order, so that they are sure of their meanings.

Starter

- Put this list of numbers on the board: 1, 2, 4, 9, 9. Tell the students that the mode is 9, the median is 4 and the mean is 5. Ask which of the three averages best represents the data.
- The mode can easily be eliminated as it is an extreme value. There is not much to choose between median and mean.
- Now add a sixth value, 71. The mode is 9, the median is 6.5 and the mean is now 16. Discuss which is the best average now. The mean can be eliminated as it is affected by an extreme value. There is not much to choose between mode and median this time.

Main lesson activity

- Ask: 'When is the mode the most useful average?' Establish that it is when dealing with non-numerical data, or trying to emphasise a certain feature of the data.
- Ask: 'When is the median the most useful average?' Establish that it is when there are extreme values that are not truly representative of the data.
- Ask: 'When is the mean the most useful average?' Establish that it is when it is important to include all the data values.
- Talk about the average age of the people in the classroom. Ask which is the best average to represent this data and establish that it depends on the method of collecting the data and the purpose for finding it. A possible response could be:
    - the mode, if the ages are given in integers
    - the median, if the ages are given in years and months
    - the mean, if the ages of the teacher and any classroom assistants are included.
- Go through the table in the Student's Book. Make sure that students understand when one average is more representative in a situation than the others.
- Students who need more **support** may struggle with some of the concepts of choice here, but as long as they are given the opportunity of revising how to find each average and start to realise that sometimes one average is better than others, then that is a success.
- Extend the discussion for students who need more of a **challenge**, encouraging them to discuss the advantages and disadvantages of each type of average and explain why one might be more suitable than others in some situations.
- The students can now do Exercise 32E.

Plenary

- Give the following data about average pocket money for a class:

  Mode = \$6

  Median = \$8

  Mean = \$12
- Which would be the best average to use in each situation?
    - You are the head teacher wanting to work out if a vending machine will be profitable.
    - You are a student after a rise in pocket money.
    - You are a parent trying to resist giving a rise in pocket money.

Homework and answers

Homework for this lesson is available in the homework folder. Consolidation and extension exercises are available in the further practice folder. Answers to the exercises can be found in the answers folder.

32.6 Stem-and-leaf diagrams

Syllabus references

Statement	Core
C9.4/E9.4	Draw and interpret stem-and-leaf diagrams.

Collins references

- Student's Book pages 660–664

Learning aims

- Construct and interpret stem-and-leaf diagrams.

Key words

- stem-and-leaf

Prior knowledge

Students need to have worked through sections 32.1 to 32.4 so that they are aware how to calculate statistical measures.

Common mistakes and remediation

The most common mistake is that students do not put a key on their diagram.

When interpreting stem-and-leaf diagrams to find the median, some students count in the wrong direction. If they look for the median value and start counting from the lowest value, then they should count from the stem each time they move to the next row. If they start counting from the highest value, then they should count from the leaf that is furthest away from the stem each time they move to the next row.

Useful tips

To help students keep the leaves neatly arranged, it helps if they work on squared paper.

Starter

- Ask students to think of a whole number between 10 and 49. Then ask them to call out their number, one by one, as you write their number on the board. When that's done ask them how easy it is to identify the mode, median and range from the jumbled-up data.
- Now ask the class to write the numbers into the following groups:
  10 – 19, 20 – 29, 30 – 39, 40 – 49. Mention that you do not want them to write down the frequency as that loses the individual data. Is it any easier to identify the mode, median and range now?

Main lesson activity

- Let the class know that they will now be drawing a new type of statistical diagram to show the data they generated called an ordered stem-and-leaf diagram, or stem-and-leaf diagram for short. In most cases the stem is the first digit of the data (the most significant digit) and the leaf is the second digit. The following data was generated by a scientific calculator – you might want to show the students how to use the 'rand' button to produce their own data.
  37, 35, 21, 38, 36, 22, 12, 20, 33, 17, 33, 14, 36, 30, 28, 28, 49, 32, 23, 12, 22, 35, 25, 33, 33, 16, 21, 24, 46, 23
- Draw a vertical line on the board and on the left-hand side put the digits 4, 3, 2 and 1 from the top to the bottom. Explain that these represent the tens digit from the data they have provided.

- Now ask the students to put the leaves (the units digits) next to the stem in the order they put them into the groups in the starter activity.

4	9 6
3	7 5 8 6 3 3 6 0 2 5 3 3
2	1 2 0 8 8 3 2 5 1 4 3
1	2 7 4 2 6

  This is known as an **unordered** stem-and-leaf diagram.

- Now ask the class to put the leaves in order with the lowest units digit next to the stem.

  key: 4/6 represents 46

4	6 9
3	0 2 3 3 3 3 5 5 6 6 7 8
2	0 1 1 2 2 3 3 4 5 8 8
1	2 2 4 6 7

  Finally they need to add a key to show what each piece of data shows.
  By putting the leaves in order it makes the data easier to analyse.

- Some students may question the need to do an unordered stem-and-leaf diagram – it takes time, but helps avoid errors in ordering the data as without it they may find they miss an item of data and then it is hard to squeeze it in the ordered diagram as the digits should be neatly placed underneath each other.
- Ask how they can use this to find the range (49 – 17 = 32), the mode (33), the median (since there are 30 values, the median will be between the 15th and 16th values, which in this case is 28 – note this is between the two values of 28).
- Students who need more **support** may have difficulty finding the median when they count down from the highest value: it is better to have them count up from the lowest value.
- Show the students that this looks like a horizontal bar chart, the advantage being that the individual items of data have not been lost.
- Go through Example 6.
- The class can now do Exercise 32F.

Plenary

- Go over Exercise 32F questions 10 and 11 and explain to the students that this type of diagram is very useful in comparing two data sets, as one set of data can be shown to the right of the stem, the other set to the left of the stem.

Homework and answers

Homework for this lesson is available in the homework folder. Consolidation and extension exercises are available in the further practice folder. Answers to the exercises can be found in the answers folder.

32.7 Using frequency tables

Syllabus references

Statement	Core
C9.3/E9.3	Calculate the mean, median and mode for individual data and distinguish between the purposes for which these are used.
C9.4/E9.4	Draw and interpret simple frequency distributions.

Collins references

- Student's Book pages 664–667

Learning aims

- Revise finding the mode and median from a frequency table.
- Calculate the mean from a frequency table.

Key words

- frequency table

Prior knowledge

Students need to know how to find the mode and median from a frequency table.

Common mistakes and remediation

The most common mistake is simply adding up the numbers in the frequency column (or the data class column) and then dividing by the number of rows in the column. This often leads to a mean that is well outside the range of data values and is, therefore, clearly nonsense. Show an example of this to the students, so that they can appreciate why it is nonsense.

When using a calculator, there are two common mistakes. Firstly, students often forget that, if they make a mistake and start again, they must be very sure that they have cleared the calculator's memory before re-entering any data. Secondly, they must be sure which way round *their* calculator accepts the data value and frequency inputs. It may be necessary to help students individually with calculator work to reduce these errors.

Useful tips

Make sure all the class have calculators for use in this exercise. Students who need more of a **challenge** should be encouraged to use the statistical functionality on their calculators but only after they have completely understood the method.

Starter

- Ask for definitions of each of the three types of average, and ways of remembering which is which.
- Choose some small sets of numbers and ask for all the averages of them.
- Select a set of straightforward numbers, with some repeated numbers (for example, 2, 2, 3, 3, 5) and ask pupils to calculate the mean. Do this with a few similar sets, leaving each set and each answer visible, then put each set into table form and ask for a quick method.

Main lesson activity

- Ask: 'What is the average number of children in your family homes?' They will suggest various numbers, but look for an accurate average.
- Draw a table on the board, and record the number of children in homes in the class by asking students to put their hands up. If there is an obvious mode, talk about that as an average.
- Ask the class where the median will be (the $\frac{n+1}{2}$ position). Ask the students for the median and explain that this is also an average.
- Ask which is the only average that takes all of the values into account (the mean). Explain how to use the table to help to calculate the mean.
- Rewrite the table vertically, with column headings: Number of children, Frequency, $N \times f$.
- Explain that multiplying the number of children by the frequency gives the total number of children. Complete this part of the table with help from the students.
- Students who need more **support** may have difficulty understanding what the numbers in the $N \times f$ column represent and what the total number of children is, so it might be helpful to list the data in full before using the table. The completed table should be similar to this table:

Number of children	Frequency	$N \times f$
1	4	4
2	9	18
3	7	21
4	3	12
5	2	10

- Ask: 'What two amounts do you need to calculate the mean?' These are the total number of children in all the families and the total number of students in the class. Make sure that the students do not confuse these two totals.
- Now ask if they can see how they can get this information from the table.
- Lead the students to see that the total frequency gives the number of students in the class, 25. Also, by adding up the $N \times f$ column, they find the total number of children in all the families, 65. Add these totals to the table.
- Now ask: 'What do you now do to calculate the mean?' (Divide 65 by 25.) Ask a student to do this on their calculator (2.6), and then ask the class to check that the result seems sensible. Explain that they may have many decimal figures in the answer and will need to round to 1 decimal place.
- Remind the class that it is often difficult to calculate the mean, so a table like this often helps. Go through the process of using a table again on the board, ensuring that each student is aware of what is being done and why.
- Go through Example 7, making sure students understand the reasons for the process.
- Statistical calculators are not helpful for students who need more **support** while they are learning the process. Such a calculator may be used by students who need more of a **challenge** once they have mastered the method.
- The class can now do Exercise 32G.

Plenary

- Show the students some frequency tables and ask them to estimate the mean for each. Ensure that students understand that the mean lies somewhere between the lowest and highest data values.

Homework and answers

Homework for this lesson is available in the homework folder. Consolidation and extension exercises are available in the further practice folder. Answers to the exercises can be found in the answers folder.

32.8 Grouped data

Syllabus references

Statement	Extended
E9.3	Calculate an estimate of the mean for grouped discrete or grouped continuous data. Identify the modal class from a grouped frequency distribution.

Collins references

- Student's Book pages 668–671

Learning aims

- Identify the modal class.
- Calculate an estimate of the mean from a grouped table.

Key words

- grouped data, estimated, modal class, continuous data, discrete data

Prior knowledge

Students should be familiar with using grouped frequency tables. They will also need to be able to understand and use inequality symbols.

Common mistakes and remediation

The most common mistake here is that students misunderstand the instruction 'find an estimate for the mean', which they understand as being 'guess the mean'. Emphasise that estimating the mean involves doing a calculation.

Students often make mistakes finding the midpoints of the groups. Encourage them to use the method of adding the end points and then dividing by 2.

Useful tips

Encourage students to do a quick check to see if their answers are sensible.

Starter

- Give students some inequalities and ask them to suggest numbers that are included in them. Make sure that some of the inequalities are inclusive and that some of the suggestions are the extreme values.
- Use a non-inclusive inequality, such as $x < 5$, and ask for the largest possible number that could be included. This may need some discussion.
- Use inequalities such as $5 \leqslant x < 10$ and $10 < x \leqslant 20$. Discuss what they mean.

Main lesson activity

- Put on the board some data similar to, but not the same as, Example 8, such as:

Pocket money, p (\$)	$2 < p \leqslant 3$	$3 < p \leqslant 4$	$4 < p \leqslant 5$
No. of students	3	10	12

- Remind students who need more **support** what the inequality symbols mean, especially the difference between the symbols, and ask why they are used. Students must understand that the class boundaries need to be clearly defined so that each person can only be in one class.
- Ask the class for the modal group ($4–$5). If necessary remind them that this is the range of values with the highest frequency.
- Ask for the median. It relates to the $\frac{26}{2}$ = 13th student, in order, and is in the $3–$4 group, but the exact amount isn't known. Explain to students who need more of a **challenge** that it is possible to estimate this amount by estimating how far into this class interval it is.
- Ask what two pieces of information are needed to calculate the mean. (The total amount of pocket money and the total of students.) Ask the students if they can find the total amount of pocket money. (No, because they don't know exactly how much each child in the group got.) Emphasise that the exact figures are not known, but that each figure lies in a certain range. This means that only an estimate for the mean can be calculated – a *calculated estimate*, not a guess.
- Explain that as they do not know exact values they can assume each student got the midpoint value of the group they are in. Some may have got more and some less. Take some time explaining this as it is important that students understand why the midpoint is used. Using the $2–$3 group of three students, ask for three suggestions of possible amounts and work out their mean. Repeat this with another three suggestions and then compare these with the mean, if all three got $2.50.
- Draw a new vertical table and add new columns – first the midpoint column. Ask students for the midpoints of the groups. Students who need more **support** may need help finding the midpoints. Explain that they can add the end points of the group and divide by 2.

Pocket money, p ($)	**Frequency, f**	**Midpoint value, m**
$2 < p \leqslant 3$	3	2.50
$3 < p \leqslant 4$	10	3.50
$4 < p \leqslant 5$	12	4.50

- Ask how to estimate the total amount of pocket money for all the students. (Multiply the frequency by the midpoint value.) Emphasise this is an estimate, not a guess.
- Add the final column heading: $f \times m$, and with the help of the students, complete the column. Then ask what they need to do next. (Find the total of $f \times m$ and find the total of the frequency.)
- Ask the students to work out these totals (96.5 and 25). Students can now work out the estimated mean (96.5 ÷ 25 = 3.86). Ask what 3.86 means. As it is money, it represents $3.86. Explain that this value does not need rounding but that as it is an estimate they could give the mean as $3.90 as a rounded figure. Go through Example 8 to reinforce the method, explaining each step carefully.
- Explain the difference between discrete and continuous data. Discrete data consists only of separate numbers – number of students, shoe sizes, goals scored – whereas continuous data can take any value within a certain range – height, mass, time, capacity.
- Students can now do Exercise 32H.

Plenary

- Give students some pairs of numbers and ask them for the midpoint of each pair.

Homework and answers

Homework for this lesson is available in the homework folder. Consolidation and extension exercises are available in the further practice folder. Answers to the exercises can be found in the answers folder.

32.9 Cumulative frequency diagrams

Syllabus references

Statement	Extended
E9.6	Draw and interpret cumulative frequency tables and diagrams. Estimate and interpret the median, percentiles, quartiles and inter-quartile range from cumulative frequency diagrams.

Collins references

- Student's Book pages 671–679

Learning aims

- Find a measure of dispersion (the inter-quartile range) and a measure of location (the median), using a graph.

Key words

- quartile, inter-quartile range, dispersion, cumulative frequency, lower quartile, upper quartile

Prior knowledge

Students should be familiar with estimating the mean from grouped continuous data. In particular, they should know the distinction between lower and upper class boundaries and mid-interval values.

Common mistakes and remediation

Students who need more **support** may write the median as $\frac{n}{2}$ rather than the $\frac{n}{2}$th value: in other words, they do not make use of the cumulative frequency curve.

Some students also plot mid-class intervals along the x-axis rather than the correct upper class boundaries, confusing their work with 'estimating the mean'. If so, work through several examples with them, emphasising how to plot and read from the cumulative frequency curve.

Useful tips

Point out to students who need more of a **challenge** that a cumulative frequency curve should give a more accurate picture as it follows the trend.

Starter

- Say that this lesson returns to grouped continuous data.
- Write various class intervals on the board, e.g. $0 \leqslant h < 10$, $10 \leqslant h < 20$, and ask students to name the lower class boundary, the interval midpoint and the upper class boundary.
- Go through a quick example of estimating the mean from a grouped distribution. Emphasise that to estimate the mean from grouped data they must plot the mid-interval values, whereas in contrast, estimating the median for that data involves using the upper class boundaries.

Main lesson activity

- On the board, copy the first two columns (mark and number of students who got various marks in a mathematics test) of the table in the Student's Book. Discuss cumulative frequency – a useful way to find how many students have scored up to a certain mark, or of finding the median.
- Demonstrate how to plot the cumulative frequency diagram, using the upper bound of each class, and point out the shape of the diagram. Explain to students that the points can be joined with straight lines or with a smooth curve, giving a cumulative frequency curve. Both are called cumulative frequency diagrams.

The median

- Show students how to find the median from the diagram, using the middle of the frequency, reading along to the line and following the vertical line down to the horizontal scale. Mention again, especially to students who need more of a **challenge**, that as this is an estimate and the frequency scale is being treated as continuous, it is quite valid to use the halfway mark. Demonstrate this on your diagram, giving a median of 65 marks.

The inter-quartile range

- Revise quartiles. The bottom quarter is bounded by the lower quartile and found as the value of one-quarter of the frequency, or $\frac{1}{4}n$. Again, point out that this is an estimate, so even if $\frac{1}{4}n$ has a decimal value it should still be used. Use your diagram to show that the lower quartile is about 49 marks.
- Repeat for the upper quartile, finding three-quarters of the frequency, and reading its value on the horizontal scale – about 83 marks. Label the median, lower quartile and upper quartile on the diagram.
- Define the inter-quartile range and explain that this can be a useful measure of the dispersion of the data around the median, as it shows the range of the middle 50%. A small inter-quartile range indicates that the middle 50% of the data are all close to the median; a larger inter-quartile range indicates a far wider spread.
- Ask for the inter-quartile range in this case. (Upper quartile – lower quartile = 83 – 49 = 34 marks)
- Now ask: 'If the top 10% scored a grade A, what is the cut off mark for a grade A?' Discuss this, pointing out that the mark for the 90% frequency must be read – to give a mark of 95.
- Help students who need more **support** by working through Example 9. Meanwhile, let students who need more of a **challenge** complete Exercise 32I. Students who need more **support** can work in pairs to complete the questions. Provide support as needed.

Plenary

- Which is better, a high inter-quartile range or a low inter-quartile range? Ask students their opinions. Can they think of practical instances where it would be preferable to obtain:
  - a low inter-quartile range
  - a high inter-quartile range?

Homework and answers

Homework for this lesson is available in the homework folder. Consolidation and extension exercises are available in the further practice folder. Answers to the exercises can be found in the answers folder.

33.1 The probability scale

Syllabus references

Statement	Core
C8.1/E8.1	Understand and use the probability scale from 0 to 1.

Collins references

- Student's Book pages 682–683

Learning aims

- Use the probability scale and the basic language of probability.

Key words

- chance, outcome, event, probability, impossible, certain

Prior knowledge

Very little mathematical knowledge is required for this lesson, as it simply involves students using their judgement to decide how likely it is that a variety of different events will happen. The results of tossing coins is a useful way of demonstrating probability and you may wish to check that students know which is the 'head' and which the 'tail' of a coin.

Common mistakes and remediation

Students sometimes believe that just because an event can either happen or not happen, that this is an even chance. If this arises, discuss with the students events that are more or less likely, such as the chance of it being rainy or windy in June.

Useful tips

- Aspects of the weather, the calendar or the time of day provide useful examples of events for illustrating probabilities.

Starter

- Discuss the idea of an event that may have several outcomes, such as tossing a coin or spinning a spinner. The event is tossing the coin, the outcomes are 'heads' or 'tails'.
- Ask students to suggest events with outcomes that they consider to be impossible. Probe to check that their suggestions really should be considered impossible and not just very unlikely.
- Do the same for events with outcomes that they consider certain.
- Suggest a list of events and outcomes and ask the students to rank them in order of likeliness. It is useful to have a numbered list of the events and a line on which the relevant numbers for the outcomes can be placed, in order.

Main lesson activity

- Students should understand the use of the five main terms associated with likeliness: impossible, unlikely, even chance, likely and certain.
- Ask: 'What is the chance of your eating rice today?' This should provoke discussion from those who say, 'It's impossible because ...' to those who say, 'Certain because ...'. Write the key words: *impossible*, *unlikely*, *evens*, *likely* and *certain* on the board as they appear from the class. Having this list on the board helps to focus on these key words.
- Draw a probability scale from 0 to 1 on the board and ask the students to suggest three to five events and outcomes that are in each category. Students who need more of a **challenge** should be able to provide more examples.
- The impossible ones must be exactly that and not just very unlikely; it is important to draw out the difference between, for instance, 'You will become famous one day', which is very unlikely, and 'You will suddenly grow another arm', which is impossible.
- Students who need more **support** will need a lot more help with the idea of evens as, surprisingly, they often do not grasp this idea well. Talk about even chances of, say, tossing a coin, or a new-born baby being a boy or girl. Emphasise that even chances do not just occur when there are only two possible options,
for example, tripping down the stairs tonight or not tripping down the stairs tonight.
- Work through Example 1 with the students.
- Students can now do Exercise 33A.

Plenary

- Ask students to suggest things that might happen to them when they get home from school.
Ask for five things that are impossible, then five that are unlikely, five that are even chances, five that are likely and finally just one that is certain. It is this last one that might actually cause more discussion as there is a fine line between certain and very likely.

Homework and answers

Homework for this lesson is available in the homework folder. Consolidation and extension exercises are available in the further practice folder. Answers to the exercises can be found in the answers folder.

33.2 Calculating probabilities

Syllabus references

Statement	Core
C8.1/E8.1	Calculate the probability of a single event.

Collins references

- Student's Book pages 684–687

Learning aims

- Calculate the probability of outcomes of events.

Key words

- outcome, equally likely, probability fraction, random

Prior knowledge

Students must know how to express one number as a fraction of another, how to simplify fractions and how to add fractions with the same denominator. They should also be able to convert between fractions, decimals and percentages.

Common mistakes and remediation

Students often express a probability incorrectly by writing an answer as 1 out of 5 instead of $\frac{1}{5}$. Correct this as soon as you see students making this error so that they realise their mistake.

Useful tips

It may be useful to keep a probability scale on the board throughout this chapter so that students are reminded that probabilities always lie in the range from 0 to 1.

Starter

- Prepare a bag containing a number of different coloured balls, or similar.
- Ask students to select a ball without looking in the bag, check the colour and put it back. Have a student collect the data as it is generated.
- At different stages, ask for suggestions about what is in the bag. Ensure that students give reasons for their suggestions.

Main lesson activity

- Take the balls out of the bag and discuss the activity and the results now that students can see the contents.
- Explain the ways in which a probability can be expressed. For example, if there were six blue balls and four red balls then the chance of getting a blue is six out of 10. Write this on the board as a fraction and ask where each number has come from. The top number (numerator) is the number of blue balls and the bottom number (denominator) is the total number of balls.

- For students who need more of a **challenge** introduce the idea of cancelling this fraction to $\frac{3}{5}$, so the chance of getting a blue ball is $\frac{3}{5}$.
- Ask: 'What is the chance of getting a red ball?' ($\frac{4}{10}$ or $\frac{2}{5}$ – accept also 0.4 or 40%)
  Introduce the idea that there are only two possibilities: students are bound to pick either a blue or a red. Picking one of these two colours is a certainty.
- The probability of picking blue or red is 1. Write this clearly on the board:
  - for students who need more **support**, $\frac{6}{10} + \frac{4}{10} = 1$
  - for students who need more of a **challenge**, $\frac{3}{5} + \frac{2}{5} = 1$
- In summary, the probability of picking one of all the possible alternatives is 1.
- Next, introduce a six-sided spinner with integers from 1 to 6 and ask: 'What is the probability of rolling this spinner and getting a 5?' $\left(\frac{1}{6}\right)$.
- Discuss the various answers from the class. Ask whether this is a good chance or not a good chance, using the vocabulary from the previous lesson.
- Discuss the chance of getting each number on the spinner. They all have the same probability, $\frac{1}{6}$, and the total of all these probabilities is $\frac{6}{6}$ or 1.
- Each time, write the fraction correctly on the board.
- If students use phrases such as 'one in four', explain that probabilities are not expressed in this way. In mathematics, probabilities are always written as either a fraction or a decimal. If students mention percentages, remind them that percentages are fractions out of 100.
- For students who need more of a **challenge**, introduce the shorthand notation P(blue), which means the probability of getting blue. You will need to explain this notation to students who need more **support** in the examples and a few of the questions in the exercise, but students who need more of a **challenge** should use it.
- Look at Examples 2 and 3.
- Explain that 'at random' means without having a choice in the outcome.
- Students can now do Exercise 33B.

Plenary

- Use a target board with whole numbers on it and ask for probabilities of different outcomes of picking a number at random, for example, P(prime number), P(less than 10) or P(multiple of 4). The choice of outcomes will depend on the target board. Be sure to choose some outcomes that are impossible.

Homework and answers

Homework for this lesson is available in the homework folder. Consolidation and extension exercises are available in the further practice folder. Answers to the exercises can be found in the answers folder.

33.3 Probability that an event will not happen

Syllabus references

Statement	Core
C8.1/E8.1	Understand that the probability of an event not occurring = 1 – the probability of the event occurring.

Collins references

- Student's Book pages 687–688

Learning aims

- Calculate the probability of an outcome of an event not happening when you know the probability of the outcome happening.

Key words

- outcome

Prior knowledge

Students must know how to find the probability of an event, such as those covered in Exercise 33B.

Common mistakes and remediation

It is most common for students not to take sufficient care in identifying mutual exclusivity of events. For example, the events 'picking a prime number' and 'picking an even number' are not mutually exclusive.

Another common mistake is not to take sufficient care in identifying the exhaustiveness of events. For example, the events 'scoring less than 2' and 'scoring more than 2' when spinning a spinner are not exhaustive.

In both cases, encourage students to list the outcomes for each event and then check for exclusivity or exhaustiveness to help to avoid possible errors.

Starter

- Practise complements to one; for example, ask students to subtract $1 - \frac{4}{5}$
- Ask if they can see an easy way to work these problems out.

Main lesson activity

- Show students a six-sided spinner numbered from 1 to 6 and ask: 'If I spin this spinner, what numbers could I get?' As they answer, write the numbers 1 to 6 in a list on the board.
- Ask: 'What is the probability of getting a 1?' Put the answer $\frac{1}{6}$ on the board. (Remind students who need more **support** of the work in the previous lesson.) Then ask about getting a 2, then a 3 and so on, putting the fraction $\frac{1}{6}$ next to the number in each case.

- Now ask: 'What is the probability of spinning a 1 or a 2 or a 3 or a 4 or a 5 or a 6?' For students who need more **support** use the wording: *spinning a number bigger than 0*. The answer is 1 as it is certain. Remind students that a probability of 1 means that an outcome is certain to happen.
- Ask the students what they get if they add up all the probabilities of all these numbers, i.e. $\frac{1}{6} + \frac{1}{6} + \frac{1}{6} + \frac{1}{6} + \frac{1}{6} + \frac{1}{6}$ leading to the answer 1.
- Now ask: 'What is the probability that I do *not* spin a 6?' This may bring various answers, with students who need more **support** just guessing an answer. Ask for explanations.
- Eventually, lead the students to recognise that to find the probability of not getting a 6 is simply 1 minus the chance of getting a 6, i.e. $1 - \frac{1}{6} = \frac{5}{6}$. Students who need more **support** may need to count that there are five other numbers to choose from, also giving $\frac{5}{6}$.
- Now ask about the chance of not spinning a 4 and expect an answer of $\frac{5}{6}$ again. Continue this for all the numbers of the spinner until the students have understood this concept.
- The subtraction of fractions may cause problems for students who need more **support**, so give plenty of practice to those who need it.
- Go through Example 4 with the students.
- Say: 'I heard this morning that the chance of rain today is 0.3. What is the chance that there is no rain?' (0.7) Put 0.3 on the board to help students focus and again ask them to justify their answers. Students who need more **support** will need extra help with the decimal subtraction, although this should simply be a reminder.
- Explain to students who need more of a **challenge** that they need to be very careful with questions such as, 'If the probability of it raining today is 0.5, what is the probability of it being sunny?' This can lead to a good discussion with many students assuming the answer is 0.5, but of course just because it doesn't rain, doesn't mean it will be sunny. The probability of an outcome is only equal to 1 – the probability of another outcome if the two outcomes cover all possibilities.
- This is a challenging concept for many students and so should not be rushed. Allow discussion of students' answers as attempting to justify their incorrect thinking often helps them to understand their errors.
- Once the students understand this concept they can do Exercise 33C.

Plenary

- Use a target board with whole numbers on it. Ask a variety of 'not' probability questions, e.g. P(not a multiple of 3), P(not less than 8) and P(not more than one digit).

Homework and answers

Homework for this lesson is available in the homework folder. Consolidation and extension exercises are available in the further practice folder. Answers to the exercises can be found in the answers folder.

33.4 Probability in practice

Syllabus references

Statement	Core
C8.2/E8.2	Understand relative frequency as an estimate of probability. Calculate expected frequencies.

Collins references

- Student's Book pages 688–691

Learning aims

- Calculate experimental probabilities and relative frequencies from experiments.
- Recognise different methods for estimating probabilities.

Key words

- experimental probability, trials, relative frequency, expected frequency

Prior knowledge

Students should already know how to calculate a probability in simple cases using equally likely outcomes.

Students must know how to express one number as a fraction of another.

Common mistakes and remediation

Students often have difficulty in identifying the relevant information from data presented in tabular form. It should be made clear to them that they need to use 'number of correct results' and 'number of trials' in all cases but they need to be able to interpret the table correctly to find these values. In some tables the values will be cumulative frequencies of the required outcome and in others they will be the frequencies of all the results. Ask students a few questions on aspects of the information given in any of the tables in section 33.4, to help them feel more confident about using them.

Useful tips

Make sure that there are plenty of calculators available for this lesson as it will often be necessary to change fractions into decimals to compare them.

Starter

- Prepare a bag or box containing a number of different coloured balls, or similar.
- Ask students to select a ball without looking in the bag, note its colour and put it back. Have a student collect the data as it is generated.
- At different stages, e.g. after four selections and again after 10, ask for suggestions about what is in the bag.
- Ensure that students give reasons for their suggestions.

Main lesson activity

- Ask: 'What is the probability of tossing a coin and getting heads?' (1/2 or 0.5) Most students will know the answer. Say that this is the theoretical probability, but ask: 'Does it actually work out in practice?'
- Ask all students to toss a coin. Ask how many students tossed a head and make a note of this on the board in a table like the one shown here. Show them how to find the probability and explain that it will be necessary to change fractions to decimals for comparison as more coins are thrown. Remind students who need more **support** how to do this, dividing the numerator by the denominator and, if necessary, simply writing down the first four or five decimal places.

Number of throws	Number of heads	Probability
30	18	$\frac{18}{30} = 0.6$

- Ask the class to toss their coins again, count again how many heads and complete the next entry in the table.

Number of throws	Number of heads	Probability
30	18	$\frac{18}{30} = 0.6$
$30 + 30 = 60$	$18 + 14 = 32$	$\frac{32}{60} = 0.5333$

- Continue to build up the table in this way. Lead students to notice that the probability gets closer and closer to 0.5. Ensure that students who need more of a **challenge** understand that using the decimal notation is nearly always the best way to compare fractions.
- Explain that this is *experimental probability*, also called *relative frequency*. Often this is the only way to find a probability, such as that of dropping a slice of bread on the floor and it landing butter side down! Students should also realise that the more trials are made, the more accurate the results should be.
- Work through Example 5.
- The syllabus mentions expected frequency of occurrences. Students should be made aware that they can use the relative frequency of an event of a sample when estimating the outcome of that event in a larger sample or population. For example, if you ask your class how many have a birthday in May and hence work out the relative frequency of a birthday in May then you can use that relative frequency to estimate the number of students who have a birthday in May for the whole school.
- Explain the use of historical data to find probabilities of events where experiments are impossible, such as the probability of an asteroid hitting the Earth, or a volcano erupting.
- Students can now do Exercise 33D.

Plenary

- Ask the students to pick a number from 0 to 9, inclusive, and write it down. Ask them how many 0s, 1s, 2s, etc. they would expect to get. They should understand that, for a class of 30, for example, there should be about three of each.
- Now collate the results. Are they as expected? It is unlikely, as 7 tends to get chosen more than any other number. Using the relative frequency of 7 for the class, estimate how many students in your school would choose 7. Discuss why the actual results did not meet expected results, or why they did.

Homework and answers

Homework for this lesson is available in the homework folder. Consolidation and extension exercises are available in the further practice folder. Answers to the exercises can be found in the answers folder.

33.5 Using Venn diagrams

Syllabus references

Statement	Core
C8.3/E8.3	Calculate the probability of combined events using Venn diagrams.

Collins references

- Student's Book pages 692–695

Learning aims

- Work out the probabilities for two outcomes occurring at the same time using Venn diagrams.

Key words

- Venn diagram

Prior knowledge

Students should already know how to calculate a probability in simple cases using equally likely outcomes.

They should also be familiar with a) the set notation used in Chapter 6 and Venn diagrams in section 6.3, and b) coordinates and straight line graphs from Chapter 15, sections 15.1 and 15.2.

Common mistakes and remediation

The most common mistake with this type of problem is that students do not pay sufficient attention to the instructions and misinterpret the language. If a class is surveyed and two questions are asked, 'Do you drink water?' and 'Do you drink milk?', then there will be some who drink both (and possibly some who drink neither), yet students forget that there are a number who will drink both, misinterpreting the question 'Do you drink water?' as 'Do you drink water only?' when interpreting the Venn diagram. Make students aware that when they interpret Venn diagrams they should remember that unless the question includes the work 'only', they are concerned with everything in the appropriate set.

Sometimes the numbers in a set represent the elements in the set, sometimes they represent the number of elements in a set and this can be confusing. Students need to take care when dealing with sets of numbers as opposed to the number in a set.

Useful tips

Emphasise that they should give their answer as a fraction in most cases.

Starter

- Ask students to draw a Venn diagram where the universal set, $\mathscr{E}$, is the numbers 1 to 10 inclusive and A = {prime numbers}, B = {odd numbers}.

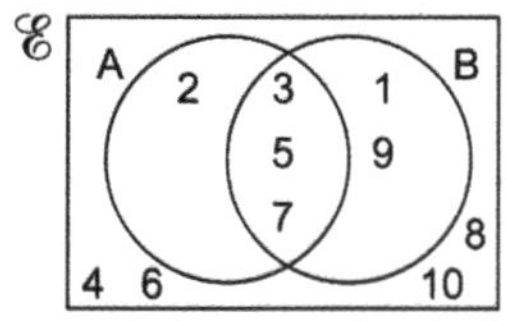

- What is n(%)? (10) What is n(*A*)? (4) What is n(*B*)? (5) What is n($A \cap B$)? (3) What is n($A \cup B$)? (6) What is n($A \cup B$)? (4)
- These results are needed for the main lesson activity.

Main lesson activity

- Using the Venn diagram from the starter, ask: What is the probability that a number chosen at random from % is a prime number? (n(*A*)/n(%) = $\frac{4}{10}$ = $\frac{2}{5}$ or 0.4 or 40%)
- Now ask: What is the probability that a number chosen at random from % is an odd prime number? (n($A \cap B$)/n(%) = $\frac{3}{10}$ or 0.3 or 30%)
- Now ask: What is the probability that a number chosen at random from % is neither odd nor prime? (n($A \cup B$)'/n(%) = $\frac{4}{10}$ or 0.4 or 40%)
- Go through the Example 6 in this section of the Student's Book.
- Students can now do Exercise 33E.

Plenary

- Go through question 5 of Exercise 33E as this is slightly different from the other questions because it requires coordinates to be put into the Venn diagram.

Homework and answers

Homework for this lesson is available in the homework folder. Consolidation and extension exercises are available in the further practice folder. Answers to the exercises can be found in the answers folder.

33.6 Probability notation

Syllabus references

Statement	Extended
E8.1	Understand and use probability notation

Collins references

- Student's Book pages 695–697

Learning aims

- Use capital letter notation for events and probabilities.

Key words

- No new key words

Prior knowledge

Students should be able to work out the probabilities of single events

Common mistakes and remediation

Students can miss the significance of the dash (e.g. A') to show an event is not happening. Emphasise the importance of this.

Useful tips

Venn diagrams can be useful when working out the probability of two other events. These will be indicated by the use of the word "and" or "or".

Starter

- Say that you have 10 cards numbered from 11 to 20. (It is a good idea to make a set as a visual aid.)
- Take a card at random. Ask for the probability that it is: an even number (0.5); a prime number (0.4); not a prime number (0.6); a multiple of 3 (0.3); not a multiple of 3 (0.7); either even or a multiple of 3 (0.6).
- Ask students to describe an event with a probability of: 0.1; 0.9; 0.2; 0.8

Main lesson activity

- Say that we can use capital letters to stand for events.
- Suppose a computer generates an integer from 1 to 15 at random. Here are some events.
- E = an even number; M = 12 or more; X = a prime number
- The probability of E is written as $P(E) = \frac{7}{15}$
- Ask students to work out P(M) and P(X). The answers are $\frac{4}{15}$ and $\frac{6}{15} = \frac{2}{5}$
- The event that it is **not** even is written as E'. This is the same as saying that it is odd.

  $P(E') = 1 - \frac{7}{15} = \frac{8}{15}$

- Ask students to explain the meaning of M' and X' and to work out P(M') and P(X').
  The answers are $\frac{11}{15}$ and $\frac{3}{5}$

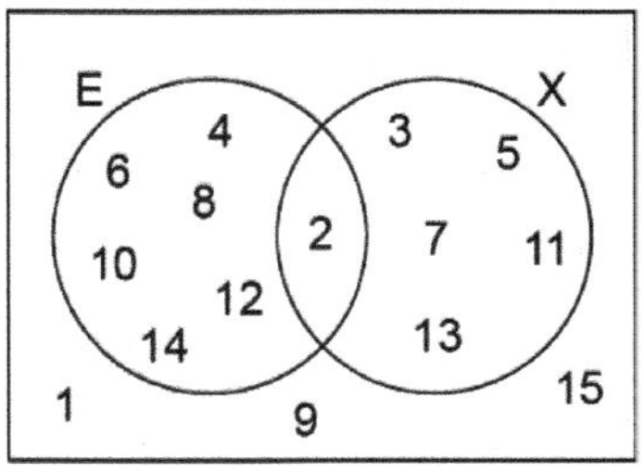

- Draw this Venn diagram to show E and X but do not put any numbers in it.
- Ask students to copy it and put the numbers from 1 to 15 in the correct places.
- Show the completed diagram for checking accuracy.
  Point out that E' means the numbers outside E.
- Ask students to describe $E \cap X$ and find $P(E \cap X)$. It means getting an even number that is prime and the probability is $\frac{1}{15}$
- Do the same for $E \cup X$ and $P(E \cup X)$.
  It means getting a number that is even or at least 12 and the probability is $\frac{12}{15} = \frac{4}{5}$

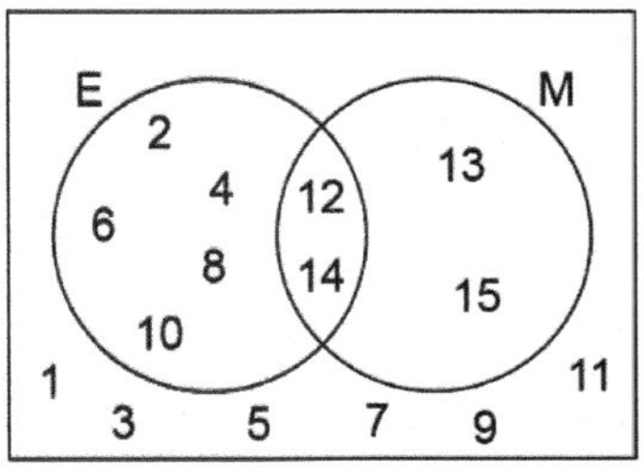

- Ask students to draw this Venn diagram to show E and M
- Ask them to use it to find $P(E \cap M)$ and $P(E \cup M)$.
  The answers are $\frac{2}{15}$ and $\frac{3}{5}$
- Students can now complete Exercise 33F.

Plenary

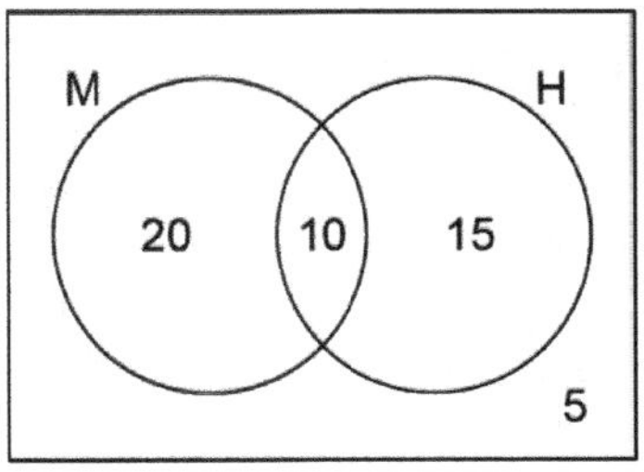

- Show this Venn diagram.
- Say that it shows the number of students in a group of 50 who study Maths (M) or History (H).
- A student is chosen at random. Ask for the probability that the student is studying Maths, P(M). the answer is $\frac{3}{5}$ or 0.6
- Ask students to describe some other events and find their probabilities. Make sure to include something not happening, such as A'.

Homework and answers

Homework for this lesson is available in the homework folder. Consolidation and extension exercises are available in the further practice folder. Answers to the exercises can be found in the answers folder.

33.7 Sample space diagrams

Syllabus references

Statement	Core
C8.3/E8.3	Calculate the probability of combined events using sample space diagrams.

Collins references

- Student's Book pages 697–700

Learning aims

- Calculate the probability of combined events using sample space diagrams.

Key words

- sample space diagram

Prior knowledge

Students should already know how to calculate a probability in simple cases using equally likely outcomes.

Common mistakes and remediation

The most common mistake with this type of problem is choosing the best format. A sample space diagram can either record two separate categories in each direction (horizontal or vertical) or can be used to combine outcomes, for example adding them or finding the difference between them. Students should decide what to record themselves, although a brief discussion with them should help them decide what is needed.

Useful tips

Emphasise that the method for finding a theoretical probability for combined events is no different from that for single events.

Starter

- Split the class into pairs. Provide each pair with two dice. Set a time limit of two minutes. Ask each pair to roll the dice, add the scores for each roll and record each total score.
- While they are doing this, prepare a tally chart on the board. Then collate all the results from the class when they have finished.
- Ask for the relative frequencies, or experimental probabilities, for each score.

Main lesson activity

- Ask: 'Which total came up the most?' Each group may have a different total, but the answer 7 is likely to occur most frequently.
- Now ask: 'Why did these numbers seem to occur more than others?' (There are more ways to get some totals.) For example, the totals 12, 11, 3 and 2 are unlikely to have occurred very often, since they can only be achieved by double 6, 5 and 6, 2 and 1, double 1, respectively. A total of 6 may be achieved by 1 and 5, 2 and 4, double 3.
- Ask: 'How could you find the number of ways of getting each total?' After a discussion lead into describing a sample space diagram.
- Draw the sample space diagram for two dice on the board. There is an example of this at the start of section 33.7.
- Show them that there are six ways to get a total of 7, but only two ways of getting a total of 11, so they should expect to get more 7s than 11s.
- Go through Examples 7 and 8 in section 33.7 of the Student's Book and point out the different ways in which sample space diagrams can be completed – either as a grid of crosses showing the outcomes, or as grid with the outcome shown as the result of a given condition.
- Explain in detail to students who need more **support** that each entry represents one possible outcome, so the diagram provides information on the total number of possibilities as well as the number that satisfy the required outcome.
- Students can now do Exercise 33G.

Plenary

- Ask the students if there is any difference between their theoretical probabilities in question 1 and the experimental probabilities they obtained in the starter.
- Should they expect any differences?
- This should again emphasise that experimental results should eventually be similar to the theoretical results when a large amount of experimental data is collected. Students who need more of a **challenge** may well note that in practice they will hardly ever be the same.

Homework and answers

Homework for this lesson is available in the homework folder. Consolidation and extension exercises are available in the further practice folder. Answers to the exercises can be found in the answers folder.

33.8 Tree diagrams

Syllabus references

Statement	Core
C8.3/E8.3	Calculate the probability of combined events using tree diagrams.

Collins references

- Student's Book pages 701–705

Learning aims

- Use tree diagrams to work out the probability of combined events.

Key words

- tree diagram

Prior knowledge

Students should be able to calculate the probability of two events happening at the same time through the use of sample space diagrams. They should also be confident in adding and multiplying simple fractions and, for probabilities that are given in decimal form, adding and multiplying decimals.

Common mistakes and remediation

Some students may decide to add probabilities along the branches, rather than multiply them. Make sure students draw the initial tree diagram to a reasonable size and label each branch with its corresponding probability, irrespective of what the question requires. Checking that a probability is $\leqslant 1$ will help to reduce this error.

Useful tips

In addition to the starter activity, provide extra practice for students who need more **support** by giving them more calculations involving sums and products of fractions. Encourage them to work in pairs so that they can help each other.

Starter

- Set students some typical calculations involving sums and products of fractions. The pace and focus of the lesson can be needlessly derailed if confidence is lacking in this area.

Main lesson activity

- So that students do not have to deal with a new situation as well as a new type of diagram, the start of section 33.8 deals with the same situation as Example 7 in section 33.7; work through it with the students.
- On the board, draw the first part of the tree diagram to show the two branches of 3 and 'not 3'. Put the probabilities on the branches.
- Add the next set of branches, for the second spinner, at the end of each first branch, and again put the probabilities on the branches.
- Explain that to find a combined probability of the first event (result on first spinner) and then a second event (result on the second spinner) happening, students must multiply the two probabilities.
- Use the bottom branches (first event being 'not 3' and second event 'not 3') to illustrate that the probability can be seen as five sixths of five sixths, so is found by multiplying the two probabilities.
- Do this for the remaining branches on the tree diagram, finishing with a set of probabilities all with denominator 36.
- Then show that these probabilities are the same as those obtained from the sample space diagram, reinforcing the rule for multiplying.
- Spend time explaining this multiplication rule. To help students who need more **support**, talk through question 1 in Exercise 33H as another example of using tree diagrams.
- All students can complete Exercise 33H. Students who need more **support** might benefit from working with a partner, if they find the later questions difficult.

Plenary

- Compare tree diagrams with sample space diagrams, which students have already learned about. Ask students what advantages or disadvantages one might have over the other. (For example, sample space diagrams give good visual presentations.)
- Ask students to suggest scenarios where it is clearly preferable to use tree diagrams instead of sample space diagrams, and vice versa.

Homework and answers

Homework for this lesson is available in the homework folder. Consolidation and extension exercises are available in the further practice folder. Answers to the exercises can be found in the answers folder.

33.9 Conditional probability

Syllabus references

Statement	Extended
E8.4	Calculate conditional probability using Venn diagrams, tree diagrams and tables.

Collins references

- Student's Book pages 705–711

Learning aims

- Calculate conditional probability from Venn diagrams and tree diagrams.

Key words

- conditional probability

Prior knowledge

Students should have worked through all the previous sections of Chapter 33 and be able to draw and interpret possibility diagrams such as Venn diagrams and tree diagrams. They should also be confident in adding and multiplying simple fractions and, for probabilities that are given in decimal form, adding and multiplying decimals.

Common mistakes and remediation

When working with tree diagrams, some students may decide to add probabilities along the branches, rather than multiply them. Make sure students draw the initial tree diagram to a reasonable size and label each branch with its corresponding probability, irrespective of what the question requires. Checking that a probability is $\leqslant 1$ will help to reduce this error, as will checking the sum of all the probabilities add to 1.

Useful tips

Encourage students who need more **support** to work in pairs so that they can help each other.

Starter

- Use the same starter as in lesson 33.5: Ask students to draw a Venn diagram where the universal set, %, is the numbers 1 to 10 inclusive and A = {prime numbers}, B = {odd numbers}.

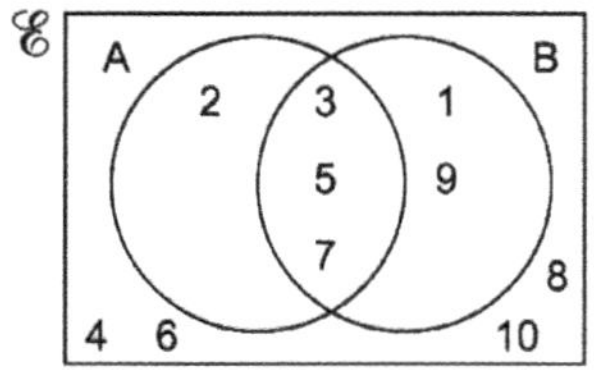

- Develop what was done earlier by asking: What is the probability that an odd number chosen at random from % is prime? ($n(A \cap B)/n(B)$ = 3/5 or 0.6 or 60%)
- Ask your students how this question differs from what was asked in lesson 33.5: they should answer that previously the event came from the universal set, not a subset of the universal set, so this time the denominator of the fraction is different.

Main lesson activity

- Inform students that this section deals with probabilities that involve a condition, so that the universal set is reduced in some way.
- On the board, work through the first part and emphasise the fact that although questions – *What is the conditional probability that the number is a factor of 12 given that it is a multiple of 3?* and *What is the conditional probability that the number is a multiple of 3 given that it is a factor of 12?* – look very similar, they are very different questions. This is because in the first instance the factor of 12 is being chosen from the multiples of 3 (of which there are five) and in the second instance the multiple of 3 is being chosen from the factors of 12 (of which there are six).
- When you work through Example 9, it is useful to have three red pens and two blue pens in a box with you so that you can show students exactly how the probabilities change after a pen is taken out of the box. The visual stimulation is a lot more impressive than a mental one.
- All students can start Exercise 33I. Students who need more **support** might benefit from working with a partner if they find the later questions difficult.

Plenary

- Go over question 9 of Exercise 33I
- Ask students to pose a conditional probability question for a neighbour, who should then answer it and have it checked by the proposer.

Homework and answers

Homework for this lesson is available in the homework folder. Consolidation and extension exercises are available in the further practice folder. Answers to the exercises can be found in the answers folder.

Collins

William Collins' dream of knowledge for all began with the publication of his first book in 1819.

A self-educated mill worker, he not only enriched millions of lives, but also founded a flourishing publishing house. Today, staying true to this spirit, Collins books are packed with inspiration, innovation and practical expertise. They place you at the centre of a world of possibility and give you exactly what you need to explore it.

Collins. Freedom to teach.

Published by Collins

An imprint of HarperCollins*Publishers*

The News Building, 1 London Bridge Street, London, SE1 9GF, UK

Macken House, 39/40 Mayor Street Upper, Dublin 1, D01 C9W8, Ireland

Browse the complete Collins catalogue at **collins.co.uk**

10 9 8 7 6 5 4

ISBN 978-0-00-854606-9

British Library Cataloguing-in-Publication Data

A catalogue record for this publication is available from the British Library.

Authors: Chris Pearce and Isabel Marsden
Updates for this edition: Chris Pearce
Publisher: Elaine Higgleton
Product manager: Letitia Luff
Copyeditor: Mark Gadd/4Science
Cover designer: Kevin Robbins and Gordon MacGilp
Cover illustrator: Ann Paganuzzi
Typesetter and illustrator: Jouve India Pvt. Ltd.
Additional illustrations: Ann Paganuzzi
Production controller: Lyndsey Rogers
Printed and bound by Ashford Colour Ltd.

MIX
Paper | Supporting responsible forestry
FSC™ C007454

This book contains FSC™ certified paper and other controlled sources to ensure responsible forest management.

For more information visit: www.harpercollins.co.uk/green

Acknowledgements

The publishers gratefully acknowledge the permission granted to reproduce the copyright material in this book. Every effort has been made to trace copyright holders and to obtain their permission for the use of copyright material. The publishers will gladly receive any information enabling them to rectify any error or omission at the first opportunity.

Cambridge International copyright material in this publication is reproduced under licence and remains the intellectual property of Cambridge Assessment International Education.

Endorsement indicates that a resource has passed Cambridge International's rigorous quality-assurance process and is suitable to support the delivery of a Cambridge International syllabus. However, endorsed resources are not the only suitable materials available to support teaching and learning, and are not essential to be used to achieve the qualification. Resource lists found on the Cambridge International website will include this resource and other endorsed resources. Any example answers to questions taken from past question papers, practice questions, accompanying marks and mark schemes included in this resource have been written by the authors and are for guidance only. They do not replicate examination papers. In examinations the way marks are awarded may be different. Any references to assessment and/or assessment preparation are the publisher's interpretation of the syllabus requirements. Examiners will not use endorsed resources as a source of material for any assessment set by Cambridge International. While the publishers have made every attempt to ensure that advice on the qualification and its assessment is accurate, the official syllabus, specimen assessment materials and any associated assessment guidance materials produced by the awarding body are the only authoritative source of information and should always be referred to for definitive guidance. Cambridge International recommends that teachers consider using a range of teaching and learning resources based on their own professional judgement of their students' needs. Cambridge International has not paid for the production of this resource, nor does Cambridge International receive any royalties from its sale. For more information about the endorsement process, please visit www.cambridgeinternational.org/endorsed-resources